EVOLUTION,
LEARNING
AND
COGNITION

EVOLUTION,

LEARNING

AND

COGNITION

Editor
Y. C. LEE
Los Alamos National Laboratory

World Scientific
Singapore • New Jersey • London • Hong Kong

Published by

World Scientific Publishing Co. Pte. Ltd.
P O Box 128, Farrer Road, Singapore 9128

USA office: World Scientific Publishing Co., Inc.
687 Hartwell Street, Teaneck, NJ 07666, USA

UK office: World Scientific Publishing Co. Pte. Ltd.
73 Lynton Mead, Totteridge, London N20 8DH, England

Library of Congress Cataloging-in-Publication Data

Evolution, learning & cognition.
1. Neural computers.   2. Artificial intelligence.   3. Cognitive science.
I. Lee, Y. C. (Yee Chun)
QA76.5.E944   1988   006.3   88-33806
ISBN 9971-50-529-0
ISBN 9971-50-530-4 (pbk.)

Printed in Singapore by Utopia Press.

PREFACE

Significant progress has been made in such diverse areas as neural networks, classifier systems, adaptive signal processing, and nonlinear prediction theory in recent years, all of which pertain to a common goal of understanding the adaptive and computational capabilities of natural and artificial complex intelligent systems. The main driving forces behind the current resurgence of interest in computational adaptation and self-organizational techniques are the ready accessibility of inexpensive, fast massively parallel computing devices which permits the modeling of large scale neural and other adaptive networks suitable for practical real world applications, and the realization, by researchers in artificial intelligent systems, of the need to incorporate automatic learning capability into knowledge-based systems in order to deal with the inherent imprecise, incomplete, and ever-changing nature of the real world knowledge base.

The publication of this volume reflects the urgent need for a global overview of this emerging interdisciplinary science. The extraordinarily rapid growth of research effort in the areas of neural networks and genetic algorithm in particular, with the attendant proliferation of research papers and conference proceedings has increasingly forced the researchers to specialize in narrow domains. The speed and magnitude of private companies jumping the bandwagon in their attempt to capitalize on the potential of these powerful techniques for commercial applications have helped to generate the recent wave of public awareness in this new discipline, but at the same time, also have helped to create on media hype surroundings the promise of the new techniques that sometimes laymen and experts alike "find it difficult" to judge whether real progress has been made by reading the frequent press releases and conference papers.

To further compound the problem, along with the big explosion of the R&D effort in both academia and the industrial and commercial sector, came a minor explosion of a confusing array of new products and terminologies. In addition, by and large, this young discipline of learning intelligent systems can still be regarded as a hacker's paradise, with a hodge-podge list of algorithms and tricks, the majorities of which are empirical and were developed with specific applications in mind. As the technology slowly inches toward maturity, it becomes increasingly important to provide a comprehensive yet coherent treatment of the field.

A major issue in the practical application of the new self-organization techniques is the speed at which the intelligent systems can be trained for each task. The learning speed not only depends on the specific system architecture, the learning algorithm employed, but is also strongly dependent on the particular task at hand which defines the shape of the error (or objective) surface. Since the adaptation process of the intelligent system can be essentially described as a kind of optimization process which seeks to improve the performance (and hence reducing the error rate) of the system with each new observation, it is reasonable to expect that the complexity of the landscape (of the error surface) is a direct reflection of the computational complexity of the task given.

The majority of the intelligent problems the systems are expected to solve is most likely to be of the NP-complete type, or at the very least, to lack efficient deterministic algorithm. For those tasks, the way learning complexities scales as the task size is of great concern. However, the potential complexities of the error landscape cannot shoulder all the blame for the slowness of the present learning algorithms. Even tasks which are simple from the algorithmic point of view oftentimes require unacceptably long training sequences. The three dominant learning strategies, i.e., the correlational (or Hebbian) learning, the gradient descent learning, and the Darwinian strategy of random mutation and crossover, are simply not sufficiently "guided" to solve the "ravine tracking" problem due to large eigenvalue spread of the Hessian (i.e., second order) matrix which frequently occurs for large dimensional tasks, even though the mathematical problems associated with the tasks are essentially linear (and therefore simple) in nature.

Another major theoretical issue is the expressive power of the self-modifying intelligent systems. One of the early attempts to build a learning machine was the "perceptron" machine popularized by Rosenblatt. Unfortunately the expressive power of the perceptron was put into question by Minsky and Papert and was found to be inadequate even for certain classes of "easy" problems. Modern neural net architectures are vastly more powerful than their perceptron predecessors. Similarly, the nonneural adaptive mapping network architectures presently being investigated are capable of approximating a large class of smooth and/or hierarchical mappings. Even more impressive are the classifier systems of Holland, described in this volume, since the expressive power of such systems is fully equivalent to that of a Turing machine. However, there seems to be a trade-off between expressive power and the speed of adaptation, as the more expressive systems tend to have more complicated architectures.

Perhaps the least understood aspect of the learning systems is the capabilities of the systems to generalize from learned examples. Some of the generalization capabilities of the adaptive mapping networks such as the locally linear and higher order mappers of Farmer (this volume), the default hierarchy net (also this volume), and the single layer perceptron can be attributed either to either the smoothness hypothesis assumed by these systems which allows explicit interpolation algorithms to be used for generalization, or to the built-in hierarchical memory organization and the sequential learning algorithm which favors generalization through hierarchy formation. The generalization that seems to be provided by classifier systems and neural nets with hidden processing units is much more difficult to comprehend.

For some investigators and industrial users, the hidden neural nets and classifier systems hold a forbidding aura of deep mystery. A few people even have gone so far as to claim this to be a major virtue of the systems and evidence of the supposed superiority of the so-called "brain metaphor". While this assertion may delight neural modelers and thrill the public media, it does nothing to clarify the matter. There is simply no substitute for a sound mathematical investigation of the characteristics of neural generalization, even if conducted in relatively circumscribed domains.

In order to partially address each of the above issues, we invited active researchers who are leaders and pioneers in their respective fields to contribute to this volume. Even though numerous research papers have already appeared in widely disparate forums, with the bulk being in the form of conference and workshop proceedings, there has been no single volume which provides access to state-of-the-art research in the broad discipline of self-organizing intelligent systems. We have tried to include the applications of one methodology in several, different domains as well as the applications of distinct methodologies to the same problems so long as it is feasible. This should allow the comparison of different methodologies and hence promote cross-fertilization.

The book can be divided roughly into three almost equal parts. In the first part, the papers selected are of a more general, mathematical nature, and as such, they provide a mathematical introduction to the general subject. The second part of the book comprises description and formulation of various adaptive architectures, whereas the last part of the book is mostly devoted to applications. Such division, however, is only approximate, since all papers selected for this volume are essentially self-contained, each with its own architectural description, mathematical formulation, and application results or suggestions.

Throughout this volume, the aim was to provide the most up-to-date account of the present status in learning intelligent system methodologies in diverse areas, and to suggest directions for future research. If this book can convey the excitement experienced by those active in this new discipline and can provide stimulus to beginning readers to participate in the advancement of the subject, then the purpose of this volume will be amply served.

The author would like to thank Dr. David K. Campbell, Director of the Center for Nonlinear Studies at Los Alamos National Laboratory, for suggesting this project to me and for his continuous support and encouragment. I also wish to thank the editorial staff of World Scientific for their valuable expert technical help.

Los Alamos
1988

Y. C. Lee

CONTENTS

x

EVOLUTION,
LEARNING
AND
COGNITION

Part One
MATHEMATICAL THEORY

Connectionist Learning
Through Gradient Following

Ronald J. Williams
College of Computer Science
Northeastern University
Boston, MA 02115

INTRODUCTION

Consider the two questions: (1) What are the processing principles, learned or innate, used by the brain to compute a given sensorimotor or cognitive function, such as visual recognition of one's grandmother, auditory recognition of a familiar melody, motor commands to control the swing of a bat at a 90 m.p.h. fastball, or a decision to make a particular chess move? (2) What are the principles used by the brain to adapt itself to meet the needs of the particular environment it finds itself in at any particular stage of its existence, so that, for example, it can improve at any of the above tasks?

One may seek answers to these questions for their own sake or as a means of identifying techniques for use in artificial systems having similar capabilities. Ultimately, the answers to these two difficult questions will depend on empirical studies of the brain itself. In the meantime, however, one can try to approach them by studying simplified formal models. The difficulty, of course, is to decide what constitutes a valid model of processing in the brain, of various sensorimotor and cognitive functions, and of adaptation and learning. Perhaps an even more fundamental difficulty is to resolve the philosophical problem of what constitutes processing *principles*, as opposed to *details*. Because of the wide latitude possible in any of these areas, it is not surprising that a wide variety of approaches to these questions have been investigated by various researchers.

There are those psychologists and artificial intelligence researchers who believe that the principles of brain-like processing are best expressed in the language of computational symbol manipulation (e.g., Newell, 1980)—at least for those specifically high-level cognitive functions, as distinguished from the more low-level sensorimotor functions. At another extreme are neurophysiologists, who would like to explain brain functioning in terms of the biochemical details of synaptic communication and neural cell growth.

Somewhere between these two extremes in the issue of what constitutes processing principles versus mere details is the study of what are variously called

---

Preparation of this article was supported by the National Science Foundation under grant IRI-8703566.

connectionist systems, parallel distributed processing networks, or *artificial neural systems.* The unifying feature of these systems is that they consist of highly interconnected networks of relatively simple processing units, the computational properties of the system being a result of the collective dynamics of the network. This approach is distinct from that of modeling biological neural networks because the individual processing units are not constrained to match in any but the most superficial way the details of biological neuron functioning. A common disclaimer is to use the term *neuron-like* to describe the individual units and *neurally inspired* to describe the resulting models.

There are many reasons why this approach is considered worthwhile for helping to provide valuable insights into brain functioning as well as suggesting useful approaches to the design of artificial systems having brain-like capabilities. Among the attractive features of such networks are: (1) their high degree of parallelism, with computational processing broadly distributed across possibly very many units; (2) their powerful associative memory properties, including best-match generalization, content addressability, and graceful degradation; and (3) their ability to rapidly compute "near-optimal" solutions to highly constrained optimization problems. These networks can form nonlinear mappings (such as Boolean functions) and are often constructed to manifest interesting nonlinear dynamics. Many of these properties are explored and discussed in, e.g., Hinton and Anderson (1981), Hopfield (1982), Hinton and Sejnowski (1983), Kohonen (1984), Feldman (1985), Hopfield and Tank (1985), Rumelhart and McClelland (1986), and McClelland and Rumelhart (1986).

This article will describe two particular approaches to arriving at answers to the questions posed above which are appropriate to a connectionist view of brain-like processing. In particular, we will examine two classes of learning algorithms for such networks, where the term "learning" is intended to be interpreted quite generally as something that can be applied either on-line, as in its usual sense, or off-line. Thus the learning algorithms to be described here may be thought of as possible answers to the second question posed above, or, alternatively, as automated techniques for proposing candidate answers to the first question.

The learning algorithms considered here are appropriate to two particular formalizations of the learning problem for a connectionist system. While these two paradigms are quite different and make different assumptions about the nature of the computation performed by the units in the net, the common thread is that algorithms for each case can be derived mathematically by first formulating the learning problem as an optimization problem and then using the simple but powerful principle of stochastic hill-climbing in this criterion function. Specifically, algorithms are presented here for each of these learning paradigms which follow the gradient—statistically, at least—of appropriate performance measures and have the further important property of being implementable locally.

CONNECTIONIST SYSTEMS

A connectionist system is simply a network of computational nodes, called *units*, and a collection of one-way signal paths, or *connections*, between them. It is assumed that this network interacts with an environment, so that some of the units, called *input units*, receive signals from the environment, and other units, called *output units*, transmit signals to the environment. In general, there may be units in the network which are neither input nor output units, and these are called *hidden units*. Hidden units provide a particular challenge for certain types of learning task because neither their actual or desired states are specified by the particular task.

There are a variety of assumptions which can be made concerning the nature of the computation performed by the individual units within a connectionist network. Each unit computes an output signal as some function of that unit's several input signals, and these input signals are themselves either equal to the outputs of units in the net or signals received from the environment. In general, these input and output signals are time-varying, but in certain restricted cases it may not be necessary to make this time dependence explicit. Input and output values of units in the net may be assumed to be discrete (Hopfield, 1982; Hinton & Sejnowski, 1983; Rosenblatt, 1962; Barto & Anderson, 1985) or continuous (Hopfield & Tank, 1985; Kohonen, 1984; Widrow & Hoff, 1960; Rumelhart et al., 1986), and the input/output function of units may be assumed to be deterministic (Hopfield, 1982; Hopfield & Tank, 1985; Kohonen, 1984; Rosenblatt, 1962; Widrow & Hoff, 1960; Rumelhart et al., 1986) or stochastic (Hinton & Sejnowski, 1983; Ackley et al., 1985; Barto & Anderson, 1985). In addition, when the time-varying nature of these signals propagating through the net is important, time may be modeled as discrete (Ackley et al., 1985; Barto & Anderson, 1985; Rumelhart et al., 1986) or continuous (Hopfield & Tank, 1985; Kohonen, 1984), with updating of output values performed synchronously (Barto & Anderson, 1985; Rumelhart et al., 1986) or asynchronously (Hopfield, 1982; Hinton & Sejnowski, 1983; Ackley et al., 1985). Still another point of variation is whether the network is assumed to have feedback loops (Hopfield, 1982; Hopfield & Tank, 1985; Hinton & Sejnowski, 1983; Kohonen, 1984, Ackley et al., 1985) or be acyclic (Barto & Anderson, 1985; Rumelhart et al., 1986).

Throughout all these variations is the common pair of assumptions, intended to capture the idea expressed in describing the computation as *neuron-like*, that: (1) signals transmitted along the connections are (time-varying) *scalars*; and (2) the computation performed at each unit is relatively simple. This second assumption is vague, but intended to rule out, for example, sophisticated encoding/decoding schemes as would be used for communication between two digital computing devices. Weighted analog summation combined with some simple nonlinearity is a typical example of a computation which is considered to satisfy this second criterion. Below we will consider some specific examples of computational units for connectionist networks.

LEARNING

There are a number of possible formulations of the learning problem for a connectionist system. The two particular learning paradigms of interest in this article are *supervised learning* and *associative reinforcement learning*, both of which involve learning on the basis of experience with a finite set of examples. The main distinction between these is the nature of the feedback provided to the system in the two cases. Figures 1 and 2 illustrate networks facing the two types of learning problem. For supervised learning the system is presented with the desired output for each training instance, while for reinforcement learning the system produces a response which is then evaluated using a scalar value indicating the appropriateness of the response. The objective in the supervised learning problem is to find network parameters which minimize some measure of the difference between actual and desired response, while the objective in the associative reinforcement learning problem is to find network parameters maximizing some function of the evaluation signal. Since the training examples for supervised learning consist of input/desired-output pairs, supervised learning might also be thought of as storage of such pairs (albeit in a way designed to permit efficient retrieval and generalization).

It is interesting to note that while there is a long history of attempts to develop what have been called self-organizing procedures for connectionist networks, it is only recently that certain obstacles faced by earlier approaches have been satisfactorily overcome. In particular, a major difficulty for the supervised learning problem has been in devising learning algorithms capable of providing effective adjustment of the parameters associated with hidden units in the network. For this reason, earlier research efforts (e.g., Rosenblatt, 1962; Widrow & Hoff, 1960) generally contented themselves with restricting learning in such networks to certain limited portions which excluded the hidden units.

It should be noted that other formulations of the learning problem are possible. One leading competitor in connectionist circles to those discussed here is that of *unsupervised learning*, in which learning occurs in the absence of any performance feedback. In this paradigm, the objective is for the network to discover statistical regularities or *clusters* in the stream of input patterns. Although we do not consider such learning procedures here, it is worth pointing out why such techniques have been (and continue to be) of interest. One reason is that, until fairly recently, there appeared to be no alternative for training the hidden units in multilayer nets in supervised or associative reinforcement learning tasks. By not depending on performance feedback of any sort, such techniques allow the independent self-organization of individual portions (typically single layers) of a network. Of course, there can thus be no assurance that the resulting performance is desirable (much less optimal) for a given task. With the recent development of promising algorithms for supervised and associative reinforcement learning in multilayer networks (Ackley, Hinton, & Sejnowski, 1985; Barto & Anandan, 1985; Rumelhart, Hinton, & Williams, 1986), the importance of this use for unsupervised learning procedures has diminished. Another source of the appeal of such procedures is their simplicity and biological plausibility. Much of the work of Grossberg (e.g., 1976) makes use of this class of algorithm, and Kohonen (1984) has demonstrated some interesting properties of certain algorithms of this type. Discussion of this general approach to learning may be found in Rumelhart and Zipser (1985).

The specific algorithms to be described here together with their gradient-following properties are the *back-propagation* algorithm (Rumelhart, Hinton, & Williams, 1986; Parker, 1982, 1985; Werbos, 1974) for supervised learning in networks of deterministic units and the *REINFORCE* class of algorithms (Williams, 1986, 1987) for associative reinforcement learning in networks of stochastic units. These latter algorithms are closely related to that investigated by Barto and Anderson (1985). Another recently developed stochastic hill-climbing algorithm which will not be discussed here is the *Boltzmann machine* learning algorithm of Ackley, Hinton, & Sejnowski (1985).

Supervised Learning vs. Associative Reinforcement Learning

Since this article discusses two different formulations of the learning problem and describes algorithms for each, it is useful to clarify the distinctions between the two and discuss briefly the question of their appropriateness.

In the associative reinforcement learning paradigm a network and its training environment interact in the following manner: The network receives a time-varying vector of inputs from the environment and sends a time-varying vector of outputs (also called *actions*) to the environment. In addition, it receives a time-varying scalar signal, called *reinforcement*, from the environment. The objective of learning is for the network to try to maximize some function of this reinforcement signal, such as the expectation of its value on the upcoming time step or the expectation of some integral of its values over all future time, as appropriate for the particular task. The precise nature of the computation of reinforcement by the environment can be anything appropriate for the particular problem and is assumed to be unknown to the learning system. In general, it is some function, deterministic or stochastic, of the input patterns produced by the environment and the output patterns it receives from the network. Figure 1 depicts the interaction between a network and its environment in an associative reinforcement learning situation.

This formulation should be contrasted with the *supervised learning* paradigm, in which the network receives a time-varying vector signal, indicating *desired output*, from the environment, rather than the scalar reinforcement signal, and the objective is for the network's output to match the desired output as closely as possible. This distinction is sometimes summarized by saying that the feedback provided to the network is *instructive* in the case of supervised learning and *evaluative* in the case of reinforcement learning. Figure 2 depicts the interaction between a network and its environment in a supervised learning situation.

We do not concern ourselves here with which is the more appropriate formalization in general, but simply note that each seems to have its place. The idea of matching a specified output pattern seems appropriate for certain problems dealing

8

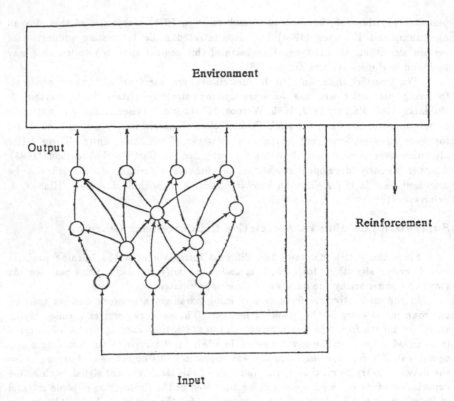

Figure 1. A connectionist network and its training environment for the associative reinforcement learning problem. The precise operation of this system consists of the following four phases:

1. The environment picks an input pattern for the network randomly (the distribution of which is assumed to be independent of prior events within the network/environment system).
2. As the input pattern to each unit becomes available, it computes its output. Thus "activation" passes through the network from the input side to the output side.
3. After all the units at the output side of the network have computed their outputs the environment evaluates the result as a (possibly stochastic) function of the given input and output patterns.
4. Each unit changes its internal parameters according to some specified function of the current value of those parameters, the input it received, the output it produced, and the environment's evaluation. The precise manner in which the evaluation, or *reinforcement*, signal is used by the individual units depends on the learning algorithm to be applied. In the simplest case, the reinforcement signal is simply broadcast to all units, but the use of additional units or interconnections designed to help in the learning process is also possible.

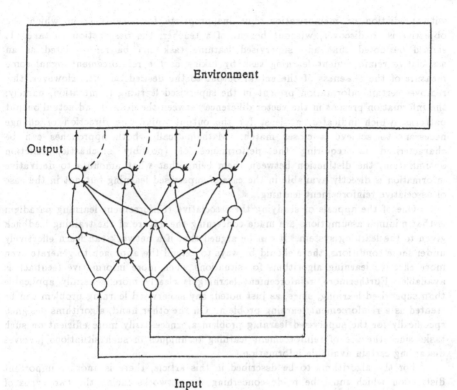

Figure 2. A connectionist network and its training environment for the supervised learning problem. Information flowing along the dashed arrows is the desired output. The precise operation of this system consists of the following four phases:

1. The environment picks an input pattern for the network randomly (the distribution of which is assumed to be independent of prior events within the network/environment system).
2. As the input pattern to each unit becomes available, it computes its output. Thus "activation" passes through the network from the input side to the output side.
3. After all the units at the output side of the network have computed their outputs the environment specifies the desired output pattern corresponding to the given input pattern.
4. Each unit changes its internal parameters according to some specified function of the current value of those parameters, the input it received, the output it produced, and additional information based on the network's actual output pattern and the desired output pattern. Not shown is the precise manner in which this additional information is made available to each unit, which may require further communication links. For example, *back-propagation* requires certain error-correction information to flow in the reverse direction along each (forward) link in the network.

with prediction or memorization, but inappropriate for situations in which the objective is to discover, without benefit of a teacher, the best action to take. It should be noted that any supervised learning task can be reformulated as an associative reinforcement learning task by taking as the reinforcement signal some measure of the closeness of the actual output to the desired output. However, this removes certain information present in the supervised learning formulation, namely, the information present in the vector difference between the desired and actual output patterns, which indicates, at least for the output units, the direction of change necessary to achieve a closer match. Mathematically, both approaches can be characterized as requiring the performance of (possibly stochastic) function optimization, the distinction between them being that what amounts to derivative information is directly available in the case of supervised learning but not in the case of associative reinforcement learning.

One of the appeals of studying the associative reinforcement learning paradigm is that minimal assumptions are made concerning the nature of the training feedback given to the learning system. It can be argued that if a network can learn effectively under these conditions, there should be ways to extend the approach to generate even more effective learning algorithms for situations when more informative feedback is available. Furthermore, reinforcement learning is clearly more generally applicable than supervised learning, since, as just noted, any supervised learning problem can be treated as a reinforcement learning problem. On the other hand, algorithms designed specifically for the supervised learning problem are necessarily more efficient on such tasks since the use of reinforcement learning techniques in such situations involves discarding certain available information.

For the algorithms to be described in this article, there is another important distinction which must be made concerning the networks facing the two types of learning task. The back-propagation supervised learning algorithm requires the use of deterministic computing units, while the REINFORCE algorithm for associative reinforcement learning is based on the use of stochastic computing units. The use of stochastic units in associative reinforcement learning networks can be seen as a necessary source of variation in the manner in which actions are chosen, allowing exploration of alternative actions. While other more sophisticated search strategies are possible, this approach is simpler to deal with mathematically and is consistent with the theory of *stochastic learning automata* (Lakshmivarahan, 1981; Narendra & Thathatchar, 1974; Narendra & Thathatchar, to appear). Barto and colleagues (Barto, Sutton, & Anderson, 1983; Barto, Sutton, & Brouwer, 1981), emphasizing this role for such stochastic units, have called them *search elements*.

FORMAL ASSUMPTIONS AND NOTATION

Now we introduce notation which will be used throughout this article. In general, for the moment suppressing explicit denotation of time dependence, let y_i denote the output of the ith unit in the network, and let $\mathbf{x}^i$ denote the pattern of input to that unit. As just described, $\mathbf{x}^i$ is a tuple whose individual elements are either the outputs of certain units in the network (those sending their output directly to the ith unit) or certain inputs from the environment (if that unit happens to be connected so that it receives input directly from the environment). If the unit is deterministic, the output y_i is computed as a function of $\mathbf{x}^i$ and a set of parameters $\{w_{ij}\}$, where j ranges over an appropriate index set. If the unit is stochastic, the output y_i is drawn from a distribution depending on $\mathbf{x}^i$ and a corresponding set of parameters $\{w_{ij}\}$. In either case, for each i, let $\mathbf{w}^i$ denote the tuple consisting of all the parameters w_{ij}. Then let $\mathbf{W}$ denote the tuple consisting of all the parameters w_{ij} for the network.

For example, suppose that for each i the output y_i is computed deterministically by

$$y_i = f(s_i),\tag{1}$$

where f is some given function (called a *squashing* function) and

$$s_i = \mathbf{w}^{i\,T}\mathbf{x}^i,\tag{2}$$

the inner product of $\mathbf{w}^i$ and $\mathbf{x}^i$. In this case, w_{ij} is called the *weight* on the connection from the jth unit to the ith unit, $\mathbf{w}^i$ is the *weight vector* for the ith unit, and $\mathbf{W}$ is the *weight matrix* for the network.

When f is the threshold (unit step) function

$$f(s) = \begin{cases} 0 & \text{if } s \le 0 \\ 1 & \text{if } s > 0, \end{cases}$$

this unit is called a *threshold logic unit*. Such units are the components of the *simple perceptron* (Rosenblatt, 1962; Minsky & Papert, 1969), and can be viewed as Boolean logic gates. Boolean functions computable by such units are called *linearly separable*. At another extreme, if f is the identity function, such a unit corresponds to the *adaline* of Widrow and Hoff (1960). An interesting choice of f lying between these two extremes is the *logistic* function

$$f(s) = \frac{1}{1+e^{-s}}.\tag{3}$$

This sigmoidal function can be viewed as a differentiable "approximation" to the threshold function. Any unit whose input/output function is computed according to (1) and (2) and having differentiable squashing function f is called a *semilinear unit*. The computation performed by a semilinear unit is depicted schematically in Figure 3.

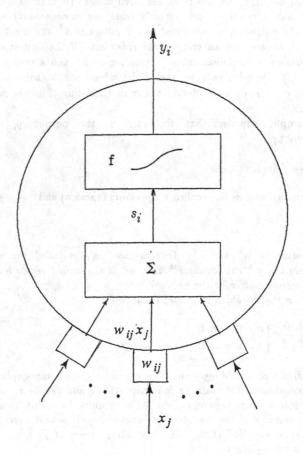

Figure 3. The manner in which a semilinear unit computes its output as a function of its input. The squashing function f is differentiable. Not shown are additional computational mechanisms required to support learning, which causes changes in the weights w_{ij}.

As an example of a stochastic unit, let

$$y_i = f\left(\mathbf{w}^{i\,T}\mathbf{x}^i + \eta\right), \tag{4}$$

where η is a deviate chosen from a given distribution with mean zero and f is a threshold function. Such units are used in the work of Barto and colleagues (Barto & Anandan, 1985; Barto & Anderson, 1985; Barto, 1985; Barto et al., 1983, Sutton, 1984). It is shown in Williams (1986) that an equivalent formulation of such a computation is obtained by using (1) and (2) to compute not the output itself, but the probability that the output is 1, with f chosen to be essentially the cumulative distribution function of η. More generally, consistent with this alternative formulation, a *stochastic semilinear unit* is a unit whose computation is performed as depicted in Figure 4, where f is a differentiable function and p_i is a parameter governing the distribution from which the output is selected. If there are just two possible output values, 0 and 1, and p_i represents the probability that the output is 1, such a unit is called a Bernoulli semilinear unit.

BACK-PROPAGATION ALGORITHM FOR SUPERVISED LEARNING

Assume that each unit in an acyclic network computes deterministically, and that the network is to be trained using supervised learning. This means that for every input pattern presented to the network, a desired output pattern is available. For each output unit there is a desired response d_k which is to be compared with the actual response y_k. Suppose that J is a criterion function depending on all the values y_k and d_k, where k ranges over output units. For example, the mean-square error criterion is given by

$$J = E\left\{\frac{1}{2}\sum_k (y_k - d_k)^2 \,|\, \mathbf{W}\right\}, \tag{5}$$

the expected value of the error conditioned on the particular choice of network parameters $\mathbf{W}$. We would like to find $\mathbf{W}$ minimizing this criterion function. Let J_p denote the value of the function J when input pattern p is presented to the network. Assuming p is randomly drawn from a given distribution of possible input patterns, J_p represents an unbiased estimate of J. Because of the linearity of the expectation operator, it follows that $\nabla_{\mathbf{W}} J_p$ is an unbiased estimate of $\nabla_{\mathbf{W}} J$. We can thus try to minimize this function by the technique of adjusting weights along these sample gradients of J, using

14

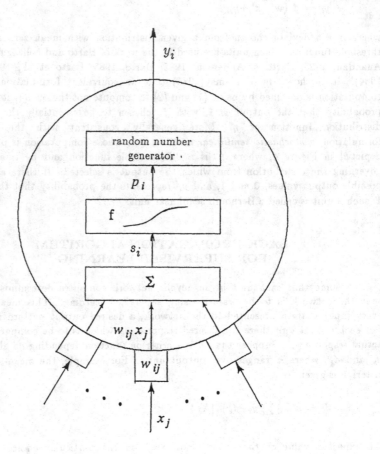

Figure 4. The manner in which a stochastic semilinear unit computes its output as a function of its input. The squashing function f is differentiable and the random number generator produces a value. according to a distribution determined by the single parameter p_i. Not shown are additional computational mechanisms required to support learning, which causes changes in the weights w_{ij}.

$$\Delta W = -\alpha \nabla_W J_p , \tag{6}$$

where α is a positive rate factor.

But interestingly, there is a straightforward computational technique for computing $\nabla_W J_p$ as long as the transfer functions of all units in the net are differentiable. Rumelhart et al. (1986) give a derivation for the case of semilinear units, but the argument readily generalizes. This straightforward computation amounts to propagating partial derivative information backward through the network, reversing information flow along every connection. That the computation has this form is a direct consequence of the chain rule for partial derivatives.

For example, suppose that the network consists of semilinear units, which compute according to equations (1) and (2). If the mean-square error function (5) is to be minimized, the algorithm then has the particular form

$$\Delta w_{ij} = -\alpha \delta_i z_j , \tag{7}$$

where δ_i is a quantity associated with the ith unit and is computed by

$$\delta_i = f'(s_i)(d_i - y_i) \tag{8}$$

for output units and

$$\delta_i = f'(s_i) \sum_k w_{ki} \delta_k \tag{9}$$

for all other units.

There are some special cases worth noting. When all units in the network are output units and the function f is the identity function, the resulting algorithm is identical to the *least-mean-square* algorithm of Widrow and Hoff (1960; Widrow & Stearns, 1985) used in adaptive signal processing and adaptive control applications. At the more nonlinear extreme, letting f be the logistic function (3) in order to obtain approximately Boolean computation, the algorithm uses

$$f'(s_i) = y_i(1 - y_i). \tag{10}$$

This *back-propagation* algorithm, given by (7), (8), (9), and (10), has been simulated on many examples and found to work quite effectively in general.[1] It has

---

1. In practice, a slight modification of this algorithm designed to accelerate convergence has often been used, as described in Rumelhart et al. (1986). Discussion of this or other such techniques will be omitted here.

been shown to be capable of setting the weights in multilayer networks to compute abstract Boolean functions which are not linearly separable, and it has also been applied with some success to more practical problems. While local minima can trap this algorithm, simulation experience suggests that it is quite rare for the weights to get stuck in local minima for which the performance is significantly worse than for the global minimum. One particularly interesting application of this algorithm was the training of a multilayer network having approximately 300 units and 18,000 weights to perform a text-to-speech mapping based on a limited sample of speech and its corresponding transcription (Sejnowski & Rosenburg, 1986). Others have begun to investigate its possible applicability to problems of speech recognition (Plaut, Nowlan, & Hinton, 1986; Burr, 1986; Elman & Zipser, 1987), with promising preliminary results.

Extended Back-Propagation

There is an interesting extension of this algorithm which has been developed to allow the training of desired behaviors which are extended through time in networks containing feedback loops. It is applicable to a situation in which a given net is to perform according to some measure through a given fixed time interval. In the case of discrete time and synchronous updating of the units in the net, there is a simple formal mapping between such learning problems and certain problems in corresponding layered acyclic nets as illustrated in Figure 5. The interesting result is that there is a simple formula for computing the sample gradients $\nabla_{\mathbf{W}} J_p$, as described in Rumelhart et al. (1986) and derived more completely in Williams (1987). This computation, however, requires "backward propagation through time," which can only be accomplished through the storage of the history of activity of all the units during the network's (forward) operation. This algorithm is thus difficult to imagine being used in an on-line manner. Relatively little simulation study of this *extended back-propagation* algorithm has been undertaken, but limited experience suggests that it can work to create networks which store sequences or carry out desired sequential processes. A study of the use of this algorithm for developing networks for speech recognition is reported by Watrous & Shastri (1986), who also derive a generalization allowing for variable link delays.

An interesting special case of the use of this extended back-propagation algorithm arises when a network's actual and desired dynamics consist of settling to a fixed equilibrium state. In this case, one can perturb the equilibrium states to desired values by use of equation (7) above, where now δ_i is computed for each unit by means of (8) and infinitely repeated use of (9) in what amounts to a settling computation. This amounts to the use of extended back-propagation applied to the problem of perturbing one constant trajectory of the system into another. Storage of the activity history is not required in this case because all past states along a constant trajectory are equal to the current one. Almeida (1987), Pineda (1987), and Rohwer and Forrest (1987) have all derived various versions of this algorithm, and Lapedes and Farber (1986) have derived a similar algorithm for this situation. In addition, Almeida has

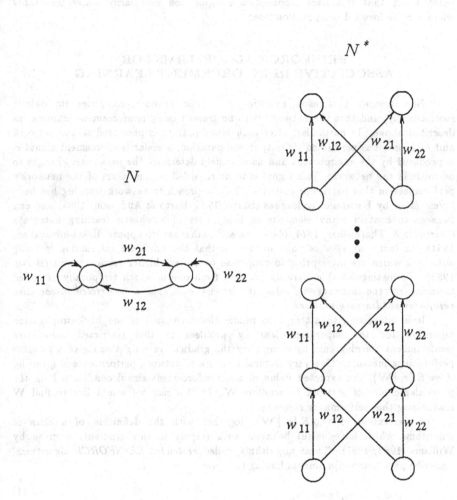

Figure 5. A network N containing cycles to be run through a certain number of time steps and its unfolded-in-time acyclic equivalent N^*. Weights in the two networks with identical labels have identical values.

pointed out that this back-propagation computation necessarily converges stably whenever the forward computation does.

REINFORCE ALGORITHMS FOR
ASSOCIATIVE REINFORCEMENT LEARNING

Now assume that each unit in an acyclic network computes its output stochastically, and that the network is to be trained using reinforcement learning, as described above. In particular, after each input pattern is presented to the network and each unit has performed its output computation, a scalar reinforcement signal r is produced by the environment and used to help determine the parameter changes to be made in the network. This signal is to be regarded as a measure of the network's performance in that particular instance. This approach to network learning has been investigated by Barto and colleagues (Barto, 1985; Barto & Anderson, 1985) and can be seen to contain many elements of the theory of stochastic learning automata (Narendra & Thathachar, 1974; Narendra & Thathachar, to appear; Lakshmivarahan, 1981). In fact, one view of this problem is that the units are interacting learning automata which are attempting to cooperate to achieve higher reinforcement (Barto, 1985). To distinguish this learning problem from one in which temporally extended behavior of the network is relevant, let us call this a *restricted associative reinforcement learning problem*.

It is interesting to attempt to mimic the derivation of the back-propagation algorithm for the supervised learning problem in this restricted associative reinforcement learning case by computing the gradient in weight space of a suitable performance measure. One very natural measure of network performance is given by $J = E\{r \mid \mathbf{W}\}$, the expected value of the reinforcement signal conditioned on the particular choice of network parameters $\mathbf{W}$. In this case we would like to find $\mathbf{W}$ maximizing this performance measure.

A derivation of $\nabla_{\mathbf{W}} E\{r \mid \mathbf{W}\}$, together with the definition of a class of algorithms which have useful behavior with respect to this gradient, is given by Williams (1986, 1987). These algorithms, called *restricted REINFORCE* algorithms, prescribe parameter-adjustments having the form

$$\Delta w_{ij} = \alpha_{ij}(r - b_{ij})e_{ij}, \tag{11}$$

where α_{ij} is a *learning rate factor* and b_{ij} is termed the *reinforcement baseline*. The quantity e_{ij} is called the *canonical eligibility* of w_{ij}, and is defined below. The factor $r - b_{ij}$ is called a *reinforcement offset*. The reinforcement baseline b_{ij} is assumed to be conditionally independent of y_i, given $\mathbf{W}$ and $\mathbf{x}^i$, and the rate factor α_{ij} is assumed to be nonnegative and essentially constant, except for possible dependence on i and j. In particular, α_{ij} is assumed not to depend on the input $\mathbf{x}^i$ to the unit.

The canonical eligibility is defined to be $e_{ij} = \dfrac{\partial \ln g_i}{\partial w_{ij}}$, where g_i is the probability mass function determining the value of y_i as a function of the parameters of the unit and its input, so that $g_i(\xi, \mathbf{w}^i, \mathbf{x}^i) = Pr\{y_i = \xi \mid \mathbf{w}^i, \mathbf{x}^i\}$ for any ξ.

The name *REINFORCE* is an acronym for "*RE*ward *I*ncrement $=$ *N*onnegative *F*actor $\times$ *O*ffset *R*einforcement $\times$ *C*anonical *E*ligibility," which describes the particular form of the algorithm. The following result is proved in Williams (1987):

Theorem. For any restricted REINFORCE algorithm, the inner product of $E\{\Delta \mathbf{W} \mid \mathbf{W}\}$ and $\nabla_{\mathbf{W}} E\{r \mid \mathbf{W}\}$ is nonnegative. Furthermore, this inner product is zero if and only if $\alpha_{ij} = 0$ for all i and j or $\nabla_{\mathbf{W}} E\{r \mid \mathbf{W}\} = 0$. Also, if $\alpha_{ij} = \alpha$ is independent of i and j, then $E\{\Delta \mathbf{W} \mid \mathbf{W}\} = \alpha \nabla_{\mathbf{W}} E\{r \mid \mathbf{W}\}$.

This means that for any such algorithm the expected update direction in weight space lies in a direction for which the performance measure $E\{r \mid \mathbf{W}\}$ is increasing.

Simulation experience with such algorithms is limited but results so far suggest that they, like back-propagation, can be used to set the weights in multilayer networks to compute abstract Boolean functions which are not linearly separable. In addition to providing algorithms which may be usefully applied, introduction of this class may also provide theoretical benefit. For example, a number of algorithms which have been studied by others can be shown to be examples of this class, and some other algorithms are closely related. This may be of value in helping provide insight into the behavior of these other algorithms as well as suggesting a unified approach to the development of such algorithms. The following algorithms are all either examples of REINFORCE algorithms or closely related:

Two-action linear reward-inaction stochastic learning automaton. Let the network consist of a single unit, with no (non-reinforcement) input from the environment. In this case the indices i and j may be suppressed. Let the possible output values y of the unit be 0 or 1, and let the possible values of the reinforcement signal likewise be 0 or 1. Let p denote a single parameter determining the distribution of y according to the Bernoulli distribution with parameter p. Then the probability mass function g satisfies $g(\xi, p) = Pr\{y = \xi \mid p\}$, where $g(0, p) = 1 - p$ and $g(1, p) = p$. If the REINFORCE algorithm is used with $\alpha = \rho p(1-p)$ and $b = 0$, the resulting algorithm is $\Delta p = \rho r(y-p)$, which is the much-studied two-action linear reward-inaction (L_{R-I}) stochastic learning automaton (Lakshmivarahan, 1981; Narendra & Thathatchar, 1974; Narendra & Thathatchar, to appear).

Team of two-action linear reward-inaction stochastic learning automata. Let the network consist of several units, with no (non-reinforcement) input from the environment and no interconnections between these units. This reinforcement learning problem corresponds to what is referred to in the standard learning automata literature as a *team* problem, in which a collection of individual participants are independently trying to determine actions which maximize their common payoff r in the absence of any communication among them. In this case we retain the index i but suppress the index j. Just as in the previous example, each unit has as its possible output values 0 or 1, determined for each unit i by a Bernoulli parameter p_i. Likewise, r takes on the two values 0 or 1. A REINFORCE algorithm for adjusting the parameters p_i is obtained by taking $\alpha_i = \rho_i \, p_i \, (1-p_i)$ and $b_i = 0$ for each i; the resulting algorithm has the form $\Delta p_i = \rho_i \, r \, (y-p_i)$ for each i. This amounts to having each member of the team use the L_{R-I} algorithm with its own individual learning rate, an algorithm investigated by Narendra and Wheeler (1983) and Wheeler and Narendra (1986).

Associative reward-penalty algorithm. Barto and colleagues (Barto, 1985; Barto & Anandan, 1985; Barto & Anderson, 1985; Anderson, 1986) have investigated a network learning algorithm they have called the *associative reward-penalty* (A_{R-P}) algorithm. This algorithm is designed for training a multilayer network of Bernoulli semilinear units facing a restricted reinforcement learning task in which there are two possible reinforcement values, which we take here to be 0 and 1. With each learning trial, the A_{R-P} algorithm prescribes weight changes having the form

$$\Delta w_{ij} = \begin{cases} \alpha(y_i - p_i)x_j & \text{if } r = 1 \\ \alpha\lambda(1 - y_i - p_i)x_j & \text{if } r = 0 , \end{cases}$$

where α is a positive learning rate parameter and $0 < \lambda \leq 1$. The corresponding algorithm with $\lambda = 0$ is called the *associative reward-inaction* (A_{R-I}) algorithm. It is shown in Williams (1986) that taking $b_{ij} = 0$ and $\alpha_{ij} = \alpha$ for all i and j in the REINFORCE algorithm yields an algorithm which coincides with the A_{R-I} algorithm for such networks in which the units have as their squashing function the logistic function.

Reinforcement comparison. In all the previous examples, the reinforcement baseline b_{ij} was fixed at 0. An interesting variant of A_{R-I}, still a REINFORCE algorithm, is obtained by taking b_{ij} to be some reasonable *a priori* prediction of what r will be on the current trial based on results obtained on earlier trials. In this scheme, the reinforcement baseline is adaptively updated, with one value used for the entire net or even a separate value used for each unit, based on that unit's inputs. This general notion of offsetting the currently obtained value of reinforcement by subtracting an adaptively determined prediction of that value was proposed and studied by Sutton (1984), who termed it *reinforcement comparison*. His simulation results suggest that superior speed of convergence is obtained when reinforcement comparison is used.

Extended REINFORCE Algorithms

As with the back-propagation algorithm, there is a way to apply the ideas behind the derivation of the restricted REINFORCE algorithm to the development of a learning algorithm for temporally extended behaviors in networks containing feedback loops. Specifically, assume a net is to be trained on an episode-by-episode basis, where each episode consists of k discrete time steps. At each time step, units compute their output synchronously in terms of their input at the previous time step (which may depend on the output of units at that previous time step). The environment may also alter its non-reinforcement input to the system at each time step. A single reinforcement value r is delivered to the net at the end of each episode.

An *extended REINFORCE* algorithm for this situation works as follows: At the conclusion of each episode, each parameter w_{ij} is incremented by

$$\Delta w_{ij} = \alpha_{ij} (r - b_{ij}) \sum_{t=1}^{k} e_{ij}(t) \tag{12}$$

where all notation is the same as that defined earlier, with $e_{ij}(t)$ representing the canonical eligibility for w_{ij} evaluated at the particular time t. This canonical eligibility is a function of the input to the unit at time $t-1$ and the output of the unit at time t. In addition, all quantities are assumed to satisfy the same conditions required for the restricted REINFORCE algorithm, where, in particular, for each i and j, the reinforcement baseline b_{ij} is independent of any of the output values $y_i(t)$ and the rate factor α_{ij} is a nonnegative constant.

It is shown in Williams (1987) that the following result holds:

Theorem. For any extended REINFORCE algorithm, the inner product of $E\{\Delta W \mid W\}$ and $\nabla_W E\{r \mid W\}$ is nonnegative. Furthermore, this inner product is zero if and only if $\alpha_{ij} = 0$ for all i and j or $\nabla_W E\{r \mid W\} = 0$. Also, if $\alpha_{ij} = \alpha$ is independent of i and j, then $E\{\Delta W \mid W\} = \alpha \nabla_W E\{r \mid W\}$.

What is noteworthy about this algorithm is that it has a plausible on-line implementation using a single accumulator for each parameter w_{ij} in the network. The purpose of this accumulator is to form the eligibility sum, each term of which depends only on the operation of the network as it runs in real time and not on the reinforcement signal eventually received.

It is expected that further analysis along these lines should help contribute to the development and understanding of more sophisticated associative reinforcement learning algorithms designed for problems having an important temporal component, such as control problems with environmental feedback delays of unknown duration, problems involving recognition or production of time-varying signals, and problems in which the network contains internal feedback connections. An example of such an algorithm is the *adaptive heuristic critic* algorithm proposed by Sutton and colleagues

(Barto, Sutton & Anderson, 1983; Sutton, 1984) for certain control problems with delayed reinforcement. This algorithm combines features similar to the extended REINFORCE algorithm with additional adaptive behavior designed to provide predictions of eventual reinforcement throughout an episode, using an approach based on the method of *temporal differences* (Sutton, 1987).

DISCUSSION

Now that we have considered the details of the back-propagation and REINFORCE algorithms and observed that they have in common the fact that they perform statistical hill-climbing, it is worthwhile to gain an intuitive feel for just how these two algorithms work and why they are different. The use of back-propagation amounts to being able to reason from effects back to their causes. It thus requires that detailed information be available on the causal chains generating the network behavior. In contrast, REINFORCE algorithms deal with unknown chains of cause and effect by the technique of perturbing the actions of individual units in the net and observing the effects on the reinforcement signal by means of what is essentially a cross-correlation. It is thus clear that in any situation where either algorithm is applicable, back-propagation must necessarily· be more efficient because it takes advantage of knowledge of the network structure.

At the same time, it is interesting to note that a certain compatibility holds between back-propagation and REINFORCE algorithms. In particular, suppose given a network containing a mixture of stochastic and deterministic units. A gradient-estimating REINFORCE algorithm for this network can be obtained by using an appropriate mix of cross-correlation and back-propagation. One simply cross-correlates at stochastic units, using back-propagation through deterministic units to compute the eligibilities of the individual parameters. This technique works because back-propagation is an entirely general computational technique for computing partial derivatives whenever a particular variable is a deterministic differentiable function of other variables. In fact, this technique can be seen in the derivation of the algorithm for training a Bernoulli semilinear unit. One simply computes the eligibility of the parameter p_i and then uses the chain rule to compute the eligibility of w_{ij}, one of the parameters on which p_i deterministically depends. This use of the chain rule corresponds to the idea of back-propagating through the sub-units consisting of the "squasher" and the "summer" as depicted in Figure 4.

At another extreme, consider the semilinear unit depicted in Figure 3 and suppose that all the individual "synaptic junctions" (each containing its individual weight w_{ij}) are to be treated as individual stochastic sub-units. We can write down a REINFORCE algorithm for networks of these larger units once we specify the manner in which the stochastic behavior is generated. Since the randomly generated weights are not influenced by any deterministic computation occurring elsewhere in the net, there is no back-propagation (i.e., no use of the chain rule) required. This gives rise to an algorithm based on random weight variations which is similar to one studied by

Hinton (personal communication, 1986). The interesting point is that this idea of having the weights rather than the output be randomly generated can simply be considered a special case of this general approach. Presumably, correlating weight variations with the reinforcement signal is less efficient than correlating unit output variations with the reinforcement signal and back-propagating this information to the weights since the latter technique takes advantage of the knowledge of how weight variations influence the output of the unit while the former technique does not.

Finally, it is important to note that REINFORCE algorithms are certainly not the only learning algorithms that perform statistical hill-climbing for the associative reinforcement learning problem. Other more complex algorithms having this property are certainly possible and will very likely prove to be useful for such problems. At this point much of the value of their introduction has been to contribute a hill-climbing perspective to the associative reinforcement learning problem and provide some new insights on existing algorithms.

SUMMARY

We have seen how posing the learning problem faced by a connectionist network as an optimization problem whose solution is to be obtained by gradient following leads to useful and surprisingly effective algorithms for supervised learning as well as promising and theoretically interesting algorithms for associative reinforcement learning. Future work is needed to discover ways to combine the two approaches when appropriate in order to obtain algorithms having the best properties of both types of algorithm described here. In particular, it would be valuable to develop algorithms approaching the superior speed of back-propagation but also having the simplicity of the on-line implementation of the extended REINFORCE algorithm. The ability to incorporate more sophisticated features into the architecture, such as adaptive predictors (Sutton, 1984, 1987) or models (Sutton & Pinette, 1985) may play an important role. It is hoped that this general approach will bring us closer to finding answers to the two questions posed at the beginning of this article.

REFERENCES

Ackley, D. H., Hinton, G. H., & Sejnowski, T. J. (1985). A learning algorithm for Boltzmann machines. *Cognitive Science, 9*, 147-169.

Almeida, L. B. (1987). A learning rule for asynchronous perceptrons with feedback in a combinatorial environment. *Proceedings of the IEEE First International Conference on Neural Networks, San Diego, CA.*

Anderson, C. W. (1986). *Learning and problem solving with multilayer connectionist systems.* Ph.D. Dissertation, Dept. of Computer and Information Science, University of Massachusetts, Amherst, MA.

Barto, A. G. (1985). Learning by statistical cooperation of self-interested neuron-like computing elements. *Human Neurobiology, 4*, 229-256.

Barto, A. G. & Anandan, P. (1985). Pattern recognizing stochastic learning automata. *IEEE Transactions on Systems, Man, and Cybernetics, 15*, 360-374.

Barto, A. G. & Anderson, C. W. (1985). Structural learning in connectionist systems. *Proceedings of the Seventh Annual Conference of the Cognitive Science Society*, Irvine, CA, 43-53.

Barto, A. G., Sutton, R. S., & Anderson, C. W. (1983). Neuronlike elements that can solve difficult learning control problems. *IEEE Transactions on Systems, Man, and Cybernetics, 13*, 835-846.

Barto, A. G., Sutton, R. S., & Brouwer, P. S. (1981). Associative search network: a reinforcement learning associative memory. *Biological Cybernetics, 40*, 201-211.

Burr, D. J. (1986). A neural network digit recognizer. *Proc. IEEE Conference on Systems, Man, and Cybernetics*, Atlanta, GA, 1621-1625.

Elman, J. L. & Zipser, D. (1987). *Learning the hidden structure of speech.* (Tech. Rep. 8701). University of California, San Diego, Institute for Cognitive Science.

Feldman, J. A. (Ed.) (1985). Special issue on connectionist models and their applications. *Cognitive Science, 9*.

Grossberg, S. (1986). Adaptive pattern classification and universal recoding: I. Parallel development and coding of neural feature detectors. *Biological Cybernetics 23*, 121-134.

Hinton, G. E. & Anderson, J. A. (Eds.) (1981). *Parallel models of associative memory*, Hillsdale, NJ: Erlbaum.

Hinton, G. E. & Sejnowski, T. J. (1983). Analyzing cooperative computation. *Proceedings of the Fifth Annual Conference of the Cognitive Science Society*, Rochester, NY.

Hopfield, J. J. (1982). Neural networks and physical systems with emergent collective computational abilities. *Proceedings of the National Academy of Sciences, 79*, 2554-2558.

Hopfield, J. J. & Tank, D. W. (1985). Neural computation of decisions in optimization problems. *Biological Cybernetics, 52*, 141-152.

Kohonen, T. (1984). *Self-organization and associative memory.* New York: Springer.

Lakshmivarahan, S. (1981). *Learning algorithms theory and applications.* New York: Springer.

Lapedes, A. & Farber, R. (1986). A self-optimizing, nonsymmetrical neural net for content addressable memory and pattern recognition. *Physica 22D*, 247-259.

McClelland, J. L. & Rumelhart, D. E. (Eds.) (1986). *Parallel Distributed Processing: Explorations in the Microstructure of Cognition. Vol. 2: Psychological and Biological Models.* Cambridge: MIT Press/Bradford Books.

Minsky, M. L. & Papert, S. (1969). *Perceptrons: an introduction to computational geometry.* Cambridge: MIT Press.

Narendra, K. S. & Thathatchar, M. A. L. (1974). Learning automata—a survey. *IEEE Transactions on Systems, Man, and Cybernetics, 4*, 323-334.

Narendra, K. S. & Thathatchar, M. A. L. (To appear). *Introduction to learning automata*. Englewood Cliffs, NJ: Prentice Hall.

Narendra, K. S. & Wheeler, R. M., Jr. (1983). An n-player sequential stochastic game with identical payoffs. *IEEE Transactions on Systems, Man, and Cybernetics, 13*, 1154-1158.

Newell, A. (1980). Physical symbol systems. *Cognitive Science 4*, 135-183.

Parker, D. B. (1982). Learning logic. (Invention Report S81-64, File 1). Stanford, CA: Office of Technology Licensing, Stanford University.

Parker, D. B. (1985). Learning-logic. (Tech. Rep. 47). Cambridge, MA: MIT Center for Computational Research in Economics and Management Science.

Pineda, F. J. (1987). Generalization of backpropagation to recurrent neural networks. (Tech. Memo S1A-63-87). Applied Physics Laboratory, Johns Hopkins University, Baltimore, MD.

Plaut, D. C., Nowlan, S. J., & Hinton, G. E. (1986). *Experiments on learning by back propagation*. (Tech. Rep. CMU-CS-86-126). Department of Computer Science, Carnegie-Mellon University.

Rohwer, R. & Forrest, B. (1987). Training time-dependence in neural networks. *Proceedings of the IEEE First International Conference on Neural Networks*, San Diego, CA.

Rosenblatt, F. (1962). *Principles of neurodynamics*. New York: Spartan.

Rumelhart, D. E., Hinton, G. E., & Williams, R. J. (1986). Learning internal representations by error propagation. In: Rumelhart, D. E. & McClelland, J. L. (Eds.) *Parallel Distributed Processing: Explorations in the Microstructure of Cognition. Vol. 1: Foundations.* Cambridge: MIT Press/Bradford Books,

Rumelhart, D. E. & McClelland, J. L. (Eds.) (1986). *Parallel Distributed Processing: Explorations in the Microstructure of Cognition. Vol. 1: Foundations.* Cambridge: MIT Press/Bradford Books.

Rumelhart, D. E. & Zipser, D. (1985). Competitive learning. *Cognitive Science 9*, 75-112.

Sejnowski, T. J. & Rosenberg, C. R. (1986). *NETtalk: a parallel network that learns to read aloud*. (Tech. Rept. EECS-8601). Baltimore: Johns Hopkins University, Department of Electrical Engineering and Computer Science.

Sutton, R. S. (1984). *Temporal credit assignment in reinforcement learning*. Ph.D. Dissertation (available as COINS Tech. Rept. 84-02), Dept. of Computer and Information Science, University of Massachusetts, Amherst, MA.

Sutton, R. S. (1987). *Learning to predict by the methods of temporal differences*. (Tech. Rept. TR87-509.1). GTE Labs, Waltham, MA.

Sutton, R. S. & Pinette, B. (1985). The learning of world models by connectionist networks. *Proceedings of the Seventh Annual Conference of the Cognitive Science Society*, Irvine, CA, 54-64.

Watrous, R. L. & Shastri, L. (1986). Learning phonetic features using connectionist networks: an experiment in speech recognition. (Tech. Rep. MS-CIS-86-78). University of Pennsylvania.

Werbos, P. J. (1974). Beyond regression: new tools for prediction and analysis in the behavioral sciences. Unpublished doctoral dissertation, Harvard University, Cambridge, MA.

Wheeler, R. M. & Narendra, K. S. (1986). Decentralized learning in finite Markov chains. *IEEE Transactions on Automatic Control, 31,* 519-526.

Widrow, B. & Hoff, M. E. (1960). Adaptive switching circuits. *IRE WESCON Convention Record, Part IV,* 96-104.

Widrow, B. & Stearns, S. D. (1985) *Adaptive Signal Processing,* Englewood Cliffs, NJ: Prentice-Hall.

Williams, R. J. (1986). *Reinforcement learning in connectionist networks: a mathematical analysis.* (Tech. Rept. 8605). La Jolla: University of California, San Diego, Institute for Cognitive Science.

Williams, R. J. (1987). *Reinforcement-learning connectionist systems* (Tech. Rept. NU-CCS-87-3). College of Computer Science, Northeastern University, Boston, MA.

Efficient Stochastic Gradient Learning Algorithm for Neural Network

Y. C. Lee

University of Maryland Institute for Advanced Computer Studies

and

Department of Physics and Astronomy
University of Maryland
College Park, MD 20742

Center for Nonlinear studies
Los Alamos National Laboratory
Los Alamos, NM 87545

December 14, 1987

Abstract

Efficient first and second order adaptive learning algorithms of stochastic gradient descent variety are described. Both algorithms can automatically adjust step sizes to achieve optimum convergence rates. Various theorems concerning the convergence properties of these algorithms are discussed.

1 Introduction

Despite much promise, progress in connectionist networks has been slowed by rather inefficient learning algorithms which, in effect, force researchers in this field to confine themselves to mostly small scale "toy" problems. The "slow" learning problem is particularly severe for a class of feedforward connectionist networks[1] that have intermediate or "hidden" layers of nonlinear neuron-like elements. The main difficulty associated with such network systems is in the apportionment of credit for the hidden units since the latter are not directly connected with input or output and therefore must somehow construct internal representations in response to the external signals. Present learning procedures for neural networks invariably are different realizations of steepest descent algorithms in the space of "connection weights". Given a performance measure $J(w)$, where w denotes the weight vector, the steepest-descent strategy is simply

$$\Delta w = -\alpha \nabla J(w). \tag{1}$$

where $\Delta w = w_{n+1} - w_n$, n denotes the time-step, α is a positive "learning" parameter, and $\nabla J(w)$ is just the gradient vector of J with respect to the weight vector.

2 Learning as Stochastic Gradient Descents

Present learning procedures are not truly gradient descent procedures since usually $J(w)$ is defined over the entire training set whereas the gradient used in those training procedures is based on a single training vector in the training set. Therefore practical procedures are only probabilitic approximations of true steepest-descent procedures. The hope here is that, given small enough step size (i.e., sufficiently small α), the individual gradient vectors will be summed to approximate the true gradient descent direction. Denoting $J(w, t_n)$ the partial performance measure with respect to the training vector t_n, the total performance measure can be expressed as

$$J(w|T) = \frac{1}{N} \sum_{i=1}^{N} J(w, t_i), \tag{2}$$

where N is the total number of training vectors in the training set $T \subseteq S$ and S is the possibly nondenumerable set of all input-output vectors. Clearly if T_i were picked at random, then as $N \to \infty$, we expect

$$\lim_{N \to \infty} J(w|T_N) = E\{J|w\} \tag{3}$$

where $E\{J|w\}$ denotes the expectation value of $J(w, t)$. The current learning procedures are generally of the form

$$w_{n+1} = w_n - \alpha_n \nabla J(w_n, t_n), \tag{4}$$

where $\nabla J(w, t_n)$ again denotes gradient with respect to the weight vector W only. For cases that $J(w, t)$ is a quadratic function of W, (4) is known as the "stochastic approximaton" or "Robbins-Monro" procedure[2]. One of the most important theorem concerning the convergence of this procedure is that, with relatively mild conditions on the statistics of $J(w, t)$, and with α_n chosen such that

$$\sum_{n=1}^{\infty} \alpha_n < \infty$$

$$\sum_{n=1}^{\infty} \alpha_n^2 = \infty, \tag{5}$$

then w_n converges to the true optimun w^* in the mean as well as in mean square with probability 1.

The existence of a convergence alone, however, does not guarantee the efficiency of the stochastic learning procedure (4). On the contrary, since Eq. (5) implies that α_n has to decrease sufficiently rapidly with n, the learning rate may decrease to zero too fast for large n, possibly making (4) a slow learning procedure. Most current learning procedures use constant $\alpha_n = \alpha$, independent of n, which in general only can guarantee convergence in the mean for sufficiently small values of α. However, this does not guarantee that the iteration will converge to the local minimum. This can be seen from the following theorems.

Definition 1. We say the vector v is "downhill" for the function $J(w)$ at the point w if

$$V^T \cdot \nabla J(w) < 0. \tag{6}$$

Definition 2. A stochastic vector V is said to be "downhill in the mean" for the stochastic function J at w if

$$E\{V^T\} \cdot \nabla E\{J(w)\} < 0. \tag{7}$$

Proposition 1. $E\{\Delta w_n | w_n\} = -\alpha \nabla E\{J|w_n\}$ where $\Delta w_n \equiv w_{n+1} - w_n$. (8)

(8) follows directly from (4) with $\alpha = \alpha_n$. Note also that

$$E\{J|w_n\} = \sum_{\xi \epsilon T} E\{J|w_n, t = \xi\} P_r\{t = \xi\}. \tag{9}$$

Lemma 1. The stochastic vector Δw_n is "downhill in the mean" mean at w_n for $J(w_n, t_n)$ provided $\nabla E\{J|w_n\} \neq 0$.

Proof.

$$E\{J|w_n\} \cdot \nabla E\{J|w_n\} = -\alpha \|\nabla E\{J|w_n\}\|^2 < 0 \qquad (10)$$

Note here that the property of "downhill in the mean" does not necessarily imply convergence even in the unimodal case. In fact, the "convergence in the mean", whose definitions are given below, is in general not assured.

Definition 3. The sequence of vectors W_n is said to "convergence in the mean" in the weak sense if

$$\lim_{n \to \infty} \|\nabla E\{J(w_n)\}\| = 0 \qquad (11)$$

Definition 4. The sequence of vectors w_n $n = 1, 2, ...,$ is said to be "convergent in the mean" in the strong sense if

$$\lim_{n \to \infty} E\{w_n\} = w^*, \qquad (12)$$

where $\nabla E\{J(w^*)\} = 0$.
The following proposition is clearly true.

Proposition 2. (12) implies (11).

3   Convergence Theorems for First Order Schemes

The following theorems establish conditions for which convergence can be obtained.

Theorem 2.
Let w^* be the unimodal minima of $E\{J(w, t)|w\}$ in a bounded region R, and let $A(w) \equiv E\{\nabla\nabla^T J(w, t)|w\}$ be the Hessian matrix which is bounded and positive definite in R and let $r(w, t)$ and d be defined by
(i) $r(w, t) = \nabla J(w, t) - (w - w^*) \cdot \nabla\nabla^T J(w, t) - \nabla J(w^*, t)$,
(ii) $d = Inf_{w \in R}\{\lambda[A(w)]\}$, where $\lambda[A(w)]$ is the minimum eigenvalue for $A(w)$.
Assume also there exists a number M such that

(iii) $||r(w,t)|| \leq ||w - w^*||^2 M \quad \forall w \epsilon R$, $\forall t \epsilon T$.

Finally, assume that $t_i's$ are statistically independent: Then the sequence of vectors W_n is convergent in the mean in the strong sense for sufficient; small values of α and δ where $||w_n - w^*|| < \delta$.

Proof.

From (4) and (i), we have

$$
\begin{aligned}
E\{w_{n+1} - w^*\} &= E\{w_n - w^*\} - \alpha E\{w_n - w^*\} \cdot \nabla\nabla^T E\{J(w^*, t_n)\} + \alpha E\{r(w_n, t_n)\} \\
&= E\{w_n - w^*\} - [I - \alpha A(w^*)] + \alpha E\{r(w_n, t_n)\}, \quad (13)
\end{aligned}
$$

where use has been made of the statistical independence between w_n and $J(w^*, t_n)$. By (ii) and (iii),

$$
\begin{aligned}
||E\{w_{n+1} - w^*\}|| &\leq ||E\{w_n - w^*\}||[1 - \alpha d + \alpha||E\{w_n - w^*\}||M] \\
&\leq ||E\{w_n - w^*\}||[1 - \alpha(d - \delta)]. \quad (14)
\end{aligned}
$$

Clearly the sequence $||E\{w_{n+1} - w^*\}||$ will converge to zero if

$$
d > \delta , \quad \alpha < \frac{1}{d - \delta}, \quad (15)
$$

and this concludes the proof.

Theorem 3. If there exists a value w^* in a bounded region R, for which $\nabla J(w^*, t) = 0$ identically for all $t \epsilon T$, and if the mean Hessian matrix $A(w)$ is positive definite and bounded at both ends for all $w \epsilon R$, then there exists a δ independent of n such that $\forall \alpha \leq \delta$ and for $||w_n - w^*|| < \delta$ for some n the sequence of vectors w_n converges to w^* with probability 1.

Proof.

By the mean value theorem

$$
\nabla J(w, t) = (w - w^*) \cdot \overline{\nabla\nabla^T J} \equiv (w - w^*) \cdot H(w, w^*, t) \quad (16)
$$

where the overbar denotes the mean values of the arguments of the components of $\nabla\nabla^T J$. Hence

$$
\begin{aligned}
w_{n+1} - w^* &= w_n - w^* - \alpha(w_n - w^*) \cdot H(w_n, w^*, t_n) \\
&= (w_n - w^*) \cdot (1 - \alpha H_n). \quad (17)
\end{aligned}
$$

Repeat applications of Campbell-Hausdorf-Baker theorem leads to

$$w_{n+1} - w^* = (w_1 - w^*) \cdot e^{\alpha \sum_{j=1}^n H_j + \alpha^2 Q_n}, \tag{18}$$

where $\|Q_n\| \leq nN$ and $H_j = H(w_j, w^*, t_j)$.

Choose a sufficiently large m such that, for some $0 < \epsilon < 1$,

$$P_r\left\{\|\frac{1}{m}\sum_{j=1}^m H(w, w^*, t_{j+\bar{n}}) - A(w, w^*)\| > \frac{3\sqrt{\sigma_H}}{\sqrt{m}}\right\} < \epsilon, \tag{19}$$

where $\bar{n}$ is arbitrary, and $A(w, w^*) = E\{H(w, w^*, t)|w, w^*\}$, and σ_H is the standard deviation for $H(w, w^*, t)$, and central limit theorems have been used in arriving at (19), and for

$$\alpha \ll \frac{1}{m}\frac{1}{N\sigma_H + A}, \quad A = sup_{w,w^* \epsilon R}\|A(w, w^*)\|, \tag{20}$$

we have, with probability $1 - t$,

$$W_{n+m} - w^* = (w_{n+1} - w^*) \cdot e^{-m\alpha A(w_n, w^*) + \alpha^2 Q_{n,m}}, \tag{21}$$

where $\|Q_{n,m}\| \leq m^2 N$. Application of the mean value theorem leads to the fact that $A(w, w^*)$ is positive definite and bounded both from above and from below, thus $\exists d$ such that

$$d = Inf_{w,w^* \epsilon R}\{\lambda[A(w, w^*)]\} > 0, \tag{22}$$

where $\lambda[A(w, w^*)]$ is the minimum eigenvalues for $A[w, w^*]$. Note also that in deriving (21), we have taken advantage of the fact that $\|W_{n+\ell} - w^*\| = O(m)$ for $m \geq \ell \geq 1$ and the mean value theorem to replace $A(W_{n+\ell}, w^*)$ by $((W_n, w^*)$. By selecting an α such that $\alpha \ll d/mN$ is also satisfied, it follows that

$$\|(W_{n+1} - w^*) \cdot e^{-m\alpha A(w_n, w^*) + \alpha^2 Q_{nm}}\| \leq e^{-\alpha dm + \alpha^2 m^2 N}\|w_{n+1} - w^*\|, \tag{23}$$

with probability $1 - \epsilon$. When (19) is not true (with probability ϵ), it is nevertheless still true from (18) that

$$\|w_{n+m} - w^*\| \leq \|w_{n+1} - w^*\|e^{\alpha^2 mN}. \tag{24}$$

Now the probability that in 2p+1 successive tries, (19) is not true for more than p times can be shown to be smaller than $\epsilon[4\epsilon(1-\epsilon)]^p$. By letting $\epsilon \le 1/4$, it follows that

$$||w_{n+(wp+1)m} - w^*|| \le ||w_{n+1} - w^*||e^{-\alpha p(d-2\alpha mN)m}, \qquad (25)$$

with probability $> 1 - (3/4)^P \epsilon$. Hence

$$\lim_{p \to \infty} ||w_{n+(2p+1)m} - w^*|| = 0, \qquad (26)$$

with probability 1. It is not hard to see, since m is arbitrary provided (18) is satisfied, in fact that

$$\lim_{n \to \infty} ||w_n - w^*|| = 0, \qquad (27)$$

with probability one.

Although theorem 3 is a strong result, it nevertheless requires the existence of w^* such that $\nabla J(w^*, t) = 0$ for all t, a condition which usually is not satisfied in practice. We conjecture that for sufficiently small values of $\nabla J(w^*, t)$ for all t, $E\{||w_n - w^*||^2\}$ is bounded asymptotically. In particular, we expect that $E\{||w_n - w^*||^2\}$ should scale like α/d as $n \to \infty$, meaning that w_n should approach w^* as $\alpha \to 0$. The proof of this conjecture is still in progress.

The rather weak convergence properties displayed by the stochastic gradient method, while discouraging, nevertheless actually may turn out to be a blessing in disguise. The reason is that gradient descent techniques usually can only locate local minimum, on the other hand it can be argued that the local minima w^* which generates large residues $||\nabla J(w^*, t)||$ is not really a good minima after all. Since large residues tend to cause w_n to wander away from w^*, the effect being equivalent to that of a random perturbing force, therefore shallow minimum can usually be avoided with reasonably certainty, allowing good (which means deep) minimum to be effectively searched. What is more disconcerting is the fact that the convergent rate of the stochastic gradient method is unnervingly slow for most practical cases where the ratios of the maximum to the minimum eigenvalues for the Hessian matrices tend to be large, especially for large systems. The following theorems establish the optimal convergence rates for the gradient method. Note that it is usually sufficient to approximate the evaluation function $J(w,t)$ by a quadratic function around the optimal point since the higher order terms to not contribute to the asymptotic limit of the rates of convergence.

Theorem 4.

Assume that $E\{J(w,t)|w\} = const. + \frac{1}{2}(w - w^*)^T \cdot A \cdot (w - w^*)$, where A is a symmetric positive definite constant matrix, assume also that t_i's are statistically independent and let λ_n and λ_1 be the maximum and minimum eigenvalues of A, respectively, then the expected optimal convergence rate is linear and is given by

$$\frac{\lambda_n - \lambda_1}{\lambda_n + \lambda_1}. \tag{28}$$

Proof. From the statistical independence between w_n and $J(w^*, t_n)$ we have

$$E\{w_{n+1} - w^*\} = E\{w_n - w^*\} \cdot [I - \alpha A]. \tag{29}$$

Let the eigenvalues of A be $\lambda_1, ..., \lambda_N$; the associated eigenvectors $v_1, ..., v_n$ which are normalized, so that $v_i^T \cdot v_j = \delta_{ij}$. Let the eigenvalues be ordered so that $\lambda_1 \le \lambda_2 \le \lambda_3 \le \cdots \le \lambda_N$ and let $\lambda_1 \ne \lambda_n$ (the case $\lambda_1 = \lambda_n$ is trivial).

We now express

$$E\{w_n - w^*\} = \sum_j \alpha_n^j v_j, \tag{30}$$

$$E\{J(w_n, t)\} = \frac{1}{2} \sum_j (\alpha_n^j)^2 \lambda_n. \tag{31}$$

From (29)

$$\alpha_{n+1}^j = (1 - \lambda_j \alpha) \alpha_n^j, \quad j = 1, ..., N. \tag{32}$$

Asymptotically ($n \to \infty$), the convergence rate of $E\{w_n - w^*\}$ is determined by the slowest rates of convergence for α_n^j, unless the corresponding α_0^j happens to be zero (which is inprobable). Since the convergence rate for α_n^j is simply $|1 - \lambda_j \alpha|$, it follows that the slowest convergence rate is either $1 - \lambda_1 \alpha$ with $0 < \alpha < 1/\lambda$, or $|| - \lambda_N \alpha| = \lambda_N \alpha - 1$ with $\alpha > 1/\lambda_N$. The choice

$$\alpha = \frac{2}{\lambda_1 + \lambda_N} \tag{33}$$

is optimal in the sense that

$$Max\{|1 - \frac{1}{\lambda_1 + \lambda_N}\lambda_1|, |1 - \frac{2}{\lambda_1 + \lambda_N}\lambda_N|\} = \frac{\lambda_N - \lambda_1}{\lambda_N + \lambda_1}$$
$$= Inf_{\alpha>0}Max\{|1 - \alpha\lambda_1|, |1 - \alpha\lambda_N|\}. \qquad (34)$$

Hence the convergence rate can be expected to be no better than $\frac{\lambda_N - \lambda_1}{\lambda_N + \lambda_1}$, so that the theorem has been proved.

The above theorem deals only with the rather weak property of "convergence in the mean". A somewhat stronger property of convergence exists for the small residue case where $\nabla J(w^*, t) \simeq 0 \forall t \epsilon T$, as is indicated by the following theorem:

Theorem 5. If there exists a w^* in R for which $\nabla J(w^*, t) = 0$ for all $t \epsilon T$ and the mean Hessian matrix $A(w)$ is positive definite and bounded in R, and the Hessian matrix $H(w, t)$ is positive semidefinite and bounded for all $t \epsilon T$, then there exist an $\epsilon : 0 < \epsilon \ll 1$ such that the optimal convergence rate is better than $1 - \frac{\epsilon}{2}\lambda_1/\lambda_N$, where λ_1 and λ_n are the minimum and maximum eigenvalues of A, respectively.

Proof.

It is sufficient to consider $J(w, t)$ to be a quadratic function of W, so that $A(w)$ could be treated to be independent of w. Again we have

$$w_{n+1} - w^* = (w_n - w^*) \cdot (I - \alpha H_n). \qquad (35)$$

Clearly, in order for $\|w_n - w^*\|$ to converge, it is necessary that $\|1 - \alpha H_n\|$ not be greater than 1 with sufficiently frequency. Hence α should be chosen such that $\alpha\lambda_{max}[H_n] < 2$ for most H_n, where $\lambda_{max}[B]$ denotes the maximum eigenvalue associated with the matrix B. As a conservative estimate, however, we will choose α to be so small that $\alpha\lambda_{max} < \epsilon$ for some $\epsilon \ll 1$, where $\lambda_{max} = sup_{H_n}\lambda_N[H_n]$. With such a α, we can again apply the Campbell-Baker-Haudorf theorem to obtain

$$w_{n+1} - w^* = (w_1 - w^*) \cdot e^{-\alpha \sum_{j=1}^{N} H_j + \alpha^2 Q_n}, \qquad (36)$$

where $\|Q_n\| \leq nN\lambda_{max}^2$. Using the same technique that was employed to prove theorem 3, we arrive at

$$w_{n+m} - w^* = (w_n - w^*) \cdot e^{-\alpha m A + \alpha^2 Q_{n,m}}, \qquad (37)$$

with high probability, where $\|Q_{n,m}\| \le m^2 N \lambda_{max}^2$. Let ϵ be sufficiently small that $\alpha^2 mN\lambda_{max}^2 < \epsilon^2 mN \ll 1$. Let $\lambda_1 \le \lambda_2 \le \lambda_3 \le \cdots \le \lambda_n$ be the eigenvalues of A; the associated eigenvector $v_1, ..., v_n$ satisfying $V_i^T \cdot V_j = \delta_{ij}$. Expanding $w - w^*$ in terms of v_i's we have

$$w - w^* = \sum_j \alpha^i v_j, \tag{38}$$

thus,

$$\|\alpha_{n+m}^j\| \le e^{-\frac{1}{2}\alpha m \lambda_j} \|\alpha_{n+1}^j\|, j = 1, ..., N. \tag{39}$$

Obviously the convergence will be the slowest for $j = 1$, which has an average convergence rate no smaller than $1 - \frac{1}{2}\alpha\lambda_1$. Since $\alpha\lambda_N < \epsilon$ from $\alpha\lambda_{max} < \epsilon$, we have shown that the convergence rate is no worse than $1 - \frac{1}{2}\epsilon\lambda_1/\lambda_N$. The average convergence rate above is determined by using the following definition

$$\text{Average convergence rate} = \frac{1}{m} log \frac{\|\alpha_{n+m}^j\|}{\|\alpha_{n+1}^j\|}. \tag{40}$$

Clearly the above choice of α is not optimal and the actual optimal convergence rate can be expected to be better than $1 - \frac{1}{2}\epsilon\lambda_1/\lambda_N$. However, the optimal convergence rate is not likely to be much better than that given in theorem 4, namely $1 - 2\lambda_1/\lambda_N$, the reason being that $1 - 2\lambda_1/\lambda_N$ is the optimal convergence rate (for large λ_N/λ_1) when $\nabla J(w, t)$ is not random ($E\{\nabla J(w,t)\} = \nabla J(w,t)$) and the addition of randomness can only make the convergence worse. Additionally, the value of α for which the rate of convergence is optimal clearly has to be larger than the α values used in Theorem 5. These are made precise by the following digression:
Lemma 2.
  The matrix

$$M_{n_1 n_2} = T \prod_{j=n_1}^{n_2} (1 - \alpha H_j) \tag{41}$$

where T is the time-ordering symbol which means $T \sum_{j=1}^{\ell} O_j = O_{\ell}O_{\ell-1}\cdots O_2 O_1$, is a monotonically decreasing function of α provided that $1 - \alpha H_j, j = n_1, ..., n_2$ are all positive.
Proof.

$$\frac{dM_{n_1 n_2}}{d\alpha} = -T \sum_{k=n_1}^{n_2} [\prod_{j=n_1}^{k-1} (I - \alpha H_j)] H_k [\prod_{i=k+1}^{n_2} (I - \alpha H_i)] < 0. \tag{42}$$

The above lemma leads, in particular, to the facts that both $\lambda_{max}[M_{n,n_2}]$ and $\lambda_{min}[M_{n,n_2}]$ decrease with α.

Theorem 6.

Let w_n satisfy (17), and $\alpha < 1/\lambda_{max}$, where $\lambda_{max} = sup_{H_n} \lambda_H[H_n]$, then the rate of convergence of w_n to w^* is a monotonically increasing function of α.

The proof of the above theorem follows straightforwardly from (18) and lemma 2.

Theorem 7.

Let (19) be true for w_n's, and in addition, let T_n's be statistically independent of one another, then the optimal rate of convergence in the mean square sense of w_n to w^* is no better than $\frac{\lambda_N - \lambda_1}{\lambda_N + \lambda_1} \simeq 1 - \frac{2\lambda_1}{\lambda_N}$.

Proof.

Let us scalar multiply (17) with itself to obtain

$$\|w_{n+1} - w^*\|^2 = (w_n - w^*)^T \cdot (I - \alpha H_n(t_n))^2 \cdot (w_n - w^*). \qquad (43)$$

Taking the conditional expectation of (43), we have

$$E\{\|w_{n+1} - w^*\|^2 | w_n\} = (w_n - w^*)^T \cdot E\{(I - \alpha H_n(t_n))^2\} \cdot (w_n - w^*). \qquad (44)$$

Now it is easy to see that

$$E\{(I - \alpha H_n)^2\} = I - 2\alpha E\{H_n\} + \alpha^2 E\{H_n^2\} \le I - 2\alpha E\{H_n\} + \alpha^2 (E\{H_n\})^2. \qquad (45)$$

where use has been made of the fact that $E\{M^2\} \ge (E\{M\})^2$. Noting that $E\{H_n\} = A$ and taking expectations on either side of (45), we get

$$E\{\|w_{n+1} - w^*\|^2\} \ge E\{(w_n - w^*)^T\} : (I - 2\alpha A + \alpha^2 A^2). \qquad (46)$$

The maximum and minimum eigenvalues of $I - 2\alpha A + \alpha^2 A^2 = (I - \alpha A)^2$ are $1 - \alpha \lambda_N$ and $1 - \alpha \lambda_1$, respectively. Expanding $w_n - w^*$ in terms of the eigenvectors V_j's:

$$w_n - w^* = \sum_{j=1}^{N} a_n^j v_j \qquad (47)$$

we find

$$E\{(\alpha_{n+1}^j)^2\} \geq (1 - \alpha\lambda_j)^2 E\{|a_n^j|^2\} \ , \ j = 1, 2, ..., N. \tag{48}$$

It is easy to show that

$$Inf_\alpha Max\{(1 - \alpha\lambda_j)^2 | j = 1, ..., n\} = (\frac{\lambda_N - \lambda_1}{\lambda_N + \lambda_1})^2, \tag{49}$$

which is reached for $\alpha = \frac{2\lambda_1}{\lambda_N + \lambda_1}$. Hence the rate of convergence of $\sum_j E\{|Q_n^j|^2\} = E\{||w_n - w^*||^2\}$ cannot exceed $(\frac{\lambda_N - \lambda_1}{\lambda_N + \lambda_1})^2$.

The above theorem can easily be extended to show that $E\{||w_n - w^*||^{2m}\}, m = 1, ..., \infty$, cannot converge faster than $(\frac{\lambda_N - \lambda_1}{\lambda_N + \lambda_1})^{2m}$.

From the preceding discussion we can conclude that the optimal convergence rate is somewhere between $1 - \frac{\epsilon}{2}\frac{\lambda_1}{\lambda_N}$ and $1 - \frac{2\lambda_1}{\lambda_N}$. From the construction of the proof to theorem 5, it is clear that $1 - \frac{\epsilon\lambda_1}{2\lambda_N}$ considerably underestimates the convergence rate. Hence it seems reasonable that the optimal convergence rate is much closer to $1 - \frac{2\lambda_1}{\lambda_N}$ which is the rate of convergence for $E\{w_n\}$.

4 Convergence of the Second Order Schemes

For the rest of the paper we will examine the possibility of employing higher order schemes to improve the convergence rates. For example, consider the following second order updating scheme for w_n's:

$$w_{n+1} = (1 + \beta)w_n - \alpha\nabla J(w_n, t_n) - \beta w_{n-1} \ , \ \alpha, \beta > 0 \tag{50}$$

Since we will be interested in the asymptotic convergence rates only, we can assume $J(w, t)$ to be quadratic in w without loss of generality. Assuming also that w^* is the optimal solution in the sense that $E\{\nabla J(w^*, t)\} = 0$, as well as the statistical independence of t_n's, we have

$$\begin{aligned} w_{n+1} - w^* = \ &(1 + \beta)(w_n - w^*) - \beta(w_{n-1} - w^*) - \alpha H_n \cdot (w_n - w^*) \\ &- \alpha\eta(w^*, t_n), \end{aligned} \tag{51}$$

where $\eta(w^*, t_n) = \nabla J(w^*, t_n)$ is the residue whose mean is zero $E\{\eta(w^*, t_n)\} = 0$. Taking expectations on both sides of Eq. (51), noting $E\{H_n \cdot w_n\} = E\{H_n\} \cdot E\{w_n\} = A \cdot E\{w_n\}$, we get

$$E\{w_{n+1} - w^*\} = (1+\beta)E\{w_n - w^*\} - \beta E\{w_{n-1} - w^*\} - \alpha A \cdot E\{w_n - w^*\}. \tag{52}$$

Theorem 8.

The optimal convergence rate for $E\{w_n\}$ is $\frac{\sqrt{\lambda_n}-\sqrt{\lambda_1}}{\sqrt{\lambda_N}+\sqrt{\lambda_1}}$.

Proof.

Following the procedure used in proving theorem 4, we expand $E\{w_n - w^*\}$ in V_j's, which gives

$$
\begin{aligned}
\alpha_{n+1}^j &= (1+\beta)\alpha_n^j - \beta\alpha_{n-1}^j - \alpha\lambda_j\alpha_n^j \\
&= (1+\beta-\alpha\lambda_j)\alpha_n^j - \beta\alpha_{n-1}^j , \quad j = 1, ..., N.
\end{aligned}
\tag{53}
$$

For any given j, Eq. (53) is a finite difference equation in α_n^j with constant coefficients. Let

$$
\alpha_n^j = p^n j_0^j,
\tag{54}
$$

Eq. (53) becomes

$$
p^2 - (1+\beta-\alpha\lambda_j)p + \beta = 0,
\tag{55}
$$

which can be solved to give

$$
p_{j\pm} = \frac{1+\beta-\alpha\lambda_j}{2} \pm \frac{\sqrt{(1+\beta-\alpha\lambda_j)^2 - 4\beta}}{2}.
\tag{56}
$$

The optimum convergence rate Γ is clearly

$$
\Gamma = Inf_{\alpha,\beta}Max_{j\pm}\{|p_{j\pm}|\}.
\tag{57}
$$

Since $p_{j\pm}$ are either real numbers or complex conjugate pairs and $p_{j+} \cdot p_{j-} = \beta$, it follows that

$$
Max\{|P_{j\pm}|\} \geq \sqrt{\beta},
\tag{58}
$$

for real $p_{j\pm}$, and

$$
|p_{j\pm}| = \sqrt{\beta},
\tag{59}
$$

for complex $p_{j\pm}$. Let

$$
f_{\pm(\lambda)} = \frac{1+\beta-\alpha\lambda}{2} \pm \frac{1}{2}\sqrt{(1+\beta-\alpha\lambda)^2 - 4\beta},
\tag{60}
$$

so that $p_{j\pm} = f_{\pm}(\lambda_j)$. Since $\beta > 1$ leads to $Max\{|P_{j\pm}|\} > 1$ and the divergence of α_n^j's, we will consider the cae $\beta \leq 1$ only.

From (47), we find $F_+(0) = 1, f_-(0) = \beta < 1$. Also

$$\frac{f'_+(\lambda)}{f_+(\lambda)} = -\frac{\lambda}{\sqrt{1 + \beta - \alpha\lambda)^2 - 4\beta}}, \tag{61}$$

which is negative definite so long as $(1+\beta-\alpha\lambda)^2 > 4\beta$, hence $f_+(\lambda)$ decreases continuously until $\lambda = \lambda_{+c} = \frac{1}{\alpha}(1 + \beta - 2\sqrt{\beta})$ at which point $f_+ = f_-$ and $|f_\pm(\lambda)|$ becomes $\sqrt{\beta}$ independent of λ. After λ moves past $\lambda_{-c} = \frac{1}{\alpha}(1 + \beta + 2\sqrt{\beta})$, $f_-(\lambda)$ becomes real and less than $-\sqrt{\beta}$. Since

$$\frac{f'_-(\lambda)}{f_-(\lambda)} = \frac{\lambda}{\sqrt{(1 + \beta - \alpha\lambda)^2 - 4\beta}}, \tag{62}$$

it follows that $f_-(\lambda), \lambda > \lambda_{-c}$, is also a decreasing function of λ.

The above analysis indicates that

$$Max_{j\pm}\{|P_{j\pm}|\} = Max\{|P_{1\pm}|, |P_{N\pm}|\}, \tag{63}$$

where λ_1 and λ_N corresponds to the smallest and the largest eigenvalues of A, respectively. Consider now $\lambda = \lambda_1$. For fixed $\beta, f_+(\lambda) = 1$ at $\alpha = 0$ and decreases as α is increased until $\alpha = \alpha_+ = (1 + \beta - 2\sqrt{\beta})/\lambda_1$, at which point $f_+ = f_-$. Further increases of α does not cause any decrease of $|f_\pm(\lambda_1)|$ from the value $\sqrt{\beta}$. Similar consideration for $\lambda = \lambda_N$ shows that

$$Inf_\alpha Max\{|P_{1\pm}|, |P_{N\pm}|\} = \sqrt{\beta}, \quad \frac{1 + \beta - 2\sqrt{\beta}}{\lambda_1} \leq \frac{1 + \beta + 2\sqrt{\beta}}{\lambda_N}. \tag{64}$$

(64) is satisfied only for $\beta \geq \beta_c = (\frac{\sqrt{\lambda_N} - \sqrt{\lambda_1}}{\sqrt{\lambda_N} + \sqrt{\lambda_1}})^2$. If we lower β below β_c, the infimum is no longer determined by the complex roots of (55). Since

$$\frac{\partial}{\partial\beta}f_t = \frac{f_+ - 1}{\sqrt{1 + \beta - \alpha\lambda)^2 - 4\beta}}, \quad \frac{\partial}{\partial\beta}f_- = -\frac{f_- 1}{\sqrt{(1 + \beta - \alpha\lambda)^2 - 4\beta}}, \tag{65}$$

it is not hard to see that the infinum increases as β decreases. Thus it follows that

$$Inf_{\alpha,\beta}Max\{|P_{1\pm}|, |P_{N\pm}|\} = \frac{\sqrt{\lambda_N} - \sqrt{\lambda_1}}{\sqrt{\lambda_N} + \sqrt{\lambda_1}}, \tag{66}$$

which is reached at $\alpha = \alpha_c = \frac{4}{(\sqrt{\lambda_N} + \sqrt{\lambda_1})^2}$ and $\beta = \beta_c$. This proves the theorem.

Theorem 8 is very interesting. It tells us that by choosing α and β to be α_c and β_c, respectively, the convergence rate of the second order update scheme (50) near a quadratic minimum of $E\{J(w, t)\}$ is $1 - 2\sqrt{\frac{\lambda_1}{\lambda_N}}$; which for very large values of λ_N/λ_1, is a vast improvement of the first order gradient scheme (4). Theorem 8 by itself gives little indication of the actual convergence rate of the scheme because it only addresses the question of convergence of the expected values. In the case of zero residues $\nabla J(w^*, t) = 0$, it is possible to show, in a manner completely analogous to that used in proving theorem 5, that the optimal convergence rate cannot be much more than $1 - \frac{\sqrt{\lambda_1}}{\sqrt{\lambda_N}}$. This fact will be stated as a theorem without proof as follows:
Theorem 9.

Assume the existence of a w^* for which $\nabla J(w^*, t) = 0$ for all $t \epsilon T$, and the expected Hessian matrix A is bounded from both sides as well as being positive definite, then there exists a $\delta, 0 < \delta \ll 1$ such that the optimal convergence rate of the second order scheme near the quadratic minimum is no worse than $1 - \delta \frac{\sqrt{\lambda_1}}{\sqrt{\lambda_N}}$.

It would be nice if an analogous theorem to theorem 7 can be demonstrated for the second order scheme. For the time being, however we have not yet been able to demonstrate such a theorem. As far as the convergence of the expected values is concerned, we are nearing the completion of the proof of the following surprising extention to theorem 8 to arbitrary order:

The optimal rate of convergence in the mean sense of the ℓth order extention of the second order scheme (50) is $\frac{1-\delta}{1+\delta}$, where

$$\delta = (\frac{\lambda_1}{\lambda_N})^{\frac{1}{\ell}}, \quad \ell = 3, 4, ... \tag{67}$$

Thus even higher convergence rate can be expected of the higher order schemes. In practice, though, it is probably not worth the trouble of going beyond the third order scheme because of the greatly increased complexity associated with such schemes.

It should be noted that the second order stochatic gradient scheme (50) has already been proposed by Rumelhart, Hinton, and Williams[1], who call it the "momentum" method. However, the α and β in their test examples were chosen in a completely ad hoc manner. What we have shown above is that drastic improvement in the convergence property is indeed possible with such a scheme provided that a suitable choice of α and β are made. This brings up the question of whether the optimum choice of α and β can be determined in practice. On first sight that seems improbable since both α_c and β_c depends on the ratio λ_N/λ_1 which is not known a priori. Fortunately it is not necessary to determine λ_1 and λ_N explicitly. Instead, α and β can be adaptively adjusted in order to optimize the convergence rate. For example, α can first take a large value (β should always be taken to be below 1) until repeated successive overshooting is observed (which indicates that $\alpha\lambda_N > 1$), then α is reduced until overshooting begins to abate. Once α is fixed this way, the optimal value of β could be estimated by a one-dimensional minimization algorithm using an approximate rate of convergence obtained from the ratio of consecutive norms of temporally, weighted averages of $\nabla J(w_n, t_n)$. Such kind of empirical schemes are certainly worth testing through numerical experimentation to determine their viability. However, for now they can't be mathematically justified. What we will present in the following is an adaptive first order scheme where some partial mathematical justification is possible.

Without loss of generality let us look only at cases where $\nabla J(w, t) = H(t)w + \eta(t)$, meaning that $J(w, t)$ is strictly a positive-semidefinite quadratic function of w. We shall further stipulate that the ratio λ_1/λ_N, where λ_1 and λ_N are respectively the smallest and the largest eigenvalues of the expectation matrix $A = E\{H(t)\}$, to be a very small number. The latter condition is motivated by the consideration that the convergence rates for moderate values of λ_1/λ_N can easily be improved by more conventional techniques. To be specific, we will assume that $\lambda_1/\lambda_N \ll \frac{1}{N}$.

Recalling that in the first order scheme with $sup\{\lambda_{max}[H_n]\} < \infty$, the w_n's converge (in the mean) to w^* in such a way that $E\{w_n - w^*\}$ is dominated asymptotically the least slowly damped eigenvector of A provided $\alpha < 2/\lambda_{max}[H]$, where $\lambda_{max}[H] = sup_n\{\lambda_{max}[H_n]\}$. Now since the least damped eigenvectors are either those associated with λ_1 or those associated with λ_N, it follows that the asymptotically only those eigenvectors survive.

Lemma 3.

$$\lim_{n \to \infty} \frac{\|E\{w_{n+2} - w_n\}\|}{\|E\{w_n - w^*\}\|} < \frac{\lambda_1}{\lambda_N}. \tag{68}$$

Lemma 3 takes into account of the fact that for α close to $2/\lambda_{Max}[H]$, the eigencomponents of $w_n - w^*$ with large eigenvalues tend to oscillate with period 2 since $1 - \alpha\lambda_N$ is very nearly -1, such oscillations can be rendered invisible by looking only at time 2 maps. Thus we can conclude from lemma 3 that the sequences $\{E\{w_{2n}\}|n = 1, 2, ..., \}$ and $\{E\{w_{2n+1}\}|n = 0, 1, ...\}$ are both slowly convergent for large n's. The variances of w_n's, however, are much harder to control. Clearly when α is close to $2\lambda_{max}[H]$, the variance of w_n has to be of order unity. One might naively think that an exponentially-weighted average of w_n's of the following form

$$u_n = \delta \sum_{\ell=0}^{n} (1 - \delta)^{\ell} w_{n-\ell}, \tag{69}$$

where $| \gg \delta \gg \lambda_1/\lambda_N$, might reduce the variance a bit as long as t_i's are statistically independent. That this could not be the case can be seen from the following lemma

Lemma 4.

$$E\{\|u_n - E\{w_n\}\|^2\} = 2\delta^2 \sum_{\ell_2 > \ell_1 = 0}^{N} (1 - \delta)^{\ell_1 + \ell_2} T_r[\prod_{j=1}^{\ell_2 - \ell_1} (1 - \alpha_{n-\ell_1-j} A) E\{w_{n-\ell_2} w_{n-\ell_2}^T\}]$$

$$- 2\delta^2 \sum_{\ell_2 > \ell_1 = 0}^{n} (1 - \delta)^{\ell_1 + \ell_2} \sum_{i=1}^{\ell_2 - \ell_1} \alpha_{n-\ell_1-j} \{[(\prod_{j=1}^{\ell_2 - \ell_1} (1 - \alpha_{n-\ell_1-j} A))]\}^T E\{w_{n-\ell_2}\}$$

$$- (E\{w_n\})^2. \tag{70}$$

Proof. Iterating $w_n = w_{n-1} - \alpha_{n-1}(H_{n-1} \cdot w_{n-1} + \eta_{n-1})$ to show that

$$w_n = [T \prod_{j=1}^{\ell} (1 - \alpha_{n-j})] \cdot w_{n-\ell} - \sum_{i=1}^{\ell} \alpha_{n-j}[T \prod_{j=1}^{i-1} (1 - \alpha_{n-j} H_{n-j})]\eta_{n-i} \tag{71}$$

where T is the time-ordering symbol which means $T \prod_{j=1}^{\ell} O_j = O_{\ell} O_{\ell-1} \cdots O_2 O_1$, for any sequence of operators O_j's. Equation (70) follows from (71) by direct substitution.

From lemma 4, and the fact that $\|1 - \alpha_j A\|_{\infty}$ with uniform norm is much closer to unity then $(1 - \delta)$, we see that $E\{\|u_n - E\{w_n\}\|^2\}$ is essentially $E\{w_n^2\} - (E\{w_n\})^2$, indicating no reduction of variance!

Even though no variance reduction is expected for u_n, nevertheless time-averaging greatly reduces fluctuations between successive iterations as can be seen from the fact that Eq. (69) is equivalent to the following equation

$$u_n = (1 - \delta)u_{n-1} + \delta w_n, \tag{72}$$

together with the initial condition that $u_0 = 0$. Clearly $||u_n - u_{n-1}||$ is of the order of δ according to Eq. (72). Furthermore, the variance associated with small-eigenvalue eigenvectors (noticably λ_1-eigenvectors) is greatly reduced by the averaging procedure since for them $|\alpha_n \lambda| \ll 1$, therefore, the time-average for them is equivalent to ensemble average.

We now propose the following adaptive update scheme for α_n's in order to ensure optimal asymptotic convergence for the first order scheme denoting $v_n = H_n \cdot w_n + \eta_n$,

$$\alpha_{n+1} = \alpha_n + p_n[v_{3n}^T \cdot v_{3n+1} - \alpha_n v_{3n}^T \cdot H_{3n+2} v_{3n+1}]. \tag{73}$$

where ρ_n's are positive real numbers satisfying

$$\sum_{n=1}^{\infty} \rho_n = \infty, \tag{74}$$

$$\sum_{n=1}^{\infty} \rho_n^2 < \infty, \tag{75}$$

and the term $H_{3n+2} \cdot (H_{3n+1} \cdot w_{3n+1} + \eta_{3n+1})$ is obtained by the following operation

$$\frac{1}{\epsilon}[\nabla J(w_n + \epsilon H_n \cdot w_n + \epsilon \eta_n, \, t_{n+1}) - \nabla J(w_n, t_{n+1})] =$$
$$\frac{1}{\epsilon}H_{n+1} \cdot [w_n + \epsilon H_n \cdot w_n + \epsilon \eta_n] + \frac{1}{\epsilon}\eta_{n+1} - \frac{1}{\epsilon}[H_{n+1} \cdot w_n + \eta_{n+1}]. \tag{76}$$

Noting the staggered arrangement in (73), the following lemma is readily verified.

Lemma 5.

$$E\{\alpha_{n+1}\} = E\{\alpha_n\} + \rho_n[E\{v_{3n}^T \cdot v_{3n+1}\} - E\{\alpha_n\}Tr(A \cdot E\{v_{3n+1}v_{3n}^T\})]. \tag{77}$$

The assumption about the smallness of λ_1/λ_N together with lemma 3 allows us to treat $E\{v_{2n}^T \cdot v_{2n}\}$ and $E\{v_{3n}v_{3n+1}^T\}$ to be essentially constant quantities, independent of n for the purpose of studying the convergence of the update scheme (73). This leads us to the following lemma:

Lemma 6.
The sequence x_n's define by the recurrence relation

$$x_{n+1} = x_n + \rho_n c - \rho_n f x_n, \tag{78}$$

where f is a positive real number and ρ_n satisfies the conditions (60a,b), converges absolutely to c/f.

Proof.
Define $\tilde{x}_n = x_n - c/f$, we have

$$\tilde{x}_{n+1} = \tilde{x}_n(1 - \rho_n f). \tag{79}$$

Hence

$$\tilde{x}_{n+1} = [\prod_{\ell=p}^{n}(1 - \rho_\ell f)]\tilde{x}_p, \tag{80}$$

where p is the first number before n for which $1 \geq \rho_p f$. The convergence of $\tilde{x}_n$'s follows from the fact that

$$\lim_{n \to \infty} \prod_{\ell=\rho}^{n}(1 - \rho_\ell f) = 0 \text{ if and only if } \lim_{n \to \infty} \sum_{\ell=\rho}^{n} \rho_\ell f = \infty. \tag{81}$$

The applicability of lemma 6 to the update scheme for α_n rests on the assumption that $E\{\alpha_n\}$'s converge to $E\{v^T \cdot v\}/Tr[A \cdot E\{vv^T\}]$ much more rapidly than $E\{v_{3n+1}^T \cdot v_{3n}\}$ and $E\{v_{3n}v_{3n+1}^T\}$ do to their respective limits (which are zero's in the ideal situation).

While lemma 6 proves that the recurrence scheme converges in the mean and provides the limiting value for α_n, it does not give any indication as to whether α_n's will actually converge to the limiting value or not. Fortunately Dvoretzhy[2] has recently provided a theorem on stochastic approximation which is highly relevant here. In fact, we have

Lemma 7.

The recurrence scheme (73) for α_n's, together with the conditions (74) for ρ_n, as well as the following conditions

$$E\{Y_n\} = 0 \ , \ \sum_{n=1}^{\infty} E\{Y_n^2\} < \infty, \tag{82}$$

where $Y_n - \rho_n \cdot [v_{3n}^T \cdot v_{3n+1} - \frac{E\{v_{3n+1}^T \cdot v_{3n}\}}{Tr[A \cdot E\{v_{3n+1} v_{3n}^T\}]} v_{3n}^T \cdot H_{3n+2} \cdot v_{3n+1}]$, imply

$$Pr\{\lim_{n=\infty} \alpha_n = \frac{E\{v^T \cdot v\}}{Tr[A \cdot E\{vv^T\}]}\} = 1. \tag{83}$$

Since α_n converges to $E\{v^T \cdot v\}/Tr[A \cdot E\{vv^T\}]$ with probability 1, we can for all intended purpose, replace α_n by the latter for sufficiently large n. To show that α_n will ultimately converge to the optimal value $2/(\gamma_1 + \gamma_M)$, we need

Lemma 8.

Let $\lambda_1 \leq \lambda_2 \leq \cdots \leq \lambda_N$ be the eigenspectrum of A, if there exists a positive number γ such that

$$Min\{|\lambda_1 - \lambda_i| \ , \ |\lambda_N - \lambda_i||\lambda_1 \neq \lambda_i \ , \ \lambda_N \neq \lambda_i \ , \ i = 2, ..., N-1\} = \Gamma, \tag{84}$$

then $E\{H_n \cdot w_n + \eta_n\}$ collapses asymptotically to the linear subspace spanned by the eigenvectors corresponding to λ_1 and λ_N at the exponential rate $e^{-n\Gamma\alpha}$.

Proof.

Expand $E\{H_n \cdot w_n + \eta_n\}$ in terms of eigenvectors of A and note that the contraction factor $(1 - \alpha_n\lambda_j)$ for $\lambda_j \neq \lambda_1$ or λ_N are at least a factor of $e^{-\Gamma\alpha_n}$ smaller (for small Γ).

Lemma 9.

$$\frac{1}{\lambda_1} \geq \frac{E\{v^T \cdot v\}}{Tr[A \cdot E\{vv^T\}]} \geq \frac{1}{\lambda_N}. \tag{85}$$

Proof. Expand v in terms of eigenvectors of A.

Lemma 10.

Suppose w^* exists such that $\nabla J(w^*, t_n) = 0 \forall_n$, and $0 < \epsilon \ll 1$, for any value of $\alpha_n \leq \epsilon / \lambda_{Max}[H], n = 1, 2, ..., w_n$ converges to w^* and

$$\lim_{n=\infty} \frac{E\{v_{3n}^T \cdot v_{3n+1}\}}{Tr[A \cdot E\{v_{3n+1} v_{3n}^T\}]} = \frac{1}{\lambda_1}. \tag{86}$$

Note that the update formula (73) for α_n is not assumed in the above lemma.
Proof.

It is clear that w_n will converge to w^*. To prove that (86) is true, we note that Eq. (37) yields

$$w_{n+m} - w^* \simeq (1 - \alpha A)^m (w_n - w^*). \tag{87}$$

Since $v_{3n} = H_{3n} \cdot w_{3n} + \eta_{3n}$ and $v_{3n+1} = H_{3n+1}$ with $\eta_{3n} = -H_{3n} \cdot w^*$ and $\eta_{3n+1} = -H_{3n+1} \cdot w^*$, and noting the statistical independence between H_{3n}, H_{3n+1}, and w_{2n}, we have

$$E\{v_{3n}^T \cdot v_{3n+1} w_{3n}\} \simeq [A(w_{3n} - w^*)]^T (1 - \alpha_{3n} A) A(w_{3n} - w^*), \tag{88}$$

$$Tr[A \cdot E\{v_{3n+1} v_{3n}^T\}] \simeq [A(w_{3n} - w^*)]^T A(1 - \alpha_{3n} A) A(w_{3n} - w^*). \tag{89}$$

Now

$$Max\{|1 - \alpha_n \lambda_i| | i = 1, ...N\} = 1 - \alpha_n \lambda_1, \tag{90}$$

which means that $w_n - w^*$ should collapse asymptotically to the low-dimensional subspace spanned by λ_1 eigenvectors exponentially, thus (86) follows.

When α_n is updated using Eq. (73), it is clear that α_n will not stay below $\epsilon / \lambda_{Max}[H]$, since lemma 7 then asserts that α_n will approach $1/\lambda_1 \gg \epsilon / \lambda_{Max}[H]$ for large n, which contradicts the precondition for the validity of lemma 10. However, if α_n grows too large ($> 2/\lambda_{Max}[H]$), then the small eigenvalue eigenvectors will cease to dominate, instead the eigenvectors with the largest eigenvalues will quickly dominate, which in turn will force down α_n. Thus it is easy to see that

Theorem 10.

If the conditions for lemmas 6 through 9 are true, and in addition if α^* exists which optimizes the convergence rate, then the update scheme (73) will allow α_n to converge to α^*.

Proof.

Assume the contrary, then α_n will either converge to some value above α^* or below α^*. In the former case the second term on the RHS of Eq. (73) will be negative, in the latter case the second term will be positive, both leading to contradiction.

It thus appears that optimal performance is assured for the first order scheme with the update scheme (73) for α_n provided that the residues are either zero or small. It also seems reasonable to assume that the update scheme (73) would be useful even for the case of large residues because of the self-correcting property of (73).

For the second order scheme one can expect similar adaptive algorithms for α_n and β_n to work. One possible adaptive scheme is the following

$$f_{n+1} = f_n + \rho_n(v_{4n}^T \cdot v_{4n+1} - f_n v_{4n}^T \cdot H_{4n+3} H_{4n+2} v_{4n+1}) \,, \tag{91}$$

$$g_{n+1} = g_n + \rho_n[(v_{4n}^T \cdot H_{4n+2} v_{4n+1})^2 - g_n(v_{4n}^T \cdot v_{4n+1})(v_{4n}^T \cdot H_{4n+3} H_{4n+2} v_{4n+1})], \tag{92}$$

$$\alpha_n = 2\sqrt{f_n g_n}, \tag{93}$$

$$\beta_n = (1 - \sqrt{g_n})^2. \tag{94}$$

The above choice of updating Eqs. (91) through (94) is dictated by the fact that they approximately reproduce the optimum value (α_c, β_c) as the stable fixed point for (α_n/β_n) provided that $w_{n+1} - w_n$ ultimately becomes asymptotic to eigendirections associated to λ_1 and λ_N only. The latter property follows directly from Eq. (63). The adaptive second order scheme is still a subject of active investigation and the result will be reported elsewhere.

5 Discussion

In neural network research, the objective function is usually taken to be the mean square error

$$J = \frac{1}{2} \sum_{t \in T} (\hat{O} - O)^T \cdot (\hat{O} - O), \tag{95}$$

where $t = (I, O)$ is the input/output pattern vector, $\hat{O}$ the target output vector, and O the actual output vector. The summation is over the training set T which serves as the teacher. The output vector O is a function of the weight vector W which is simply a set of adjustable parameters, as well as a function of the input vector I. It is generally acknowledged that nonlinear neural networks are much more powerful than linear ones. That means that O is preferably a nonlinear vector function of I and/or of W, the latter precludes the possibility of using more efficient linear optimization schemes.

One question which usually comes up in the discussion of learning algorithms for neural networks concern the choice of stochastic gradient-type algorithms over more conventional gradient optimization algorithms, especially Newton, Quasi-Newton, and conjugate gradient type schemes which potentially have faster convergence rates. The answer lies in the fact that neural networks, unlike most other optimization problems, usually contain large numbers of adjustable parameters. It can even be argued that a neural net derives its power primarily from having an inordinate amount of adjustable weights in order to adequately construct internal representations consistent with training exemplars. Another characteristics of neural net systems is the use of large amount of training data. In fact, even though no neural net systems now in existence are operated this way, in order for the neural net to be a reasonable model of intelligent self-learning system, it is necessary for the neural net to be able to extract environmental regularities from a continuous stream of sensory inputs. Both characteristics pose serious problems for the conventional optimization algorithms. Newton's method as well as Quasi-Newton schemes, while quite efficient for small systems, simply requires too much storage and computation for any problems having more than 100 adjustable parameters. Algorithms of the conjugate gradient variety can handle larger problems, but storage and amount of computation is still a big problem. Certainly, there is no way for even conjugate gradient schemes to handle a continuous stream of input data. The adaptation of stochastic approximation methods for neural net research was a step in the right direction, but the excessive slowness of the learning algorithms presently in use still hinders progress. Evidences accumulated in Quantum Monte Carlo studies have indicated that the minimum eigenvalues of large positive hessian matrices decrease exponentially with the ranks of the matrices. The methods proposed in this paper may be more efficient than the methods presently employed, yet they would still fall short of beating the combinatorial explosion. The solution, it seems, is not to build larger and larger quasihomogeneous networks, but to build a large neural net intelligent system out of smaller modular neural nets which can be trained separately. The learning efficiency of heterogeneous or hierarchical network systems will be a subject of future study.

References

[1] Rumelhart, D. E., Hinton, G. E., and Williams, R. J. (1986). "Learning Internal Representations by Error Propagation,". In: Rumelhart, D. E. and McClelland, J. L. (Eds.) Parallel Distributed Processing: Explorations in the Microstructure of Cognition. Vol. 1: Foundations. Cambridge: MIT Press/Bradford Books,

Parker, D. B. (1985). "Learning-Logic." (Tech. Rep. 47). Cambridge, MA: MIT Center for Computational Research in Economics and Management Science,

Grossberg, S. (1986). "Adaptive Pattern Classification and Universal Recoding: I. Parallel Development and Coding of Neural Feature Detectors." Biological Cybernetics 23, 121-134.

Hinton, G. E. and Sejnowski, T. J. (1983). "Analyzing Cooperative Computational Proceedings of the Fifth Annual Conference of the Cognitive Science Society, Rochester, NY.

Barto, A. G., Sutton, R. S., and Brouwer, P. S. (1981). "Associative Search Network: A Reinforcement Learning Associative Memory" Biological Cybernetics, 40, 201-211.

Hopfield, J. J. and Tank, D. W. (1985). "Neural Computation of Decisions in Optimization Problems" Biological Cybernetics, 52, 141-152.

[2] Narendra, K. S. and Thathatchar, M. A. L. (1974). "Learning Automata - a Survey." IEEE Transactions on Systems, Man., and Cybernetics, 4, 323-334.

Dvoretzky, A. (1986), "Stochastic Approximation Revisited," Advances in Applied Math. 7, 220-227.

Robbins, H. and Siegmund, D. (1971). "A Convergence Theorem for Nonnegative Almost Supermartigales and some Applications," Optimizing Methods in Statistics, J. Rustagi, (ed.) Academic Press, NY, 233-257.

[3] Gill, P. E., Murray, W. and Wright, M. (1981). "Practical Optimization," Academic Press, NY, 83-154.

INFORMATION STORAGE IN FULLY CONNECTED NETWORKS

Demetri Psaltis*   Santosh S. Venkatesh[t]

1   INTRODUCTION

1.1   Neural Networks

We will consider two models of associative memory based upon ideas from neuronal nets, and characterise their performance. A neural network consists of a highly interconnected agglomerate of cells called *neurons*. The neurons generate action trains dependent upon the strengths of the *synaptic* interconnections. The instantaneous state of the system is described by the collective states of each of the individual neurons (firing or not-firing). Models of learning (the Hebbian hypothesis [6]), and associative recall based on such networks (cf. [7] for instance), have been developed, and illustrate how distributed computational properties become evident as a collective consequence of the interaction of a large number of simple elements (the neurons). The success of these biological models for memory has sparked considerable interest in developing powerful distributed processing systems utilising the neural network concept. Central features of such systems include a high degree of parallelism, distributed storage of information, robustness, and very simple basic elements performing tasks of low computational complexity.

Our focus is on the distributed computational aspects evidenced in such networks. In particular, we consider the capacity of two specific neural networks for storage of randomly specified vectors and their capability for error correction (or equivalently, nearest neighbor search). We will utilise a system theoretic analogue of a neural network so as to facilitate the analysis.

We will be concerned with a specific neural network structure in what follows. We assume a densely interconnected network with neurons communicating with each other through linear synaptic connections. Neurons change state either in concert (the synchronous mode), or independently (the asynchronous mode), depending upon the net synaptic potential present at the neuron due to the action of all the

---
*Department of Electrical Engineering, California Institute of Technology, Pasadena, CA 91125

[t]Moore School of Electrical Engineering, University of Pennsylvania, Philadelphia, PA 19104

neurons in the system. In essence then, the state vector is operated upon by a global linear operation followed by a pointwise non-linear operation. The nature of state changes, and the flow in state space is hence entirely determined by the choice of the linear transformation, and the non-linear decision rule. We will consider two specific constructive schemes for generation of the matrix of synaptic weights corresponding to the global linear transformation, and utilise a thresholding decision rule.

1.2 Organisation

In the next section, we will describe the neural network structure from a system-theoretic point of view, and define the parameters that are important for the system to function as an efficient associative memory.

In section 3 we review the archetype scheme for generation of the matrix of synaptic weights using outer-products of the prescribed datum vectors. Our focus in this section is on the presentation of a formal analysis of the storage capacity of the outer-product scheme.

In section 4 we outline a technique for generating the weight matrix by tailoring the linear transformation to obtain a desired spectrum [24]. We analyse the performance of this scheme as an associative memory in some detail, and obtain estimates of it's performance from purely theoretical considerations. We will use $\mathbf{W}$ to represent the linear operation for both schemes; in the event that it is important to discriminate between the linear operators of the two schemes, we use $\mathbf{W}^{op}$ for the outer-product scheme, and $\mathbf{W}^s$ for the spectral scheme.

Sections 5 and 6 are devoted to computer simulations of the two techniques, *ad hoc* modifications, and discussions.

Formal proofs of quoted results are delegated to the appendices. Propositions 1–5 are proved in appendix A. Theorem 1, and corollary 1 are proved in appendix B. Theorems 2 and 3 are proved in appendix C. Technical lemmas which ae needed in the proofs are stated and proved only in the relevant appendices.

1.3 Notation

Several of the arguments used in the paper involve asymptotics. Here we develop some of the notation that we utilise.

Let $\{x_n\}_{n=1}^{\infty}$ and $\{y_n\}_{n=1}^{\infty}$ be two sequences of positive real numbers. As $n \to \infty$ we shall say that:

1. $x_n \sim y_n$ if $x_n/y_n \to 1$;

2. $x_n = O(y_n)$ if x_n/y_n is bounded from above;

3. $x_n = \Omega(y_n)$ if x_n/y_n is bounded from below;

4. $x_n = o(y_n)$ if $x_n/y_n \to 0$; we also denote $x_n = o(1)$ if $x_n \to 0$.

We will also utilise the terminology *almost all* to mean all but an asymptotically negligible fraction. We denote by $\mathbb{B}$ the set $\{-1, 1\}$.

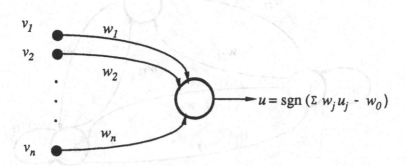

Figure 1: A model neuron as a threshold element (after McCulloch and Pitts).

2   THE MODEL OF McCULLOCH-PITTS

2.1   State-Theoretic Description

The basic computational elements that we utilise are the model neurons introduced by McCulloch and Pitts [14]. Each neuron is modelled as a threshold element as illustrated in figure 1. The neuron is characterised by n real *weights*, $w_1, w_2, \ldots, w_n$, and a real *threshold*, w_0; it accepts n real inputs, $v_1, v_2, \ldots, v_n$, and produces an output, u, of -1 or 1 according to the rule

$$u = \mathrm{sgn}(\sum_{j=1}^{n} w_j v_j - w_0) = \begin{cases} 1 & \text{if } \sum w_j v_j \geq w_0 \\ -1 & \text{if } \sum w_j v_j < w_0 \end{cases}.$$

We shall say that the output, $u \in \mathbb{B}$, of the neuron represents the neural *state*.

We consider a densely interconnected network of n model neurons with the outputs of the neurons fed back to constitute the n inputs of each neuron. The schema of figure 2 illustrates a network of three densely interconnected neurons. The neurons are labelled from 1 to n, and the instantaneous state of the system is denoted by the binary n-tuple $\mathbf{u} \in \mathbb{B}^n$, where the component u_i of $\mathbf{u}$ denotes the state of the i-th neuron, $i = 1, \ldots, n$. Each neuron, say i, has n weights, $w_{i1}, w_{i2} \ldots, w_{in}$. We will throughout fix the thresholding level at zero, $w_{i0} = 0$, for each neuron. Thus, if the system state is $\mathbf{u}$, the new state, $\hat{u}_i$, of the i-th neuron is given by

$$\hat{u}_i = \mathrm{sgn}\left(\sum_{j=1}^{n} w_{ij} u_j\right).$$

As this is a closed, feedback system, once the initial state of the system is specified,

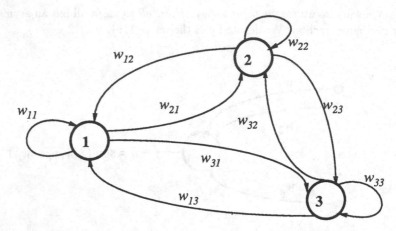

Figure 2: A densely interconnected neural network.

a trajectory is traced through the state space of binary n-tuples, $\mathbb{B}^n$, of the system. The adaptation of the system with time, or the flow in state space, is governed by two mathematical operations: (1) a globally acting linear transformation $\mathbf{W}$: $\mathbb{R}^n \to \mathbb{R}^n$, (corresponding to the *fixed* synaptic strengths w_{ij}), and (2) a pointwise thresholding operation Δ : $\mathbb{R}^n \to \mathbb{B}^n$.

Two distinct modes of operation are possible: *synchronous*, where the entire state vector updates itself, and *asynchronous*, where state changes are synonymous with bit changes and only a single randomly chosen neuron updates it's state per adaptation. For the sake of notational simplicity, we use Δ to denote both these modes of operation, and it will be clear from context to which mode we are actually referring to at any given time.

For the synchronous algorithm, the thresholding is done for each neuron, $i = 1, \ldots, n$, whereas for the asynchronous algorithm, only the i-th neuron (corresponding to some randomly chosen $i \in \{1, 2, \ldots, n\}$) is updated per adaptation, with the others held fixed. We can clearly consider just the restriction of $\mathbf{W}$ to $\mathbb{B}^n$, as wandering in the state space is confined to binary n-tubles by the nature of the algorithm. The total adaptation algorithm can hence be described as a cascade of operations $\Delta \circ \mathbf{W}$: $\mathbb{B}^n \to \mathbb{B}^n$.

Other processing modes are feasible in such neural networks. Models based on linear mappings without feedback, for instance, have been examined by Longuet-Higgins [13] and Gabor [3], while Poggio [18] has considered certain polynomial mappings.

2.2 Associative Memory

Our concern in this paper is with the information storage capability of the specified structure. In order for it to function as an associative memory we require that any prescribed set of state vectors $\{u^{(1)}, u^{(2)}, \ldots, u^{(m)}\} \in \mathbb{B}^n$ is storable in the network, and that these state vectors are invokable by any input which is sufficiently close to any of the stored vectors in some sense, i.e., these states function as *attractors*. In order to formally evaluate the functioning of a network as an associative memory, we need precise answers to the folowing intuitive notions:

- What is meant by "storing" a memory, $u^{(\alpha)}$, in a network?

- In some sense, can we evaluate the *maximum* number, m, of prescribed state vectors that can be stored in a network?

- What are the characterestics of a network that result in stored vectors functioning as attractors?

We will henceforth refer to the prescribed set of state vectors as *memories* to distinguish them from all other states of the system. Now it is a desideratum that the prescribed memories $u^{(1)}, u^{(2)}, \ldots, u^{(m)} \in \mathbb{B}^n$ are self-perpetuating, i.e., are stable. We shall say that a memory $u^{(\alpha)} \in \mathbb{B}^n$ is *strictly stable* (or simply *stable*) if $(\Delta \circ W) u^{(\alpha)} = u^{(\alpha)}$. Thus a state is strictly stable if it is a *fixed point* of the neural network. Clearly, for this definition of stability, it is immaterial whether the system is synchronous or asynchronous.

An associated idea of stability is weak stability· a memory is said to be *weakly stable* if it is mapped into a ball of small radius surrounding the memory in the state space by repeated applications of the adaptation algorithm. Strictly stable memories are weakly stable according to this notion. We will utilise in main the notion of strict stability, viz., $u^{(\alpha)} \mapsto u^{(\alpha)}$, in the analysis, and leave references to weak stability to the sections on computer simulations and discussions. Weak stability allows more systems for consideration within the framework; our opting for strict stability however yields some ease of analysis, and a strong regularisation in the algorithms.

With the nature of the thresholding operation fixed, we have the flexibility of realising different networks only by the appropriate choice of linear transformation W, or equivalently, by the choice of the synaptic strengths w_{ij}, which are the components of W in the standard basis. An algorithm for storing memories is, hence, simply a prescription for determining the weights w_{ij} (or, equivalently, the matrix W) as a function of the prescribed memories $u^{(1)}, u^{(2)}, \ldots, u^{(m)}$.

For any fixed algorithm, we define capacity to be a rate of growth rather than an exact number as in traditional channel capacity in information theory. Specifically, consider an algorithm, X, for storing *prescribed* memories, $u^{(1)}, u^{(2)}, \ldots, u^{(m)}$, in a model neural network. We specify an undelying product probability space by choosing the components of the memories from a sequence of symmetric Bernoulli trials; specifically, the memory components, $u_i^{(\alpha)}$,

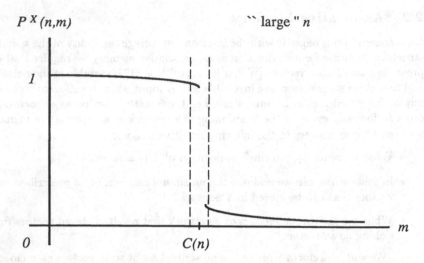

Figure 3: The 0-1 behaviour of capacity for large n: a schematic plot of the probability, $P^X(n, m)$, that all the memories are stored as fixed points by algorithm X versus the number of memories, m.

$i = 1, \ldots, n$, $\alpha = 1, \ldots, m$, are chosen from a sequence of independent trials with

$$u_i^{(\alpha)} = \begin{cases} 1 & \text{with probability } 1/2 \\ -1 & \text{with probability } 1/2 \end{cases}.$$

Definition 1 A sequence of integers $\{C(n)\}_{n=1}^{\infty}$ is a *sequence of capacities* for algorithm X iff for every $\lambda \in (0, 1)$, as $n \to \infty$ the probability that each of the memories is strictly stable approaches one whenever $m \leq (1 - \lambda)C(n)$, and approaches zero whenever $m \geq (1 + \lambda)C(n)$.

The above definition of capacity was formally arrived at by a consideration of lower and upper limits for capacity, and has been found to be particularly well suited in the present context [23]. In figure 3 we schematically illustrate the 0-1 behaviour required by the definition of capacity. A consequence of the definition is that if a sequence of capacities does exist, then it is not unique.

Proposition 1 If $\{C(n)\}$ is any sequence of capacities for algorithm X, then so is $\{C(n)(1 \pm o(1))\}$. Conversely, if $\{C(n)\}$ and $C^*(n)\}$ are any two sequences of capacity for algorithm X, then $C(n) \sim C^*(n)$ as $n \to \infty$.

We can thus consider an *equivalence class* of sequences of capacities for algorithm X, whose member sequences have the above asymptotic property. Without further ado we will refer to $C(n)$ as the *capacity of algorithm X* if $\{C(n)\}$ is any member of the equivalence class of sequences of capacities for algorithm X.

In the structure of association, we would like the stored memories to exercise a region of influence around themselves, so that if an input vector is "sufficiently close" to a memory (in the Hamming distance sense), the adaptive algorithm will cause the neural network to settle into a stable state centered at that memory. The determination of attraction behaviour in general depends on the specific structure of the linear transformation $\mathbf{W}$. For the specific case where $\mathbf{W}$ is symmetric, however, we can demonstrate Lyapunov functions for the system, and this suffices as an indicator of attraction behaviour.

Let $E : \mathbb{B}^n \to \mathbb{R}$ be the quadratic form

$$E(\mathbf{u}) = -\frac{1}{2}\langle \mathbf{u}, \mathbf{W}\mathbf{u} \rangle = -\frac{1}{2}\sum_{i=1}^{n}\sum_{j=1}^{n} w_{ij}u_i u_j . \tag{1}$$

By analogy with physical systems we refer to E as the *energy* of the current system state.

Proposition 2 The trajectories in state space follow contours of non-increasing energy if:
(a) $\mathbf{W}$ is symmetric, and has non-negative diagonal elements, and the mode of operation is asynchronous; or
(b) $\mathbf{W}$ is symmetric, non-negative definite, for *any* mode of operation.

As in physical systems, the system dynamics proceed in the direction of decreasing energy. This observation, for the case of asynchronous operation, is essentially due to Hopfield [8]. However, the result does not hold in a synchronous mode of operation for arbitrary symmetric weight matrices . For such cases an alternative Lyapunov function—the so-called *Manhattan norm*—can be demonstrated [4]. Let $F : \mathbb{B}^n \to \mathbb{R}$ be defined by

$$F(\mathbf{u}) = -\sum_{i=1}^{n}\left|\sum_{j=1}^{n} w_{ij}u_j\right| . \tag{2}$$

Proposition 3 For any choice of symmetric $\mathbf{W}$, and a synchronous mode of operation, the trajectories in state space are such that the functional F is non-increasing.

In either case, fixed-points of the system reside as minimas of either E or F for systems with symmetric weight matrices. If the memories are programmed to be fixed-points by suitable choice of symmetric $\mathbf{W}$, then under the conditions of the assertions above, trajectories in state space in the vicinity of the memories will tend to settle into strictly stable states at the memories thus establishing basins of attraction around them. (Limit cycles are possible with either E or F being identically zero for those states. This, however, has small probability in most cases.)

3   THE OUTER-PRODUCT ALGORITHM

3.1   The Model

We review here a correlation based scheme for generating the linear transformation W. The scheme is based upon the sum of the outer-products of the memory vectors, and has been well documented in the literature (cf. [7], for instance). Nakano [16] coined the term "Associatron" for the technique, and demonstrated how a linear net constructed using outer-products of prescribed state vectors could be combined with a pointwise thresholding rule to obtain a time-sequence of associations, with some ability for recall and error correction. More recent papers emphasise the role of the non-linearity in the scheme and include both synchronous and asynchronous approaches. The conditions under which long term correlations can exist in memory have been investigated by Little [11], and Little and Shaw [12] utilising a synchronous model. Using an asynchronous model, Hopfield [8] demonstrated that the flow in state space was such as to minimise a bounded "energy" functional, and that associative recall of chosen memories was hence feasible with a measure of error correction.

We now describe the model. We assume that m memories $\mathbf{u}^{(1)}$, $\mathbf{u}^{(2)}$, ..., $\mathbf{u}^{(m)} \in \mathbb{B}^n$ have been chosen randomly. The matrix of weights is constructed according to the following prescription: for $i, j = 1, \ldots, n$, set

$$w_{ij}{}^{op} = \begin{cases} \sum_{\alpha=1}^{m} u_i^{(\alpha)} u_j^{(\alpha)} & \text{if } i \neq j \\ 0 & \text{if } i = j \end{cases} . \tag{3}$$

Thus $[w_{ij}{}^{op}]$ is a symmetric, zero-diagonal matrix of weights. Let $\mathbf{U} = [\,\mathbf{u}^{(1)}\ \mathbf{u}^{(2)} \cdots \mathbf{u}^{(m)}\,]$ be the $n \times m$ matrix of memories to be stored. Then, from equation 3 the matrix of weights can be directly constructed as a sum of *Kronecker outer-products*,

$$\mathbf{W}^{op} = \mathbf{U}\,\mathbf{U}^T - m\,\mathbf{I} , \tag{4}$$

where $\mathbf{I}$ is the $n \times n$ identity matrix.

An inspection of equation 4 shows that the outer-product matrix can be recursively computed. Let $\mathbf{W}^{op}[k]$ denote the outer-product matrix of weights generated by the first k memories. We then have

$$\mathbf{W}^{op}[k] = \mathbf{W}^{op}[k-1] + \mathbf{u}^{(k)}(\mathbf{u}^{(k)})^T - \mathbf{I} , \qquad k \geq 1 , \tag{5}$$

where $\mathbf{W}^{op}[0] \overset{\text{def}}{=} 0$.

We now demonstrate that the memories are stable (at least in a probabilistic sense). Assume that one of the memories, say $\mathbf{u}^{(\alpha)}$, is the initial state of the neural network. For each $i = 1, \ldots, n$, we have

$$
\begin{aligned}
(\mathbf{W}^{op}\mathbf{u}^{(\alpha)})_i &= \sum_{j=1}^{n} w_{ij}{}^{op}u_j^{(\alpha)} \\
&= \sum_{\substack{j=1 \\ j \neq i}}^{n} \sum_{\beta=1}^{m} u_i^{(\beta)} u_j^{(\beta)} u_j^{(\alpha)} \\
&= (n-1)u_i^{(\alpha)} + \sum_{\beta \neq \alpha} \sum_{j \neq i} u_i^{(\beta)} u_j^{(\beta)} u_j^{(\alpha)} .
\end{aligned}
\tag{6}
$$

As before, we assume that the memory components, $u_i^{(\alpha)}$, $i = 1, \ldots, n$, $\alpha = 1, \ldots, m$, are generated from a sequence of symmetric Bernoulli trials; specifically, $\mathbf{P}\{u_i^{(\alpha)} = -1\} = \mathbf{P}\{u_i^{(\alpha)} = 1\} = 1/2$. It then follows that the second term of equation 6 has zero mean, and variance equal to $(n-1)(m-1)$, while the first term is simply $(n-1)$ times the sign of $u_i^{(\alpha)}$. The terms w_{ij} from equation 3 are the sum of m independent random variables, and are hence asymptotically normal *vide* the Central Limit Theorem. It can similarly be demonstrated that the summands of the double sum in equation 6 are independent, binary random variables, so that a large deviation Central Limit Theorem applies (cf. appendix B). Hence, the bit $u_i^{(\alpha)}$ will be stable only if the mean to standard deviation given by $\sqrt{n-1}/\sqrt{m-1}$ is large. Thus, as long as the storage capacity of the system is not overloaded, we expect the memories to be stable in probability. Later in the section we quote a result which yields the storage capacity of the scheme; note that the simple argument used above immediately yields that $m = o(n)$.

Stable memories tend to exhibit attraction basins for this model *vide* propositions 2 and 3. Specifically, the algorithmic flow in state space is towards the minimisation of a bounded functional (either the energy E or the Manhattan norm F, depending on the mode of operation). Trajectories hence tend to terminate in stable states which are local minima of the appropriate functional (nominally the stored memories). The algorithm thus demonstrates the characterestics required of a physical content-addressable memory provided we are working in the regime where the stored memories are stable. Overloading the storage capacity of the algorithm causes a breakdown in the associative behaviour around the memories.

3.2 Storage Capacity

A rigorous demonstration of the capacity of the outer-product scheme is quite involved (cf. [15] and [23]). Much of the technical difficulties encountered in the detailed proofs of can be bypassed, however, with the use of a simplifying Gaussian hypothesis which leads to a demonstration of the correct result with an alternate, much simpler proof which also provides a heuristic justification for the use of the *signal-to-noise ratio* as a performance yardstick. We will present this approach here, while refering the reader to the technical literature for more rigourous demonstrations of the result.

A GAUSSIAN ASSUMPTION: The elements $w_{ij}{}^{op}$ *of the weight matrix are jointly normal.*

It is clear that asymptotically each w_{ij} is individually Gaussian by virtue of the *Central Limit Theorem*, so that the conjecture is plausible. (Note that there is some abuse of terminology here, as with n approaching infinity, more and more random variables—the w_{ij}—are required to be jointly Gaussian.) However, it must be understood that the Gaussian assumption, while plausible, cannot be rigorously defended. It does, however, provide a vehicle for a simple, and intuitive, justification of the correct result.

Let the sequence $\{m_n\}_{n=1}^\infty$ denote explicitly the number of memories as a function of the number of neurons n. We define the sequence of probabilities $\{P^{op}(n)\}_{n=1}^\infty$ by

$$P^{op}(n) = \mathbf{P}\{\mathbf{u}^{(\alpha)} \text{ is a stable state, } \alpha = 1, \ldots, m_n\} . \tag{7}$$

Given the number of neurons, n, $P^{op}(n)$ is the probability that *all* the m_n chosen memories are stable. The following yields a formal estimate of the storage capacity of the scheme by estimating what is the maximum allowable rate of increase of m_n with n such that the prescribed memories are stable with high probability. All logarithms are to base e.

Theorem 1 Let δ be a parameter with $(\log n)^{-1} \leq \delta \leq \log n$. As $n \to \infty$ if

$$m_n = \frac{n}{4\log n}\left[1 + \frac{3\log\log n + \log(128\pi\delta^2)}{4\log n} + o(\frac{1}{\log n})\right] , \tag{8}$$

then the probability that each of the m_n memories is strictly stable is asymptotically $e^{-\delta}$, i.e., $P^{op}(n) \sim e^{-\delta}$ as $n \to \infty$.

Corollary 1 $C(n) = n/4\log n$ is the storage capacity of the outer-product algorithm.

The capacity result was based on the requirement that *all* of the memories be strictly stable. A recomputation of capacity based on the less stringent requirement that *one* of the memories be strictly stable with high probability yields a capacity of $n/2\log n$. (This latter requirement yields that the *expected number* of strictly stable memories is $m - o(m)$.) Thus, requiring all the memories to be stable instead of just one reduces capacity by only a factor of a half.

4 SPECTRAL ALGORITHMS

4.1 Outer-Products Revisited

We again assume that m memories $\mathbf{u}^{(1)}$, $\mathbf{u}^{(2)}$, ..., $\mathbf{u}^{(m)} \in \mathbb{B}^n$ have been chosen randomly. For strict stability, we require that $(\Delta \circ \mathbf{W})(\mathbf{u}^{(\alpha)}) = \mathbf{u}^{(\alpha)}$ for $\alpha = 1, ..., m$. Specifically, if $\mathbf{W}\mathbf{u}^{(\alpha)} = \mathbf{v}^{(\alpha)} \in \mathbb{R}^n$, we require that $\mathrm{sgn}(v_i^{(\alpha)}) = u_i^{(\alpha)}$ for each $i = 1, ..., n$.

For the outer-product scheme for generating the elements of the weight matrix, we have from equation 3

$$
\begin{aligned}
(\mathbf{W}^{op}\mathbf{u}^{(\alpha)})_i = \sum_{j=1}^{n} w_{ij}{}^{op}u_j^{(\alpha)} &= \sum_{\substack{j=1 \\ j \neq i}}^{n} \sum_{\beta=1}^{m} u_i^{(\beta)}u_j^{(\beta)}u_j^{(\alpha)} \\
&= (n-1)u_i^{(\alpha)} + \sum_{\beta \neq \alpha} \sum_{j \neq i} u_i^{(\beta)}u_j^{(\beta)}u_j^{(\alpha)} \\
&= (n-1)u_i^{(\alpha)} + \delta u_i^{(\alpha)},
\end{aligned}
\tag{9}
$$

where $\mathrm{E}(\delta u_i^{(\alpha)}) = 0$, $\mathrm{Var}(\delta u_i^{(\alpha)}) = (n-1)(m-1)$. Hence

$$
\frac{\mathrm{E}(|(n-1)u_i^{(\alpha)}|)}{\sqrt{\mathrm{Var}(\delta u_i^{(\alpha)})}} = \frac{\sqrt{n-1}}{\sqrt{m-1}} \longrightarrow \infty \qquad \text{as } n \to \infty,
$$

where we require that $m = o(n)$ from corollary 1 so that the memories are stable with high probability. We can hence write

$$
\mathbf{W}^{op}\mathbf{u}^{(\alpha)} = (n-1)\mathbf{u}^{(\alpha)} + \delta\mathbf{u}^{(\alpha)}
$$

where $\delta\mathbf{u}^{(\alpha)}$ has components $\delta u_i^{(\alpha)}$ whose contributions are small compared to $(n-1)u_i^{(\alpha)}$, at least in a probabilistic sense. In essence then, the memories $\mathbf{u}^{(\alpha)}$ are "*eigenvectors-in-mean*" or "*pseudo-eigenvectors*" of the linear operator $\mathbf{W}^{op}$, with "*pseudo-eigenvalues*" $n-1$.

4.2 Constructive Spectral Approaches

In this section we demonstrate constructive schemes for the generation of the weight matrix which yield a larger capacity than the outer-product scheme. This construction ensures that the given set of memories is stable under the algorithm; specifically, we obtain linear operators $\mathbf{W}^s$ which ensure that the conditions $\mathrm{sgn}(\mathbf{W}^s\mathbf{u}^{(\alpha)})_i = u_i^{(\alpha)}$, $i = 1, ..., n$, $\alpha = 1, ..., m$, are satisfied for $m \leq n$. The construction entails an extension of the approach outlined in the previous section so that the memories $\mathbf{u}^{(\alpha)}$ are *true eigenvectors* of the linear operator $\mathbf{W}^s$ [21,24]. Related approaches that have been considered before include those of Kohonen [9], who considers a purely linear mapping which is optimal in the mean-square sense, and Poggio's polynomial mapping technique [18]. Other schemes which are formally related to this approach are the interesting orthogonalisation techniques proposed by Amari [1] and Personnaz, et al [17].

We now utilise a result due to J. Komlós on binary n-tuples, to establish two results which have a direct bearing on the construction of the weight matrix.

Proposition 4 (a) For all randomly chosen binary (-1,1) n-tuples $\mathbf{u}^{(1)}$, $\mathbf{u}^{(2)}$, ..., $\mathbf{u}^{(m)} \in \mathbb{B}^n$ with $m \leq n$, define the $n \times m(-1,1)$ matrix $\mathbf{U} = \left[\mathbf{u}^{(1)} \ \mathbf{u}^{(2)} \cdots \mathbf{u}^{(m)}\right]$. Then $\mathbf{P}\{\text{rank } (\mathbf{U}) = m\} \to 1$ as $n \to \infty$.
(b) Let $\mathcal{B}_n$ be the family of bases for $\mathbb{R}^n$ with all basis elements constrained to be binary n-tuples; (i.e., $E = \{\mathbf{e}_1, \mathbf{e}_2, \ldots, \mathbf{e}_n\} \in \mathcal{B}_n$ iff $\mathbf{e}_1, \mathbf{e}_2, \ldots, \mathbf{e}_n \in \mathbb{B}^n$ are linearly independent). Then asymptotically as $n \to \infty$, *almost all* vectors $\mathbf{u} \in \mathbb{B}^n$ have a representation of the form

$$\mathbf{u} = \sum_{j=1}^{n} \alpha_j \mathbf{e}_j, \qquad \alpha_j \neq 0 \text{ for each } j = 1, \ldots, n, \qquad (10)$$

for *almost all* bases E in $\mathcal{B}_n$.

We use these results to establish the validity of the following schemes for constructing the weight matrix $\mathbf{W}^s$.

Fix $m \leq n$, and let $\lambda^{(1)}$, $\lambda^{(2)}$, ..., $\lambda^{(m)} \in \mathbb{R}^+$ be fixed (but arbitrary) positive real numbers. Let $\mathbf{u}^{(1)}$, $\mathbf{u}^{(2)}$, ..., $\mathbf{u}^{(m)} \in \mathbb{B}^n$ be the m randomly chosen memories to be stored in the memory. In the following we formally define two "spectral" formulations for the interconnection matrix.

Strategy 1 Define the $m \times m$ diagonal matrix

$$\Lambda = \mathbf{dg}[\lambda^{(1)}, \ldots, \lambda^{(m)},],$$

and the $n \times m(-1, 1)$ matrix of memories

$$\mathbf{U} = [\mathbf{u}^{(1)} \mathbf{u}^{(2)} \cdots \mathbf{u}^{(m)}].$$

Set

$$\mathbf{W}^s = \mathbf{U}\Lambda(\mathbf{U}^T\mathbf{U})^{-1}\mathbf{U}^T. \qquad (11)$$

Strategy 2 Choose any $(n - m)$ vectors $\mathbf{u}^{(m+1)}$, $\mathbf{u}^{(m+2)}$, ..., $\mathbf{u}^{(n)} \in \mathbb{B}^n$ such that the vectors $\mathbf{u}^{(1)}$, ..., $\mathbf{u}^{(m)}$, $\mathbf{u}^{(m+1)}$, ..., $\mathbf{u}^{(n)}$ are linearly independent. Define the augmented $n \times n$ diagonal matrix Λ_a, and the augmented $n \times n(-1, 1)$ matrix $\mathbf{U}_a$ by

$$\Lambda_a = \mathbf{dg}[\lambda^{(1)}, \ldots, \lambda^{(m)}, 0, \ldots, 0],$$

and
$$\mathbf{U}_a = [\, \mathbf{u}^{(1)} \cdots \mathbf{u}^{(m)} \, \mathbf{u}^{(m+1)} \cdots \mathbf{u}^{(n)} \,].$$

Set
$$\mathbf{W}^s = \mathbf{U}_a \mathbf{\Lambda}_a \mathbf{U}_a^{-1}. \tag{12}$$

The crucial assumption of linear independence of the memories in the formal definitions above is vindicated by proposition 4. Specifically, rank$(\mathbf{U}) = m$, and rank$(\mathbf{U}_a) = n$ for almost any choice of memories, so that the inverses are well defined.

Note that in both strategies, $\{\, \lambda^{(1)},\, \lambda^{(2)},\, \ldots,\, \lambda^{(m)} \,\}$ is the *spectrum* of the linear operator $\mathbf{W}^s$, and the memories $\mathbf{u}^{(1)}$, $\mathbf{u}^{(2)}$, ..., $\mathbf{u}^{(m)}$ are the corresponding eigenvectors. Furthermore, in strategy 2 there is considerable flexibility in the choice of the $(n - m)$ linearly independent vectors $\mathbf{u}^{(m+1)}$, $\mathbf{u}^{(m+2)}, \ldots, \mathbf{u}^{(n)}$, as almost all choices of $(n - m)$ vectors will satisfy linear independence asymptotically. Alternative schemes can also be obtained by combining the two strategies: specifically, we can choose fewer than $(n - m)$ additional linearly independent vectors, and then use the *pseudo-inverse* scheme of strategy 1 on the augmented matrix of memories. In fact, for $m = n$, the two strategies are identical.

Theorem 2 The storage capacity of all spectral strategies is linear in n; specifically, $C(n) = n$ for all spectral strategies.

Remarks: *(1) Additional stable states can be created by both strategies:* Let us for simplicity consider the eigenvalues $\lambda^{(\alpha)}$ to be equal to some value $\lambda > 0$. Let $\Gamma = \text{span}\{\mathbf{u}^{(1)},\, \mathbf{u}^{(2)},\, \ldots,\, \mathbf{u}^{(m)}\} \subset \mathbb{R}^n$. Clearly, if $\mathbf{u}$ belongs to $\Gamma \cap \mathbb{B}^n$, then $\mathbf{u}$ is also stable for both strategies. By proposition 4, however, there will not be many such stable states created if $m < n$. In addition there will be some more stable states created in more or less random fashion in both strategies. Such stable states satisfy the more general stability requirement: $\text{sgn}(\mathbf{W}^s\mathbf{u})_i = u_i$ for each $i = 1, \ldots, n$, and are not eigenvectors of the linear operator $\mathbf{W}^s$.

(2) Both strategies have some capacity for positive recognition of unfamiliar starting states: Let $\Phi \subset \mathbb{R}^n$ denote the null space of $\mathbf{W}^s$. For strategy 1, Φ is the orthogonal subspace to Γ, while for strategy 2, $\Phi = \text{span}\{\mathbf{u}^{(m+1)},\, \mathbf{u}^{(m+2)},\, \ldots,\, \mathbf{u}^{(n)}\}$. If $\mathbf{u} \in \Phi$, we have $\mathbf{W}^s\mathbf{u} = 0$. Consequently, at least for a synchronous algorithm, $(\Delta \circ \mathbf{W}^s)$ will iteratively map $\mathbf{u}$ to some vector $\mathbf{u}^{(0)} \in \mathbb{B}^n$ for all $\mathbf{u} \in \Phi$. The vector $\mathbf{u}^{(0)}$ in this case represents a positive indication that the starting state was not familiar.

In light of the above, we would expect all spectral strategies to share similar characterestics as associative memories. Note, however, that there is a computational advantage in choosing strategy 1 as it involves just a $m \times m$ matrix inversion as opposed to the $n \times n$ matrix inversion required in strategy 2. In what follows we assume that we construct $\mathbf{W}^s$ according to the prescription of strategy 1.

64

The linear capacity evidenced in the spectral schemes presents a considerable improvement over the (inverse-) logarithmic capacity of the outer-product algorithm. The improvement in capacity, however, is at the cost of increased complexity in the construction of the weight matrix. In general, this increased complexity implies that simple update rules like equation 5 cannot be found. For the particular, but important, special case where the spectrum is chosen to be m-fold degenerate, however, some simplicity attains. Let $\mathbf{W}^s[k]$ denote the matrix of weights generated by strategy 1 given the first k memories, $k \geq 1$. Then $\mathbf{W}^s[k]$ can be recursively constructed as follows.

Theorem 3 Let the eigenvalues, $\lambda^{(j)}$, $j = 1, \ldots, m$, in strategy 1 be chosen to be m-fold degenerate, i.e., $\lambda^{(j)} = \lambda > 0$, $j \geq 1$. For each $j \geq 1$, let $\mathbf{e}^{(j)}$ be the n-vector defined by

$$\mathbf{e}^{(j)} \stackrel{\text{def}}{=} (\lambda \mathbf{I} - \mathbf{W}^s[j-1])\mathbf{u}^{(j)} ,$$

where we define $\mathbf{W}^s[0] \equiv \mathbf{0}$. Then

$$\mathbf{W}^s[k] = \mathbf{W}^s[k-1] + \frac{\mathbf{e}^{(k)}(\mathbf{e}^{(k)})^T}{(\mathbf{u}^{(k)})^T \mathbf{e}^{(k)}} , \qquad k \geq 1 . \tag{13}$$

4.3 Basins of Attraction

We now probe the question of whether there exists a region of attraction around each memory. We will restrict ourselves to the case where $\mathbf{W}$ has an m-fold degenerate spectrum. For definiteness, we consider variants of the matrix $\mathbf{W}$ chosen according to the pseudo-inverse scheme of strategy 1.

As in the case of the outer-product algorithm, the *signal-to-noise ratio* (SNR) serves as a good *ad hoc* measure of attraction capability. Specifically, let $\lambda > 0$ be the m-fold degenerate eigenvalue of $\mathbf{W}$. Then we claim that

$$\|W\mathbf{x}\| \leq \lambda \|\mathbf{x}\| \qquad \text{for all } \mathbf{x} \in \mathbb{R}^n .$$

To see this we write $\mathbf{x}$ in the form $\mathbf{x} = \mathbf{x}_\| + \mathbf{x}_\perp$, where $\mathbf{x}_\| \in \Gamma$, and $\mathbf{x}_\perp \in \Phi$. (Recall that we defined $\Gamma = \text{span}\{\mathbf{u}^{(1)}, \mathbf{u}^{(2)}, \ldots, \mathbf{u}^{(m)}\}$, and Φ was the orthogonal subspace to Γ.) Then $\mathbf{W}\mathbf{x} = \mathbf{W}\mathbf{x}_\| = \lambda\mathbf{x}_\|$. Also $\|\mathbf{x}\|^2 = \|\mathbf{x}_\|\|^2 + \|\mathbf{x}_\perp\|^2 \geq \|\mathbf{x}_\|\|^2$. Hence, $\|\mathbf{W}\mathbf{x}\| = \lambda\|\mathbf{x}_\|\| \leq \lambda\|\mathbf{x}\|$.

Now, if $\mathbf{u}^{(\alpha)}$ is a stable state of the system, and $\mathbf{u} = \mathbf{u}^{(\alpha)} + \delta\mathbf{u}$, then $\mathbf{W}\mathbf{u} = \lambda\mathbf{u}^{(\alpha)} + \mathbf{W}\delta\mathbf{u}$, so that $\mathbf{u}$ will be mapped into $\mathbf{u}^{(\alpha)}$ by the adaptation algorithm only if the perturbation term $\mathbf{W}\delta\mathbf{u}$ is sufficiently small. As a measure of the strength of the perturbation, we define the SNR by $\|\mathbf{W}\mathbf{u}^{(\alpha)}\|/\|\mathbf{W}\delta\mathbf{u}\| = \lambda\sqrt{n}/\|\mathbf{W}\delta\mathbf{u}\|$; a high SNR implies that the perturbation term is weak, and conversely. From the discussion in the preceding paragraph, we have that the SNR is bounded below by $\sqrt{n}/\|\delta\mathbf{u}\|$. If d denotes the Hamming distance between $\mathbf{u}$ and $\mathbf{u}^{(\alpha)}$, then $\|\delta\mathbf{u}\| = 2\sqrt{d}$. For vectors $\mathbf{u}$ in the immediate neighbourhood of $\mathbf{u}^{(\alpha)}$, we have

$d \ll n$. We hence obtain a large SNR which is lower bounded by $\sqrt{n}/2\sqrt{d}$, which is indicative of a small perturbation term (compared to the "signal" term).

The SNR argument provides a quantitative measure of the attraction radius. The existence of attraction basins is, however, ensured only in probability, insofar as we accept the SNR as an accurate barometer of attraction behaviour. For the case where $\mathbf{W}$ has an m-fold degenerate spectrum, however, a direct analytical argument can be supplied for the existence of a flow in the state space towards stable states. We use the following

Fact. If the spectrum of $\mathbf{W}$ generated by strategy 1 is m-fold degenerate, then $\mathbf{W}$ is symmetric, non-negative definite.

Proof. Let $\lambda > 0$ be the m-fold degenerate eigenvalue of $\mathbf{W}$. Then $\mathbf{W}^T = \mathbf{W} = \lambda \mathbf{U}(\mathbf{U}^T\mathbf{U})^{-1}\mathbf{U}^T$, so that $\mathbf{W}$ is symmetric. Furthermore, for any $\mathbf{u}$, set $\mathbf{u} = \mathbf{u}_{\parallel} + \mathbf{u}_{\perp}$, where $\mathbf{u}_{\parallel}$ lies in the degenerate subspace of eigenvectors Γ, and $\mathbf{u}_{\perp}$ lies in the orthogonal subspace to Γ. Then

$$\begin{aligned}\langle \mathbf{u}, \mathbf{W}\mathbf{u}\rangle &= \langle \mathbf{u}_{\parallel} + \mathbf{u}_{\perp}, \mathbf{W}\mathbf{u}_{\parallel} + \mathbf{W}\mathbf{u}_{\perp}\rangle \\ &= \langle \mathbf{u}_{\parallel} + \mathbf{u}_{\perp}, \lambda\mathbf{u}_{\parallel}\rangle \\ &= \lambda\|\mathbf{u}_{\parallel}\|^2 \geq 0. \quad \blacksquare\end{aligned}$$

It then follows by proposition 2 that model trajectories in state space follow contours of decreasing energy. As the energy attains (local) minima at stable memories, basins of atttraction are typically established. In the general case, however, this does not preclude the possibility of lower energy stable states being incidentally created close to a memory, so that the attractive flow in the region is dominated by the extraneous stable state. For the case of the m-fold degenerate spectral scheme, however, this does not happen.

Proposition 5 Global energy minima are formed at the memories for the m-fold degenerate spectral scheme of strategy 1.

This result is not true in general for the outer-product scheme. Note that as $\mathbf{W}$ is non-negative definite the energy is always non-positive. All vectors in the null space of $\mathbf{W}$ have zero energy so that the flow in state space is typically away from these vectors. Vectors in the null space hence constitute *repellor states*.

When the spectrum of $\mathbf{W}$ is not degenerate, however, the above argument does not hold, and the algorithm does not always generate flows that decrease energy. A statistical argument can be aduced instead of the analytical argument above, however, to show that such flows are typically the case.

4.4   Choice of Eigenvalues

We saw in the last section that with a choice of a degenerate spectrum, we obtain the desired flow in state space towards the memories. However, as m approaches n, the number of extraneous stable states created also increases, and for $m = n$ *all* the states become stable, as $\mathbf{W}$ is exactly a scalar multiple of the identity matrix. It is thus advantageous (even when $m < n$) to put in some scatter in the eigenvalues so as to break the degeneracy. (Specifically, linear combinations of eigenvectors will not be eigenvectors any longer as the spectrum is no longer degenerate; this results in fewer additional stable states being created.)

When the scatter of the eigenvalues is small compared to their mean value, λ, we expect the flow in state space to be still towards the minimisation of energy. Essentially, if $|\lambda^{(\alpha)} - \lambda| \leq \epsilon$, where ϵ is small compared to λ, then there is an aditional perturbation term in equation 15 (appendix A) for the change in energy, and this term is no larger than $(\epsilon/\lambda) \sum |w_{kj}|$. For small enough perturbation of eigenvalues (small ϵ), we do not expect a substantial change in the overall flow in state space towards minimising energy. At the same time, however, an improvement in the attraction radius is effected as the number of additional states created is expected to decrease, thus decreasing the probability that conflicting regions of attraction are established close to the memories. Thus we expect that *the introduction of a small amount of scatter in the eigenvalues improves overall performance*.

For large perturbations of the eigenvalues around their mean, the energy functional is no longer appropriate for the description of the flow in state space. The memories, however, are still stable, and exercise a small region of attraction around them. If $\mathbf{u} \in \mathbb{B}^n$, we again write $\mathbf{u} = \sum_{\alpha=1}^{m} c_\alpha \mathbf{u}^{(\alpha)} + \mathbf{u}_\perp$, where $\mathbf{u}_\perp$ lies in the null space of $\mathbf{W}$. We then have $\mathbf{W}\mathbf{u} = \sum_{\alpha=1}^{m} c_\alpha \lambda^{(\alpha)} \mathbf{u}^{(\alpha)}$. The memories corresponding to larger eigenvalues hence tend to dominate the flow in state space. To quantify this a little, we rewrite $\mathbf{u}$ as $\mathbf{u}^{(\alpha)} + \delta \mathbf{u}$. Let the Hamming distance between $\mathbf{u}^{(\alpha)}$ and $\mathbf{u}$ be d. As $\mathbf{W}$ is linear, there is a real number k such that $\|W\mathbf{x}\| \leq k\|\mathbf{x}\|$ for every choice of vector $\mathbf{x} \in \mathbb{R}^n$. This yields a lower bound for the signal-to-noise ratio:

$$\text{SNR} = \frac{\|\mathbf{W}\mathbf{u}^{(\alpha)}\|}{\|\mathbf{W}\delta\mathbf{u}\|} \geq \frac{\lambda^{(\alpha)}\sqrt{n}}{2k\sqrt{d}} .$$

Hence, for a given Hamming distance, increasing the eigenvalue improves the SNR. Thus in general we expect that *the radius of attraction increases as the corresponding eigenvalue increases*.

We now consider a simple *ad hoc* technique for introducing a small degree of scatter in the eigenvalues. Let $\rho_{\alpha\beta}$ denote the inner-product between the memories $\mathbf{u}^{(\alpha)}$ and $\mathbf{u}^{(\beta)}$, i.e., $\rho_{\alpha\beta} = \langle \mathbf{u}^{(\alpha)}, \mathbf{u}^{(\beta)} \rangle = \sum_{i=1}^{n} u_i^{(\alpha)} u_i^{(\beta)}$. Set

$$\lambda^{(\alpha)} = n - \frac{\sum_{\beta \neq \alpha} \rho_{\alpha\beta}}{m-1}, \qquad \alpha = 1, \ldots, m . \tag{14}$$

The rationale behind the above scheme is as follows: $\rho_{\alpha\beta}$ is a measure of the distance between the memories $\mathbf{u}^{(\alpha)}$, and $\mathbf{u}^{(\beta)}$. Specifically, $\rho_{\alpha\beta}$ achieves its maximum value of n when the memories are identical, and its minimum value of $-n$ when the memories are negations of one another; a value of $\rho_{\alpha\beta} = 0$ indicates that the memories are $n/2$ apart in Hamming distance. If two of the memories are close to each other, (i.e., $\rho_{\alpha\beta}$ is large and positive), we have from equation 14 that the corresponding eigenvalues would be roughly equal. As a consequence, neither memory will dominate the attractive flow in state space, so that one memory will not poach upon the region of attraction of the other. The choice of eigenvalues according to the above "correlation" method tends to decrease the eigenvalues corresponding to those memories that are close (in Hamming distance) to many other memories, and increase the eigenvalues corresponding to those memories which are remote from most of the other memories.

The eigenvalues of equation 14 are identically distributed random variables; we have

$$\mathbf{E}\{\lambda^{(\alpha)}\} = n - \frac{1}{m-1} \sum_{\beta \neq \alpha} \mathbf{E}\{\rho_{\alpha\beta}\} = n \,,$$

$$\mathrm{Var}\{\lambda^{(\alpha)}\} = \frac{1}{(m-1)^2} \sum_{\beta \neq \alpha} \sum_{\gamma \neq \alpha} \mathbf{E}\{\rho_{\alpha\beta}\rho_{\alpha\gamma}\} = \frac{n}{m-1} \,.$$

The mean-to-standard deviation is hence given by $\sqrt{n(m-1)}$, so that the expected scatter is small for large n. Note that the scatter in the eigenvalues decreases as the number of memories stored increases.

Equation 14 suggests a simple method of introducing scatter into the eigenvalues. Other methods are, of course, possible, whereby we could pay more attention to high correlation terms as these are potentially more damaging.

5   COMPUTER SIMULATIONS

A question at issue in determining the relative performance of the two storage algorithms is what the radius of attraction is for the two schemes, and how rapidly it dwindles with increases in the number of memories to be stored. Sharp analytical bounds for the attraction radius as a function of the number of memories stored are difficult to arrive at in the spectral case—in part because of the difficulty of appropriate statistical modelling. Trends observed in computer simulations for systems with state vectors of between 32 to 64 bits bolster the intuitive supposition that the increased storage capacity of the spectral approach (vis-à-vis the outer-product scheme) results in significantly improved performance as an associative memory.

We encapsulate some of the observed trends in the following discussion. We utilise the Hamming radius of attraction corresponding to a given number of memories as our performance measure. For comparison with the outer-product scheme, a pseudo-inverse spectral strategy with an m-fold degenerate spectrum

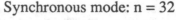

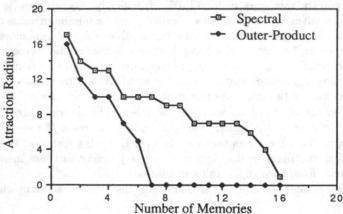

Figure 4: Attraction in the outer-product and spectral algorithms under synchronous operation.

is used, the eigenvalues, $\lambda^{(\alpha)}$, all being chosen equal to n. The memories are chosen randomly using a binomial pseudo-random number generator, and test vectors at specified Hamming distances from any specified memory are generated by reversing the signs of randomly chosen components.

For a small number of memories, m, the performance of the two schemes is roughly the same with the spectral strategy showing a slightly larger radius of attraction. For small m the observed radius of attraction is a sizeable fraction of n (roughly $n/2$ for the range of n considered). As m increases, the performance of the outer-product algorithm deteriorates much more rapidly than that of the spectral scheme. For m large enough (about $m = 6$ for $n = 32$, and $m = 12$ for $n = 64$) the outer-product scheme becomes overloaded, and the memories themselves are not stable any longer. At this point the spectral scheme still exhibits a sizeable radius of attraction around each memory. Figures 4, 5, and 6 illustrate these results for a *typical* memory. (As the plots represent a typical sample rather than a statistical average, some deviations from monotonicity are visible in the figures; however, the plots sans the fluctuations are quite representative of the average functioning of the memories as attractors under the two algorithms.)

For the range of n between 32 and 64 in the simulations, it appears that the spectral scheme exhibits some attraction behaviour for m up to about the order of $n/2$. (Of course, the memories themselves are stable up to $m = n$ for the spectral algorithm.) The linear nature of this relationship is evident in figure 7 where the largest number of memories, m, which can be stored with at least a unit

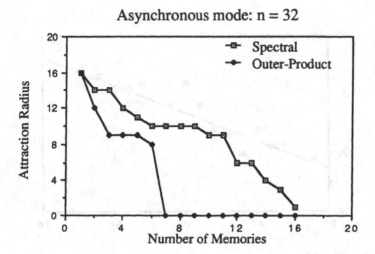

Figure 5: Attraction in the outer-product and spectral algorithms under asynchronous operation.

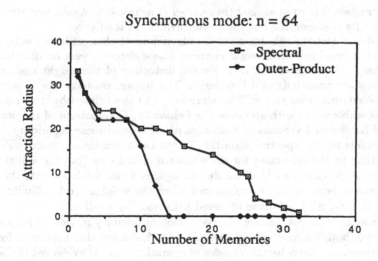

Figure 6: Attraction in the outer-product and spectral algorithms under synchronous operation.

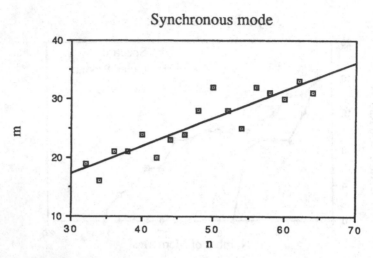

Figure 7: Number of memories that can be stored in the spectral scheme (using equal eigenvalues) with attraction over unit distance.

radius of attraction is plotted against the number of neurons, n. Again, statistical averaging results in smoothing out the scatter evident in the figure.

In order to test the robustness of the algorithms to changes in the weight matrix, we considered modified weight matrices whose elements were hardlimited to have binary values. Even for this extreme distortion of the weight matrix, the algorithms are essentially still functional. The storage capacity can be seen to decrease, but memories can still be stored—up to the (diminished) storage capacity—as stable states with attractor-like behaviour. Comparisons of the performance of hardlimited versions of both algorithms showed some superiority in attraction radius for the spectral algorithm in the cases considered, with qualitative similarity to the behaviour for the non-hardlimited case (barring a slight decrease in storage capacity) as illustrated in figures 8 and 9. It appears that sample behaviour becomes more pronounced when the weights are hardlimited, which leads to more fluctuations in observed behaviour for small n.

Some additional features are observed. Not surprisingly in view of propositions 2 and 3, both schemes exhibit virtually identical attraction behaviour for both synchronous and asynchronous modes of operation. This is evidenced in the close similarity of figures 4 and 5, and figures 8 and 9. Weak stability (where convergence is to states close to the memories rather than the memories themselves) is observed when the number of memories stored is large for both schemes. Ringing in the form of state cycles $A \to B \to A$, and similar, more complicated state cycles, are observed on occassion for the outer-product scheme, but rarely

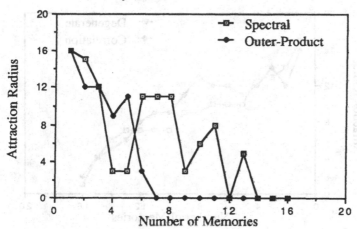

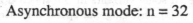

Figure 8: Attraction in the hardlimited outer-product and hardlimited spectral algorithms under synchronous operation.

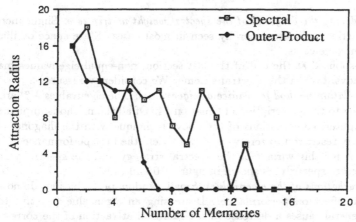

Figure 9: Attraction in the hardlimited outer-product and hardlimited spectral algorithms under asynchronous operation.

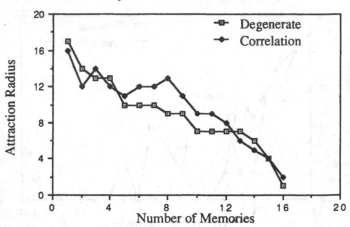

Figure 10: Attraction in the degenerate spectrum and correlation variant spectral algorithms under synchronous operation.

for the spectral scheme.

We also considered some variants in the spectral approach with a view to improving performance: (i) zero-diagonal spectral strategies, and (ii) non-equal eigenvaules.

Rendering the diagonal of the spectral weight matrix zero: Slight increases in the attraction radii are generally seen in most cases. State space oscillations are noted in some cases.

As espoused at the end of the last section, non-equal eigenvalues may be used with advantage in the spectral scheme. We consider two issues:

Correlation method for choice of eigenvalues: The eigenvalues $\lambda^{(\alpha)}$ are chosen according to the prescription of equation 14. Of all the methods implemented on the computer, combinations of the above technique with the diagonal of the weight matrix restricted to zero were seen to yield the best performance. A comparative plot for this variant of the spectral strategy and the spectral strategy with degenerate spectrum is shown in figures 10 and 11.

Perturbations of the eigenvalues: Small, random perturbations do not have a significant effect on performance. Decreasing an eigenvalue (relative to the mean) in general causes a decrease in the radius of attraction of the corresponding memory, while increasing an eigenvalue sufficiently increases the radius of attraction in general, as anticipated by the SNR argument. Figures 12 and 13 illustrate the effect of changing the eigenvalue upon the attraction radius of a single memory.

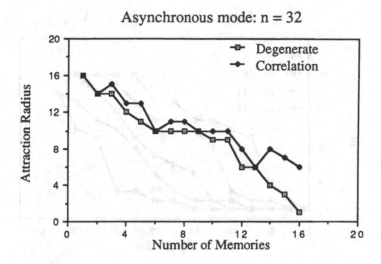

Figure 11: Attraction in the degenerate spectrum and correlation variant spectral algorithms under asynchronous operation.

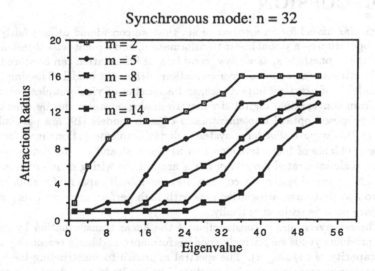

Figure 12: Attraction radius of a typical memory plotted as a function of the eigenvalue; the eigenvalues of all other memories is held fixed at n.

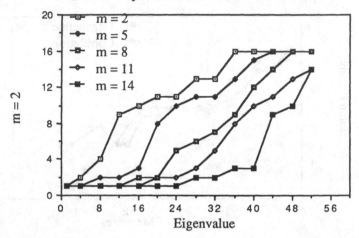

Figure 13: Attraction radius of a typical memory plotted as a function of the eigenvalue; the eigenvalues of all other memories is held fixed at n.

6   DISCUSSION

The particular model for an associative net that we considered utilises fairly elementary operations—a global linear transformation, and a point-wise thresholding decision rule. Nonetheless, as we saw, considerable information can be stored and retrieved with some measure of error correction within this model. The simplicity of the model is of practical importance, and a number of viable implementations ranging from conventional digital circuitry to analog, systems can be envisaged. Recently proposed optical implementations of such models [19] are particularly exciting in this regard. For large systems, digital implementations may founder upon the problems of full interconnection as the cost and circuit complexity of very large scale integrated circuits devolves around the wiring or interconnection problem [20]. Optical systems in contrast, have an inbuilt capacity for global communication so that large associative nets with high performance, and very rapid convergence may be realised optically.

The relatively simple construction of the linear transformation by means of outer-products yields surprisingly good performance and has a reasonably large storage capacity of $(n/4 \log n)$. The spectral approach to constructing the linear transformation is more complex in structure, but results in considerable improvement in performance, with a storage capacity of n; attraction behaviour, however, was noticed only for upto about $n/2$ datums stored in the range of n simulated. The question remains whether different (more optimal) choices of linear transformation could effect a substantial improvement in performance. The maximal fixed point storage capacity for such structures is $2n$ (cf. [23,22]) so that at most $2n$ datums can be stored as fixed points in such a neural network structure. Thus, alternative recipes for choice of linear transformation can impove on the performance of the spectral scheme by at best increasing the attraction radius of the stored datums so that a number of datums approaching $2n$ can be stored *with* exhibition of attraction behavior.

The larger storage capacity of the spectral scheme is reflected in increased pre-processing costs for computing the components of the weight matrix. The outer-product algorithm requires $mn^2/2 - O(mn)$ elementary operations for direct computation of the matrix of weights. The outer-product rule is also local, so that matrix updates to incorporate new datums can be done efficiently "in-place." In contrast, the spectral algorithm requires $mn^2 + m^2n + m^3/2 + O(n^2)$ elementary operations for direct computation of the weight matrix. For m of the order of $n/4 \log n$, the spectral algorithm hence requires about twice as many elementary operations as the outer-product algorithm; for m of the order of n, the spectral algorithm requires about five times as many elementary operations as the outer-product algorithm (the outer-product algorithm does not function well for this range of m, however, as we saw earlier). While the spectral rule is not local, when the eigenspace is degenerate, however, matrix updates can be performed relatively simply, and the previously stored datums need not be separately recorded.

A surprising result observed in the simulations was that the performance of synchronous and asynchronous processing schemes was very similar. On the surface, synchronous algorithms would seem to be more insensitive to local bit changes as they seem to function on a "signal-to-noise" ratio basis by considering all the bits in error. Changes made by asynchronously functioning schemes however, seem to reflect a more local character, and tend to follow a path of decreasing energy. It appears that error correction in the asynchronous algorithm for each of a sequence of bits in error translates to error correction as a whole to the block of error bits in the synchronous algorithm. We conjecture that asynchronous processing is somewhat more sensitive to the presence of local energy traps (incidental stable states close to the datum) than the synchronous algorithm, but that synchronous processing is less prone to recover from an initial incorrect decision. A relaxational approach with a constraint-satisfaction paradigm could be used with advantage to iron out the effect of local energy traps; this could be particularly efficacious if we are guaranteed that the datums form global energy minima (as is the case for the spectral algorithm). An alternative approach could utilise a compromise between the synchronous, and the asynchronous procedures by modifying a few bits at a time. Structures related to the relaxational approaches are the so called Boltzmann machines which incorporate random threshold levels in the algorithm. The effect again is to iron out local energy traps.

Acknowledgement

The work was supported in part by the NSF under grant EET-8709198.

A PROPOSITIONS

Proposition 1 If $\{C(n)\}$ is any sequence of capacities for algorithm X, then so is $\{C(n)[1 \pm o(1)]\}$. Conversely, if $\{C(n)\}$ and $C^*(n)\}$ are any two sequences of capacity for algorithm X, then $C(n) \sim C^*(n)$ as $n \to \infty$.

Proof. Let $\{C(n)\}$ be a sequence of capacities for algorithm X. Let p denote the probability that the neural network specified by algorithm X stores m memories. Now, for every $\lambda \in (0,1)$ we can find $\lambda^* \in (0,1)$ such that for large enough n, $[1 \pm o(1)](1 - \lambda) \leq (1 - \lambda^*)$. Fix $\lambda \in (0,1)$, and for n large, choose $m \leq (1 - \lambda)[1 \pm o(1)]C(n) \leq (1 - \lambda^*)C(n)$. As $\{C(n)\}$ is a sequence of capacities we have that $p \to 1$ as $n \to \infty$, so that the first part of definition 1 holds for the sequence $\{C(n)[1 \pm o(1)]\}$. Similarly, it can be shown that for every $\lambda \in (0,1)$, we have $p \to 0$ as $n \to \infty$ whenever $m \geq (1 + \lambda)[1 \pm o(1)]C(n)$. This completes the first part of the proof.

To prove the converse part, let $\{C(n)\}$ and $\{C(n)^*\}$ be any two sequences of capacities for algorithm X. Without loss of generality, let $C(n)^* = [1 + \alpha_n]C(n)$. We must prove that $\alpha_n = \pm o(1)$. Fix λ, $\lambda^* \in (0,1)$. For $m \leq (1 - \lambda)C(n)^* = (1 - \lambda)[1 + \alpha_n]C(n)$, we have $p \to 1$ as $n \to \infty$. Further, for $m \geq (1 + \lambda^*)[1 + \alpha_n]C(n)$, we have $p \to 0$ as $n \to \infty$. Hence, for *every* choice of scalars, λ, $\lambda^* \in (0,1)$, we require that $[1 + \alpha_n] < (1 + \lambda^*)/(1 - \lambda)$ for large enough n. It follows that $|\alpha_n| = o(1)$. ∎

Proposition 2 The trajectories in state space follow contours of non-increasing energy if:
(a) $\mathbf{W}$ is symmetric, and has non-negative diagonal elements, and the mode of operation is asynchronous; or
(b) $\mathbf{W}$ is symmetric, non-negative definite, for *any* mode of operation.

Proof. For each state $\mathbf{u} \in \mathbb{B}^n$, let the algorithm result in a flow in state space defined by $\mathbf{u} \mapsto \mathbf{u} + \delta\mathbf{u}$, where $\delta\mathbf{u}$ is an n-vector whose components take on values -2, 0, and +2 only. The change in energy $\delta E(\mathbf{u}) = E(\mathbf{u} + \delta\mathbf{u}) - E(\mathbf{u})$ is given by

$$\delta E(\mathbf{u}) = -\frac{1}{2}[\langle \delta\mathbf{u}, \mathbf{W}\mathbf{u}\rangle + \langle \mathbf{u}, \mathbf{W}\delta\mathbf{u}\rangle + \langle \delta\mathbf{u}, \mathbf{W}\delta\mathbf{u}\rangle]$$

$$= -\langle \delta\mathbf{u}, \mathbf{W}\mathbf{u}\rangle - \frac{1}{2}\langle \delta\mathbf{u}, \mathbf{W}\delta\mathbf{u}\rangle, \tag{15}$$

as W is symmetric for both part (a) and part (b). Now, regardless of the mode of operation, every non-zero component of δu has the same sign as the corresponding component of Wu by the prescription for state changes, so that $\langle \delta u, Wu \rangle \geq 0$. To show that $\delta E(u) \leq 0$ for every $u \in \mathbb{B}^n$ under the conditions of part (a) and part (b), it suffices, hence, to show that $\langle \delta u, W\delta u \rangle$ is non-negative.

Part (a): Under asynchronous operation, δu is either identically 0, or has precisely one component, say the k-th, non-zero. If $\delta u = 0$ we are done. Now assume $\delta u_k \neq 0$. Then $\langle \delta u, Wu \rangle = w_{kk}(\delta u_k)^2 \geq 0$ as the diagonal elements of W are assumed non-zero. This proves (a).

Part (b): The result follows for any mode of operation as W is non-negative definite. ∎

Proposition 3 For any choice of symmetric W, and a synchronous mode of operation, the trajectories in state space are such that the functional F is non-increasing.

Proof. Let $u[k]$, $k = 1, 2, \ldots$, denote the succession of states for some trajectory in state space under synchronous operation. We can rewrite equation 2 as

$$F(u[k]) = -\sum_{i=1}^{n} u_i[k+1] \sum_{j=1}^{n} w_{ij} u_j[k]$$

as $u_i[k+1]$ has the same sign as $\sum_{j=1}^{n} w_{ij} u_j[k]$ by the prescription for state changes. The change in F for successive states on a trajectory is given by

$$
\begin{aligned}
\delta F(u[k]) &= F(u[k+1]) - F(u[k]) \\
&= -\sum_{i=1}^{n} u_i[k+2] \sum_{j=1}^{n} w_{ij} u_j[k+1] + \sum_{j=1}^{n} u_j[k+1] \sum_{i=1}^{n} w_{ji} u_i[k] \\
&= -\sum_{i=1}^{n} (u_i[k+2] - u_i[k]) \sum_{j=1}^{n} w_{ij} u_j[k+1]
\end{aligned}
$$

where we've used the symmetry of W. Let $\mathcal{I} \subseteq \{1, 2, \ldots, n\}$ be the index set in which components $u_i[k+2]$ and $u_i[k]$ have opposite signs. Then

$$
\begin{aligned}
\delta F(u[k]) &= -2 \sum_{i \in \mathcal{I}} u_i[k+2] \sum_{j=1}^{n} w_{ij} u_j[k+1] \\
&= -2 \sum_{i \in \mathcal{I}} \left| \sum_{j=1}^{n} w_{ij} u_j[k+1] \right|.
\end{aligned}
$$

Hence $\delta F(u[k]) \leq 0$ for all states u. ∎

Proposition 4 (a) For all randomly chosen binary $(-1,1)$ n-tuples $\mathbf{u}^{(1)}$, $\mathbf{u}^{(2)}$, ...,
$\mathbf{u}^{(m)} \in \mathbb{B}^n$ with $m \leq n$, define the $n \times m(-1,1)$ matrix $\mathbf{U} = \left[\mathbf{u}^{(1)} \ \mathbf{u}^{(2)} \cdots \mathbf{u}^{(m)}\right]$.
Then $\mathbf{P}\{\text{rank }(\mathbf{U}) = m\} \to 1$ as $n \to \infty$.
(b) Let $\mathcal{B}_n$ be the family of bases for $\mathbb{R}^n$ with all basis elements constrained to
be binary n-tuples; (i.e., $E = \{\mathbf{e}_1, \mathbf{e}_2, \ldots, \mathbf{e}_n\} \in \mathcal{B}_n$ iff $\mathbf{e}_1, \mathbf{e}_2, \ldots, \mathbf{e}_n \in \mathbb{B}^n$ are
linearly independent). Then asymptotically as $n \to \infty$, *almost all* vectors $\mathbf{u} \in \mathbb{B}^n$
have a representation of the form

$$\mathbf{u} = \sum_{j=1}^{n} \alpha_j \mathbf{e}_j, \qquad \alpha_j \neq 0 \text{ for each } j = 1, \ldots, n, \tag{16}$$

for *almost all* bases E in $\mathcal{B}_n$.

Proof. (a) This is essentially Komlós' result [10]. Let A_n denote the number of
singular $n \times n$ matrices with binary elements $(-1,1)$. Then Komlós demonstrated
that

$$\lim_{n \to \infty} A_n 2^{-n^2} = 0. \tag{17}$$

(Komlós' result was for $n \times n(0,1)$ matrices, but it holds equally well for $n \times$
$n(-1,1)$ matrices.) Let $A_{n,m}$ denote the number of $n \times m(-1,1)$ matrices with
rank strictly less than m. Then we have that $A_{n,m}2^{n(n-m)} \leq A_n$, so that from
equation 17 we have that $A_{n,m}2^{-nm} \to 0$ as $n \to \infty$. It then follows that asymptotically as $n \to \infty$, almost all $n \times m(-1,1)$ matrices with $m \leq n$ are full rank.
This proves the first part of the theorem.
 (b) We first estimate the cardinality of $\mathcal{B}_n$: Let

$$\mathcal{X}_n \stackrel{\text{def}}{=} \{T = \{\mathbf{d}_1, \mathbf{d}_2, \ldots, \mathbf{d}_n\} \subset \mathbb{B}^n : T \text{ is a linearly dependent set}\}.$$

We have

$$|\mathcal{B}_n| = \binom{2^n}{n} - |\mathcal{X}_n|$$

$$= \binom{2^n}{n}\left[1 - \frac{n!\,|\mathcal{X}_n|}{2^{n^2}\left(1 - \frac{1}{2^n}\right)\left(1 - \frac{2}{2^n}\right)\cdots\left(1 - \frac{n-1}{2^n}\right)}\right]. \tag{18}$$

Let $T = \{\mathbf{d}_1, \mathbf{d}_2, \ldots, \mathbf{d}_n\} \in \mathcal{X}_n$. Then $[\mathbf{d}_1\,\mathbf{d}_2\cdots\mathbf{d}_n]$ is a singular matrix. Each
permutation of the column vectors $\mathbf{d}_1, \mathbf{d}_2, \ldots, \mathbf{d}_n$ yields another distinct singular
matrix. Since the column vectors are all distinct, we have $n!\,|\mathcal{X}_n| \leq A_n$. We
further have

$$\left(1 - \frac{1}{2^n}\right)\left(1 - \frac{2}{2^n}\right)\cdots\left(1 - \frac{n-1}{2^n}\right) > \left(1 - \frac{n-1}{2^n}\right)^n > \left(1 - \frac{n(n-1)}{2^{n-1}}\right).$$

Combining these results with equation 18 we get

$$1 - \frac{A_n}{2^{n^2}\left(1 - \frac{n(n-1)}{2^{n-1}}\right)} \leq \frac{|B_n|}{\binom{2^n}{n}} \leq 1 .$$

Define the sequence $\{\kappa_n\}$ by

$$\kappa_n = 1 - \frac{A_n}{2^{n^2}\left(1 - \frac{n(n-1)}{2^{n-1}}\right)} . \tag{19}$$

Then from equation 17 we have that $\kappa_n \to 1$ as $n \to \infty$.

Define a sequence of random variables $\{S_n\}_{n=1}^{\infty}$ such that S_n takes on the value 0 if a randomly chosen binary n-tuple $\mathbf{u} \in \mathbb{B}^n$ has the representation 16 in a randomly chosen basis $E \in B_n$, and 1 otherwise. To complete the proof it suffices to show that $\mathbf{E}\{S_n\} = \mathbf{P}\{S_n = 1\} \to 0$ as $n \to \infty$.

Fix $\mathbf{u} \in \mathbb{B}^n$, $E \in B_n$, and assume that 16 does not hold. Then there is a $j \in \{1, 2, \ldots, n\}$ such that $\alpha_j = 0$. Assume without loss of generality that $\alpha_n = 0$. Then

$$\mathbf{u} = \sum_{j=1}^{n-1} \alpha_j \mathbf{e}_j , \qquad \alpha_j \geq 0 . \tag{20}$$

We hence have that $\{\mathbf{e}_1, \mathbf{e}_2, \ldots, \mathbf{e}_{n-1}, \mathbf{u}\} \in X_n$. An overestimate for the number of choices of $\mathbf{u}$ and E such that equation 20 holds is $\binom{2^n}{n-1} 2^n$. Also, the total number of ways that we can choose $E \in B_n$, and $\mathbf{u} \in \mathbb{B}^n$ is $2^n |B_n|$. Hence, from this and equation 19, we have

$$\mathbf{P}\{S_n = 1\} \leq \frac{\binom{2^n}{n-1}}{|B_n|} \leq \frac{\binom{2^n}{n-1}}{\kappa_n \binom{2^n}{n}} \leq \frac{n 2^{-n}}{\kappa_n \left(1 - \frac{n-1}{2^n}\right)} .$$

By definition of κ_n, we then have that $\mathbf{P}\{S_n = 1\} \to 0$ as $n \to \infty$. ∎

Proposition 5 Global energy minima are formed at the memories for the m-fold degenerate spectral scheme of strategy 1.

Proof. For each memory, $\mathbf{u}^{(\alpha)}$, the energy is given by

$$E(\mathbf{u}^{(\alpha)}) = -\frac{1}{2}\langle \mathbf{u}^{(\alpha)}, \mathbf{W}\mathbf{u}^{(\alpha)}\rangle = -\frac{\lambda n}{2} .$$

Let $\mathbf{u} \in \mathbb{B}^n$ be arbitrary. We can write $\mathbf{u}$ in the form

$$\mathbf{u} = \sum_{\alpha=1}^{m} c^{(\alpha)} \mathbf{u}^{(\alpha)} + \mathbf{u}_\perp = \mathbf{u}_\| + \mathbf{u}_\perp \, ,$$

where $\mathbf{u}_\|$ is the projection of $\mathbf{u}$ into the linear span of the m memories, $\mathbf{u}^{(1)}$, $\mathbf{u}^{(2)}$, ..., $\mathbf{u}^{(m)}$, and $\mathbf{u}_\perp$ is a vector in the orthogonal subspace. Then we have $\langle \mathbf{u}_\|, \mathbf{u}_\perp \rangle = 0$, and by the Pythagorean theorem,

$$\|\mathbf{u}\|^2 = \|\mathbf{u}_\|\|^2 + \|\mathbf{u}_\perp\|^2 = \left\| \sum_{\alpha=1}^{m} c^{(\alpha)} \mathbf{u}^{(\alpha)} \right\|^2 + \|\mathbf{u}_\perp\|^2 \, .$$

The energy is then given by

$$
\begin{aligned}
E(\mathbf{u}) = -\frac{1}{2} \langle \mathbf{u}, \mathbf{W} \mathbf{u} \rangle &= -\frac{1}{2} \left\langle \sum_{\alpha=1}^{m} c^{(\alpha)} \mathbf{u}^{(\alpha)} + \mathbf{u}_\perp \, , \sum_{\alpha=1}^{m} \lambda c^{(\alpha)} \mathbf{u}^{(\alpha)} \right\rangle \\
&= -\frac{\lambda}{2} \left\| \sum_{\alpha=1}^{m} c^{(\alpha)} \mathbf{u}^{(\alpha)} \right\|^2 \\
&\geq -\frac{\lambda}{2} \|\mathbf{u}\|^2 = -\frac{\lambda n}{2} \, . \quad \blacksquare
\end{aligned}
$$

B   OUTER-PRODUCT THEOREMS

Lemma 1 Let $\{\omega_n\}_{n=1}^{\infty}$ be a positive sequence such that $\omega_n = \Omega\left((\log n)^\kappa\right)$ with $1/2 < \kappa < 1$, and let $\Phi : \mathbb{R} \to [0,1]$ be the cumulative Gaussian distribution function. Then for every $a \in \mathbb{R}$, $n^a \Phi(-\omega_n) \to 0$ as $n \to \infty$.

Proof. By the asymptotic formula for the error function (cf. [2], pp. 175—178) we have

$$\Phi(-\omega_n) \sim \frac{1}{\sqrt{2\pi}\,\omega_n} e^{-\omega_n^2/2} \qquad \text{as } n \to \infty \, .$$

Hence

$$
\begin{aligned}
\log(n^a \Phi(-\omega_n)) &\sim a \log n - \log \omega_n - \frac{\omega_n^2}{2} - \log \sqrt{2\pi} \\
&< -\frac{\omega_n^2}{2} \left(1 - \frac{a \log n}{\omega_n^2} \right) \\
&\to -\infty \qquad \text{as } n \to \infty \, . \quad \blacksquare
\end{aligned}
$$

Theorem 1 Let δ be a parameter with $(\log n)^{-1} \leq \delta \leq \log n$. As $n \to \infty$ if

$$m_n = \frac{n}{4 \log n} \left[1 + \frac{3 \log \log n + \log(128\pi\delta^2)}{4 \log n} + o(\frac{1}{\log n}) \right] , \qquad (21)$$

then the probability that each of the m_n memories is strictly stable is asymptotically $e^{-\delta}$, i.e., $P^{op}(n) \sim e^{-\delta}$ as $n \to \infty$.

Proof. We shall prove the above result using the Gaussian assumption quoted in section 3; this approach not only leads to the correct result with considerable simplification in analysis, but lends validity to the use of the easily computed signal-to-noise ratio in the analysis of these networks. The reader may rest assured that completely rigorous formulations of these proofs exist (cf. [15,23]).
Define the random variables $\{X_{\alpha i}\}_{\alpha=1, i=1}^{m_n, n}$ by

$$X_{\alpha i} = u_i^{(\alpha)} \sum_{\substack{j=1 \\ j \neq i}}^{n} w_{ij} u_j^{(\alpha)} = (n-1) + \sum_{j \neq i} \sum_{\beta \neq \alpha} u_i^{(\beta)} u_j^{(\beta)} u_i^{(\alpha)} u_j^{(\alpha)} .$$

Then

$$E(X_{\alpha i}) = n - 1 \overset{\text{def}}{=} \mu ,$$

and

$$E\{(X_{\alpha_1 i_1} - \mu)(X_{\alpha_2 i_2} - \mu)\} = \begin{cases} 1 & \text{if } \alpha_1 \neq \alpha_2 , i_1 \neq i_2 \\ m_n - 1 & \text{if } \alpha_1 = \alpha_2 , i_1 \neq i_2 \\ n - 1 & \text{if } \alpha_1 \neq \alpha_2 , i_1 = i_2 \\ (m_n - 1)(n-1) & \text{if } \alpha_1 = \alpha_2 , i_1 = i_2 . \end{cases}$$

The requirement that each of the memories $\mathbf{u}^{(\alpha)}$ is strongly stable implies that for each $\alpha = 1, \ldots, m_n$, and $i = 1, \ldots, n$, the random variables $X_{\alpha i}$ satisfy $\operatorname{sgn} X_{\alpha i} = 1$, i.e., $X_{\alpha i} \geq 0$. We will utilise the Gaussian hypothesis that the random variables $X_{\alpha i}$ are jointly normal. Now let $\{Y_{\alpha i}\}_{\alpha=0, i=0}^{m_n, n}$ be an i.i.d. set of Gaussian random variables with zero mean, and unit variance. We construct the normal random variables $\{Z_{\alpha i}\}_{\alpha=1, i=1}^{m_n, n}$ as follows:

$$Z_{\alpha i} = (n-1) - Y_{00} - Y_{\alpha 0}\sqrt{m_n - 2} - Y_{0i}\sqrt{n-2} + Y_{\alpha i}\sqrt{(m_n - 2)(n-2)} .$$

The Gaussian random variables $Z_{\alpha i}$ have the same statistics as the random variables $X_{\alpha i}$. Define

$$f_n(a, b, c) \overset{\text{def}}{=} \prod_{\alpha=1}^{m_n} \Phi\left(\frac{(n-1) - c - b_\alpha\sqrt{m_n - 2} - a\sqrt{n-2}}{\sqrt{(m_n - 2)(n-2)}} \right) \qquad (22)$$

and

$$I_n(\mathbf{b}, c) \stackrel{\text{def}}{=} (2\pi)^{-1/2} \int_{a=-\infty}^{\infty} f_n(a, \mathbf{b}, c) e^{-a^2/2} \, da \,, \tag{23}$$

where Φ is the cumulative Gaussian distribution function

$$\Phi(x) = (2\pi)^{-1/2} \int_{-\infty}^{x} e^{-a^2/2} \, da \,, \qquad \text{for all } x \in \mathbb{R} \,.$$

Now we have

$$P^{op}(n) = \mathbf{P}\{Z_{\alpha i} \geq 0 \,:\, \alpha = 1, \ldots, m_n \,, \, i = 1, \ldots, n\}$$

$$= \mathbf{P}\left\{Y_{\alpha i} \geq -\left[\frac{(n-1) - Y_{00} - \sqrt{n-2}\,Y_{0i} - \sqrt{m_n - 2}\,Y_{\alpha 0}}{\sqrt{nm_n - 2(n + m_n - 2)}}\right] \,:\right.$$

$$\alpha = 1, \ldots, m_n \,, \, i = 1, \ldots, n\right\} \,.$$

The events

$$\left\{Y_{\alpha i} \geq -\left[\frac{(n-1) - Y_{00} - \sqrt{n-2}\,Y_{0i} - \sqrt{m_n - 2}\,Y_{\alpha 0}}{\sqrt{nm_n - 2(n + m_n - 2)}}\right]\right\}$$

are conditionally independent given the random variables Y_{00}, $Y_{\alpha 0}$, and Y_{0i}. Hence, after some manipulation of integrals we have

$$P^{op}(n) = (2\pi)^{-(m_n+1)/2}$$

$$\int_{-\infty}^{\infty}\int_{-\infty}^{\infty}\cdots\int_{-\infty}^{\infty} [I_n(\mathbf{b}, c)]^n \exp\left\{-\frac{1}{2}\left(\sum_{\alpha=1}^{m_n} b_\alpha^2 + c^2\right)\right\} \, db_1 \cdots db_{m_n} \, dc \,, \tag{24}$$

where $I_n(\mathbf{b}, c)$ is as defined in equation 23.

Define the positive sequence $\{\omega_n\}_{n=1}^{\infty}$ s.t. as $n \to \infty$, $\omega_n/\log n \to 0$, and $\log n/\omega_n^2 \to 0$. (For instance, $\omega_n = (\log n)^\kappa$, with κ satisfying $1/2 < \kappa < 1$ works.) Rewrite equation 24 as

$$P^{op}(n) = I_1 + I_2 \tag{25}$$

where

$$I_1 = (2\pi)^{-(m_n+1)/2}$$

$$\int\int\cdots\int [I_n(\mathbf{b}, c)]^n \exp\left\{-\frac{1}{2}\left(\sum_{\alpha=1}^{m_n} b_\alpha^2 + c^2\right)\right\} \, db_1 \cdots db_{m_n} \, dc \,, \tag{26}$$

with all integrals having domain of integration $(-\omega_n, \omega_n)$, and I_2 is equal to a sum of terms like I_1 with each term having at least one integral with domain of integration $(-\infty, -\omega_n) \cup (\omega_n, \infty)$.

From equations 22 and 23 we have $0 \leq I_n(\mathbf{b}, c) \leq 1$. Also

$$\frac{1}{\sqrt{2\pi}} \int_{|b|>\omega_n} e^{-b^2/2} \, db = 2\Phi(-\omega_n) \,.$$

Hence

$$0 \leq I_2 \leq \sum_{j=1}^{m_n+1} \binom{m_n+1}{j} [2\Phi(-\omega_n)]^j \quad = \quad [1 + 2\Phi(-\omega_n)]^{m_n+1} - 1$$

$$\lesssim \quad 4(m_n + 1)\Phi(-\omega_n). \qquad (27)$$

Hence $I_2 \to 0$ as $n \to \infty$ by lemma 1, and with m_n as in equation 21. Now rewrite $I_n(\mathbf{b}, c)$ as

$$I_n(\mathbf{b}, c) = J_1(\mathbf{b}, c) + J_2(\mathbf{b}, c), \qquad |c|, |b_1|, \ldots, |b_{m_n}| \leq \omega_n, \qquad (28)$$

$$J_1(\mathbf{b}, c) = \frac{1}{\sqrt{2\pi}} \int_{|a| \leq \omega_n} f_n(a, \mathbf{b}, c) e^{-a^2/2} \, da, \qquad (29)$$

$$J_2(\mathbf{b}, c) = \frac{1}{\sqrt{2\pi}} \int_{|a| > \omega_n} f_n(a, \mathbf{b}, c) e^{-a^2/2} \, da.$$

Recalling that Φ is the cumulative Gaussian distribution function we have that $0 \leq f_n(a, \mathbf{b}, c) \leq 1$ from equation 22. Hence $0 \leq J_2(\mathbf{b}, c) \leq 2\Phi(-\omega_n)$, and $0 \leq J_1(\mathbf{b}, c) \leq 1$. Substituting in 28 and recalling that

$$[J_1(\mathbf{b}, c) + 2\Phi(-\omega_n)]^n \lesssim J_1(\mathbf{b}, c)^n + 4n\Phi(-\omega_n) \lesssim J_1(\mathbf{b}, c)^n$$

as $n \to \infty$ by lemma 1, we have

$$I_n(\mathbf{b}, c)^n \sim J_1(\mathbf{b}, c)^n \qquad \text{as } n \to \infty. \qquad (30)$$

Substituting 26, 27, and 30 in 25, we have

$$\inf_{\substack{|c|, |b_\alpha| \leq \omega_n \\ \alpha = 1, \ldots, m_n}} J_1(\mathbf{b}, c)^n \ [1 - 2\Phi(-\omega_n)]^{m_n+1} \lesssim P^{op}(n)$$

$$\lesssim \sup_{\substack{|c|, |b_\alpha| \leq \omega_n \\ \alpha = 1, \ldots, m_n}} J_1(\mathbf{b}, c)^n [1 - 2\Phi(-\omega_n)]^{m_n+1}.$$

With m_n as in 21, we have by lemma 1 that

$$1 \geq [1 - 2\Phi(-\omega_n)]^{m_n+1} \gtrsim 1 - 4(m_n + 1)\Phi(-\omega_n) \to 1 \qquad \text{as } n \to \infty.$$

Hence

$$\inf J_1(\mathbf{b}, c)^n \lesssim P^{op}(n) \lesssim \sup J_1(\mathbf{b}, c)^n.$$

Using the monotone increasing character of Φ, we get from 22 and 29 that

$$\Phi\left(\frac{\sqrt{n}}{\sqrt{m_n}} - \theta_n\right)^{n m_n} \lesssim \inf J_1(\mathbf{b}, c)^n \leq \sup J_1(\mathbf{b}, c)^n \lesssim \Phi\left(\frac{\sqrt{n}}{\sqrt{m_n}} + \kappa_n\right)^{n m_n},$$

where $\theta_n, \kappa_n = O(\omega_n/\sqrt{m_n})$.

As $n \to \infty$, both the asymptotic lower bound and the asymptotic upper bound approach $\Phi(\sqrt{n}/\sqrt{m_n})^{nm_n}$, as can be verifed from the asymptotic formula for the error function. Hence, as $n \to \infty$,

$$P^{op}(n) \sim \Phi\left(\frac{\sqrt{n}}{\sqrt{m_n}}\right)^{nm_n} \sim \left[1 - \frac{\sqrt{m_n}}{\sqrt{2\pi n}} e^{-\frac{n}{2m_n}}\right]^{nm_n}, \qquad (31)$$

where we again have recourse to the asymptotic form for the error function. Substituting for m_n from 21, as $n \to \infty$ we have

$$P^{op}(n) \sim \left[1 - \delta\left(\frac{4\log n}{n^2}\right) + o(\frac{1}{\log n})\right]^{\frac{n^2}{4\log n}(1+o(1))} \sim e^{-\delta}. \quad \blacksquare$$

Lemma 2 Let $P^{op}(n,m)$ explicitly represent the probability that each of m memories is strictly stable in a network of n neurons. Then as $n \to \infty$,

$$P^{op}(n,m) \lesssim P^{op}(n,m^*) \iff m > m^*.$$

Proof. The cumulative Gaussian distribution function Φ is monotone increasing, so that

$$\Phi\left(\frac{\sqrt{n}}{\sqrt{m}}\right) < \Phi\left(\frac{\sqrt{n}}{\sqrt{m^*}}\right) \iff m > m^*.$$

The lemma then follows directly from equation 31. $\quad \blacksquare$

Corollary 1 $C(n) = n/4\log n$ is the storage capacity of the outer-product algorithm.

Proof. Fix $0 < \lambda < 1$, and set

$$m = \frac{n}{4\log n}(1 - \lambda).$$

Set $\delta = 1/\log n$ in theorem 1, and let m^* be the corresponding number of memories. Clearly $m < m^*$ for large enough n, so that by lemma 2 and theorem 1

$$1 \geq P^{op}(n,m) \gtrsim P^{op}(n,m^*) \sim e^{-\frac{1}{\log n}} \longrightarrow 1 \qquad \text{as } n \to \infty.$$

Now fix $\lambda > 0$, and set

$$m = \frac{n}{4\log n}(1 + \lambda).$$

Set $\delta = \log n$ in the theorem. For large enough n, $m > m^*$, so by lemma 2 and theorem 1,

$$1 \leq P^{op}(n,m) \lesssim P^{op}(n,m^*) \sim e^{-\log n} \longrightarrow 0 \qquad \text{as } n \to \infty. \quad \blacksquare$$

C PROOFS OF SPECTRAL THEOREMS

Theorem 2 The storage capacity of all spectral strategies is linear in n; specifically, $C(n) = n$ for all spectral strategies.

Proof. Assume the memories $u^{(1)}$, $u^{(2)}$, ..., $u^{(m)}$, are linearly independent (over $\mathbb{R}$). Then, for strategy 1 we have

$$(\Delta \circ W^s)u^{(\alpha)} = (\Delta \circ (U \Lambda (U^T U)^{-1} U^T))u^{(\alpha)} = \Delta (\lambda^{(\alpha)} u^{(\alpha)}) = u^{(\alpha)} ,$$

as $\lambda^{(\alpha)} > 0$ so that $\text{sgn}(\lambda^{(\alpha)} u_i^{(\alpha)}) = u_i^{(\alpha)}$. Similarly, for strategy 2 we have

$$(\Delta \circ W^s)u^{(\alpha)} = (\Delta \circ (U_a \Lambda_a U_a^{-1}))u^{(\alpha)} = \Delta (\lambda^{(\alpha)} u^{(\alpha)}) = u^{(\alpha)} .$$

Thus the memories are stable whichever strategy is adopted.

Now note that the assumption of linear independence holds with probability one for large n by proposition 4. Hence, almost all choices of memories are stable whichever be the adopted spectral strategy. The capacity result follows because a linear transformation can have at most n linearly independent eigenvectors in n-space. ∎

Theorem 3 Let the eigenvalues, $\lambda^{(j)}$, $j = 1, \ldots, m$, in strategy 1 be chosen to be m-fold degenerate, i.e., $\lambda^{(j)} = \lambda > 0$, $j \geq 1$. For each $j \geq 1$, let $e^{(j)}$ be the n-vector defined by

$$e^{(j)} \stackrel{\text{def}}{=} (\lambda I - W^s[j-1])u^{(j)} , \tag{32}$$

where we define $W^s[0] \equiv 0$. Then

$$W^s[k] = W^s[k-1] + \frac{e^{(k)}(e^{(k)})^T}{(u^{(k)})^T e^{(k)}} , \qquad k \geq 1 . \tag{33}$$

Remarks: The construction above uses an elegant recursive computation of the pseudo-inverse of a matrix due to Greville [5]. We shall give a simple direct proof of the theorem.

Proof. Let $\{W[k]\}$ be the sequence of matrices generated by the recursion 33. We will show by induction that the sequence of matrices generated by equation 33 are indeed the weight matrices generated by strategy 1. Note that it suffices to show that for each $k = 1, 2, \ldots$,

1. The memories, $u^{(1)}$, $u^{(2)}$, ..., $u^{(k)}$, are eigenvectors of $W[k]$, and

2. All vectors in the orthogonal subspace to the linear span of $\{\mathbf{u}^{(1)}, \mathbf{u}^{(2)}, \ldots, \mathbf{u}^{(k)}\}$, are in the null space of $\mathbf{W}[k]$.

Induction Base $k = 1$: From equation 32 we have

$$\mathbf{e}^{(1)} = \lambda \mathbf{u}^{(1)} .$$

From equation 33 we then have

$$\mathbf{W}[1] = \frac{\lambda \mathbf{u}^{(1)}(\mathbf{u}^{(1)})^T}{(\mathbf{u}^{(1)})^T \mathbf{u}^{(1)}} = \frac{\lambda}{n} \mathbf{u}^{(1)}(\mathbf{u}^{(1)})^T .$$

Hence

$$\mathbf{W}[1]\mathbf{u}^{(1)} = \lambda \mathbf{u}^{(1)} .$$

Now let $\mathbf{x}_\perp \in \mathbb{R}^n$ be any vector orthogonal to $\mathbf{u}^{(1)}$. Then

$$\mathbf{W}[1]\mathbf{x}_\perp = \frac{\lambda}{n} \mathbf{u}^{(1)}\langle \mathbf{u}^{(1)}, \mathbf{x}_\perp \rangle = \mathbf{0} .$$

Inductive Hypothesis: The matrix, $\mathbf{W}[j-1]$, generated by the rule 33 is precisely the spectral matrix generated by strategy 1 to store memories $\mathbf{u}^{(1)}$, $\mathbf{u}^{(2)}$, $\ldots$, $\mathbf{u}^{(j-1)}$.

We begin with the observation (which can easily be demonstrated by induction) that the matrices, $\mathbf{W}[k]$, constructed according to the rule 33 are symmetric. From 33 we then have

$$\mathbf{W}[j] = \mathbf{W}[j-1] + \frac{(\lambda \mathbf{I} - \mathbf{W}[j-1])\mathbf{u}^{(j)}(\mathbf{u}^{(j)})^T(\lambda \mathbf{I} - \mathbf{W}[j-1])}{(\mathbf{u}^{(j)})^T(\lambda \mathbf{I} - \mathbf{W}[j-1])\mathbf{u}^{(j)}} .$$

Setting

$$\mathbf{v} \stackrel{\text{def}}{=} \mathbf{W}[j-1]\mathbf{u}^{(j)} ,$$

we have

$$\mathbf{W}[j] = \mathbf{W}[j-1] + \frac{(\lambda \mathbf{u}^{(j)} - \mathbf{v})(\mathbf{u}^{(j)})^T(\lambda \mathbf{I} - \mathbf{W}[j-1])}{\lambda n - (\mathbf{u}^{(j)})^T \mathbf{v}} .$$

Now, for $i = 1, 2, \ldots, j-1$, we have

$$
\begin{aligned}
\mathbf{W}[j]\mathbf{u}^{(i)} &= \mathbf{W}[j-1]\mathbf{u}^{(i)} + \frac{(\lambda \mathbf{u}^{(j)} - \mathbf{v})(\mathbf{u}^{(j)})^T(\lambda \mathbf{I} - \mathbf{W}[j-1])}{\lambda n - (\mathbf{u}^{(j)})^T \mathbf{v}}\mathbf{u}^{(i)} \\
&= \mathbf{W}[j-1]\mathbf{u}^{(i)} + \frac{(\lambda \mathbf{u}^{(j)} - \mathbf{v})(\mathbf{u}^{(j)})^T(\lambda \mathbf{u}^{(i)} - \mathbf{W}[j-1]\mathbf{u}^{(i)})}{\lambda n - (\mathbf{u}^{(j)})^T \mathbf{v}} . \quad (34)
\end{aligned}
$$

But by the induction hypothesis, we have

$$\mathbf{W}[j-1]\mathbf{u}^{(i)} = \lambda \mathbf{u}^{(i)} , \qquad i = 1, 2, \ldots, j-1 .$$

The entire second term in equation 34 hence vanishes, so that we are left with

$$\mathbf{W}[j]\mathbf{u}^{(i)} = \lambda\mathbf{u}^{(i)}, \qquad i = 1, 2, \ldots, j-1.$$

Now consider

$$
\begin{aligned}
\mathbf{W}[j]\mathbf{u}^{(j)} &= \mathbf{W}[j-1]\mathbf{u}^{(j)} + \frac{(\lambda\mathbf{u}^{(j)} - \mathbf{v})(\mathbf{u}^{(j)})^T(\lambda\mathbf{I} - \mathbf{W}[j-1])}{\lambda n - (\mathbf{u}^{(j)})^T\mathbf{v}}\mathbf{u}^{(j)} \\
&= \frac{(\lambda n - (\mathbf{u}^{(j)})^T\mathbf{v})\mathbf{v} + (\lambda\mathbf{u}^{(j)} - \mathbf{v})(\mathbf{u}^{(j)})^T(\lambda\mathbf{u}^{(j)} - \mathbf{v})}{\lambda n - (\mathbf{u}^{(j)})^T\mathbf{v}} \\
&= \frac{(\lambda n - (\mathbf{u}^{(j)})^T\mathbf{v})\mathbf{v} + (\lambda\mathbf{u}^{(j)} - \mathbf{v})(\lambda n - (\mathbf{u}^{(j)})^T\mathbf{v})}{\lambda n - (\mathbf{u}^{(j)})^T\mathbf{v}} \\
&= \frac{\lambda n\mathbf{v} - [(\mathbf{u}^{(j)})^T\mathbf{v}]\mathbf{v} + \lambda^2 n\mathbf{u}^{(j)} - \lambda n\mathbf{v} - \lambda[(\mathbf{u}^{(j)})^T\mathbf{v}]\mathbf{u}^{(j)} + [(\mathbf{u}^{(j)})^T\mathbf{v}]\mathbf{v}}{\lambda n - (\mathbf{u}^{(j)})^T\mathbf{v}} \\
&= \frac{\lambda[\lambda n - (\mathbf{u}^{(j)})^T\mathbf{v}]\mathbf{u}^{(j)}}{\lambda n - (\mathbf{u}^{(j)})^T\mathbf{v}} \\
&= \lambda\mathbf{u}^{(j)}.
\end{aligned}
$$

Now, let $\mathbf{x}_\perp \in \mathbb{R}^n$ be any vector in the orthogonal subspace to the linear span of $\{\mathbf{u}^{(1)}, \mathbf{u}^{(2)}, \ldots, \mathbf{u}^{(j-1)}, \mathbf{u}^{(j)}\}$. Consider

$$\mathbf{W}[j]\mathbf{x}_\perp = \mathbf{W}[j-1]\mathbf{x}_\perp + \frac{(\lambda\mathbf{u}^{(j)} - \mathbf{v})(\mathbf{u}^{(j)})^T(\lambda\mathbf{I} - \mathbf{W}[j-1])}{\lambda n - (\mathbf{u}^{(j)})^T\mathbf{v}}\mathbf{x}_\perp.$$

By inductive hypothesis $\mathbf{W}[j-1]\mathbf{x}_\perp \equiv 0$. Hence

$$\mathbf{W}[j]\mathbf{x}_\perp = \frac{(\lambda\mathbf{u}^{(j)} - \mathbf{v})(\mathbf{u}^{(j)})^T\lambda\mathbf{x}_\perp}{\lambda n - (\mathbf{u}^{(j)})^T\mathbf{v}};$$

and as $\langle\mathbf{u}^{(j)}, \mathbf{x}_\perp\rangle = 0$, we have

$$\mathbf{W}[j]\mathbf{x}_\perp = 0.$$

We have shown that under the inductive hypothesis, the memories $\mathbf{u}^{(1)}$, $\mathbf{u}^{(2)}, \ldots, \mathbf{u}^{(j-1)}, \mathbf{u}^{(j)}$, are eigenvectors of the matrix $\mathbf{W}[j]$ with common eigenvalue λ, and any vector in the orthogonal subspace to the linear span of these memories is in the null space of $\mathbf{W}[j]$. Hence, the sequence of matrices, $\{\mathbf{W}[k]\}$, generated recursively by rule 33 is precisely the sequence of spectral matrices, $\{\mathbf{W}^s[k]\}$, generated by strategy 1. ∎

References

[1] S. Amari, "Neural theory of association and concept formation," *Biol. Cybern.*, vol. 26, pp. 175–185, 1977.

[2] W. Feller, *An Introduction to Probability Theory and Its Applications*, vol. I. New York: Wiley, 1968.

[3] D. Gabor, "Associative holographic memories," *IBM J. Res. Devel.*, vol. 13, pp. 156–159, 1969.

[4] E. Goles and G. Y. Vichniac, "Lyapunov function for parallel neural networks," *Neural Networks for Computing: AIP Conf. Proc.*, vol. 151, pp. 165–181, 1986.

[5] T. N. E. Greville, "Some applications of the pseudoinverse of a matrix," *SIAM Rev.*, vol. 2, pp. 15–22, 1960.

[6] D. O. Hebb, *The Organisation of Behaviour*. New York: Wiley, 1949.

[7] G. E. Hinton and J. A. Anderson, *Parallel Models of Associative Memory*. Hillsdale: Lawrence-Erlbaum, 1981.

[8] J. J. Hopfield, "Neural networks and physical systems with emergent collective computational abilities," *Proc. Natl. Acad. Sci. USA*, vol. 79, pp. 2554–2558, 1982.

[9] T. Kohonen, *Associative Memory: A System-Theoretical Approach*. Berlin Heidelberg: Springer Verlag, 1977.

[10] J. Komlós, "On the determinant of (0,1) matrices," *Studia Scientarum Mathematicarum Hungarica*, vol. 2, pp. 7–21, 1967.

[11] W. A. Little, "The existence of persistent states in the brain," *Math. Biosci.*, vol. 19, pp. 101–120, 1974.

[12] W. A. Little and G. L. Shaw, "Analytic study of the memory storage capacity of a neural network," *Math. Biosci.*, vol. 39, pp. 281–290, 1978.

[13] H. C. Longuet-Higgins, "The non-local storage of temporal information," *Proc. Roy. Soc. B*, vol. 171, pp. 327–334, 1968.

[14] W. W. McCulloch and W. Pitts, "A logical calculus of the ideas immanent in nervous activity," *Bull. Math. Biophys.*, vol. 5, pp. 115–133, 1943.

[15] R. J. McEliece, E. C. Posner, E. R. Rodemich, and S. S. Venkatesh, "The Capacity of the Hopfield Associative Memory," *IEEE Trans. Inform. Theory*, vol. IT-333, pp. 461–482, 1987.

[16] K. Nakano, "Associatron—a model of associative memory," *IEEE Trans. Sys., Man, and Cybern.*, vol. SMC-2, no. 3, pp. 380–388, 1972.

[17] L. Personnaz, I. Guyon, and G. Dreyfus, "Information storage and retrieval in spin-glass like neural networks," *Jnl. Physique Lett.*, vol. 46, pp. L359–L365, 1985.

[18] T. Poggio, "On optimal nonlinear associative recall," *Biol. Cybern.*, vol. 19, pp. 201–209, 1975.

[19] D. Psaltis and N. Farhat, "Optical information processing based on an associative-memory model of neural nets with thresholding and feedback," *Optics Letters*, vol. 10, no. 2, pp. 98–100, 1985.

[20] I. E. Sutherland and C. A. Mead, "Microelectronics and computer science," *Scientific American*, vol. 237, pp. 210–228, 1977.

[21] S. S. Venkatesh and D. Psaltis, "Efficient strategies for information storage and retrieval in associative neural nets," *Workshop on Neural Networks for Computing*, Santa Barbara, California, April 1985.

[22] S. S. Venkatesh, "Epsilon capacity of neural networks," *Neural Networks for Computing: AIP Conf. Proc.*, vol. 151, pp. 440–445, 1986.

[23] S. S. Venkatesh, *Linear Maps with Point Rules: Applications to Pattern Classification and Associative Memory.* Ph.D. Thesis, California Institute of Technology, 1987.

[24] S. S. Venkatesh and D. Psaltis, "Linear and logarithmic capacities in associative neural networks," to appear in *IEEE Trans. Inform. Theory.*

NEURONIC EQUATIONS AND THEIR SOLUTIONS

E. R. CAIANIELLO

Dipartimento di Fisica Teorica
Università di Saleruo

1. Introduction

1.1. Reminiscing

My first contact with Cybernetics (I use this word in Norbert Wiener's sense, as used in Germany and USSR, not in the more restricted meaning attached to it in the USA) was during casual reading in Copenhagen, while working as a junior theoretical physicist at the Niels Bohr Institute, where CERN was then located, of "The Living Brain" by Grey Walter. It was love at first sight.

That was in early 1955. One year later — I had soon after won a Chair of Theoretical Physics in Naples — I started a group, funded by modest but vital USA Army and Air Force grants for research in neuroanatomy and electronics, to study the "brain".

All my colleagues at that time doubted the sanity of *my* brain. A long time had to pass before some Italian support could be obtained. I remember, in particular, two episodes that characterized the situation at that time. The first occured when I told my Ph.D. mentor in physics, soon to become the President of the American Physical Society, about my initiative: he smiled, and hinted that in some top level, top secret meeting in the USA where he had just come from, this very subject had been declared of top importance and priority; the second took place in the USSR with Gel'fand: "Cybernetics" was a very bad word there, being considered an expression of Capitalism, but he and Nobel laureate Tamm had succeeded in launching a major project in this field because, as they argued and pointed out, the existence in Italy of my little group certainly hardly had any connection with "capital"...

Of those pioneering times I also remember with gratitude my long association with Norbert Wiener, who spent in all a year and half in Naples and was the first to read and approve the manuscript of my first work[1] on the subject, the basis of all my further activity in this field, and with Warren McCulloch, a radiating vulcano of ideas and human warmth.

My indebtedness to them both, for their warm humanity and friendly encouragement, cannot be underestimated.

My greatest difficulty was to free myself of the prejudice and perhaps arrogance of a not totally unsuccessful, at the time, young physicist, until I could come to regard with due humility and respect the labours of biologists, physiologists and all others who had to study *living* matter, and to attempt to develop my own feeling for this new field, *from inside* so to say.

In studying quantum field theory I was fortunate to introduce an algorithm, the Pfaffian,[2] which soon proved to be the key to a much better understanding of Ising models of all sorts: what could have been simpler for me then, than using the obvious analogy of McCulloch's neurons with spin elements, and my own tools to transfer the plentiful know-how on the subject at hand?

Let me quote myself[1]: "a dynamical interpretation... would indeed be quite natural...; we deem it more meritorious, at this stage, to resist the temptation of adapting the available quantum-field theoretical knowledge to these problems than to yield to it".

I still feel the same way now, after a quarter century, whenever I see so many claim that problems have been "almost solved" when most basic questions cannot even be soundly formulated or even guessed.

The situation has changed somewhat in the last years: any statistical theory of neural nets has to introduce additional parameters, something like a temperature and a Hamiltonian, and there we are.

Useful as all this may be in particular instances, the problem at hand still appears to me of far greater magnitude; information, entropies of many kinds are "naturals" to our subjects, whereas standard Hamiltonian and stochastic techniques may curtail our vision and thus blind us to the real expanse of the subject. This I say not to criticize those who do so, but rather as an explanation, not an apology, for my continuing adherence to a deterministic point of view. My aim, in posing the problem as I did in 1961, was to formulate a mathematical theory and, when possible, solve the problems connected to it.

That this could be done seemed to me at first, due to my ignorance and to the utter nonlinearity of everything that I was writing, to be inconceivable. Only later did I realize that it was precisely this latter feature that permitted "exact" solutions, and hence, at least in principle, the synthesis of any neural net with fixed coefficients (or "cellular automaton": the latter term is a synonym, with added restrictions, of the first[3]).

I shall present a brief summary of this itinerary and refer to past works for clarification.

Finally, I wish to emphasize that there should be no *quérélle* among deterministic and statistic approaches: on the contrary, I am convinced that the more one knows about exact deterministic solutions, the easier it becomes to evaluate when and how statistics can be of real use in concrete situations.

1.2. The 1961 Model

I shall limit my discussion to only one of the three basic elements of my model: a set of nonlinear equations that describes the behaviour of a system of coupled binary decision elements in discrete time or "neuronic equations": NE.

Their solution is an essential preliminary to the remaining parts, "mnemonic equations" and "adiabatic learning hypothesis", to anatomical, or technological, structural information. These are closely related.

I would like to start by mentioning the following points, which appear to me now, after a quarter century of meditation, of special relevance. I sum them up in a few sentences:

1. Strategy: the study of neural models, or organizations, or complex structured systems such as natural languages and social structures needs a common approach as regards functional and relational aspects.

2. Levels of hierarchical nature are essential in this study; each level requires its specific logic, mathematics, relation to other levels. Their discrete quantization is basic to the stability of the system.

3. Quantitative information about populations of levels, also basic, cannot be derived from brain research alone, but also from natural languages and all sorts of other systems (but this is another story).

Most of present day mathematics, as applied to physical or other sciences, originated from the study of the "continuum": gravitation, fields, and so on. It carries with it an underlying notion of "space", be it Descartes's emptiness or the physicist's "vacuum", into which things happen whose description is to be sought by exploring "neighbourhoods", through (differential) linear approximations. Higher order and nonlinear terms are added as corrections when the linear description of reality is inadequate.

Everybody knows that this is done for lack of techniques capable of solving in a general and exact way even trivial nonlinear equations.

The exceptions are well documented in literature, and mostly belie the guesses that stem from such a "linearized" Weltanschauung. We find thus two concomitant elements; a historical propensity to "think linear" and the understanding that the general method available for computation is to fragment a problem into linear pieces.

There are, however, more and more instances in which nonlinearity is basic: for example Boolean algebra, computer science, decision-making, models of neural activity. This is an entirely different **Weltanschauung**. Things may become as clumsy (though elegant verbiage may act as a cover-up) as they appear to the physicist whose expansions refuse to converge. The long-term behaviour, the collective action of aggregates of interconnected yes-or-no decision elements, and many other such questions, seem unanswerable. At least the effort is not made to answer these questions.

Both these aspects are present in the model of neural activity I proposed in order to extend and put into algebra the pioneering work of McCulloch and Pitts.[4] This work describes the behaviour of a net of interconnected yes-or-no elements — the "neurons" — by means of three distinct laws:

1. those which describe the behaviour of the net with constant or frozen connections: *neuronic (or decision) equations (NE)*;

2. those which account for the change of the couplings among neurons, such as the

structure of the net, as a consequence of the activity described by 1): *mnemonic (or evolution) equations (ME)*;

3. those that fix the respective time scales appropriate for 1) and 2); ME may require a sequence of time scales , the smallest estimated for biological neurons, from retrograde amnesia, to be $\sim 10^6$ larger than that of NE. We called this *the adiabatic learning hypothesis (ALH)*. It is clearly necessary to decouple 1) from 2): N. Wiener's commented: "without it, we would be playing Alice's game of croquet, changing rules during the game".

The main challenge is presented in the first point — NE. The interested reader may find in Ref. 1 is a description of the whole model. We still consider this a sound basis for further study. We restrict our discussion here to a synopsis of results obtained since then on NE, several of which are unpublished. We shall commence by commenting on some features of special relevance.

NE describe, at time $t + \tau$ (τ is a constant delay), the state of the net as determined by its situation at time t. According to the problem at hand, it may be convenient to use values (0,1) or (- 1,1) to denote the two allowed states of a "neuron". We therefore take Heaviside or signum functions of real functions of binary variables. The discontinuous and the continuous aspect appear tied together in an essential way. Their interplay is fundamental to NE, whether the system described by them is neural or not.

Its full understanding is possible by a lucky circumstance. Many authors have preferred nonlinearities of smoother types such as the sigmoid, or quadratic, so as to use mathematics of continuum, such as standard calculus. Because we used the opposite approach, that is, totally discontinuous functions, we were able to derive the exact solution of NE and all connected problems. This will be outlined in the sequel. The breakthrough came in a most elementary manner, from the obvious property

$$\text{sgn } xy = \text{sgn } x \cdot sgn \ y$$

which may be said to express a full half of the properties of linear functions.

In concluding the introduction, I wish to emphasize that any general model of neural activity, including ours, can only express the laws of a neural medium, and certainly not the behaviour of a robot or even a brain. Likewise, physics produces laws, not automobiles or TV's: for these, rules have to be found or invented. Our search for appropriate rules (which, at another level, become laws once again) has led us along different paths such as natural languages and models of hierarchically structured system.[5] These will not concern us here. The solution of NE puts them on the same footing as linear equations for the study of exact or approximate models of general systems. In this report we focus our attention exclusively on this point.

I also wish to note that the application of so crude a model to biological situations (where a real neuron may be conceived as a VLSI made of our "mathematical" neurons) gave results far exceeding our expectations.[6]

1.3. Notation

The "neuron" is a binary decision element, whose states can be best described as $x = (0,1)$ or $\xi = (-1,1)$ according to the specific purpose. Of course

$$x = \frac{1+\xi}{2} .$$

The net has N neurons, whose interconnections determine its structure. We are not concerned here with specific structure; NE describe a general net as if it were a physical medium of which NE describe the laws.

Denote with

$$x \equiv \mathbf{x} \equiv \{x^1, x^2, \ldots, x^N\}; \quad x^h = 0, 1$$

$$\xi \equiv \boldsymbol{\xi} \equiv \{\xi^1, \xi^2, \ldots, \xi^N\}; \quad \xi^h = \pm 1$$

variables, vectors, or one-column matrices, whose components have values as specified. Let $F(\xi); \Phi(x)$ be any real function subject only to the condition

$$F(\xi) \neq 0; \quad \Phi(x) \neq 0$$

for any choice of variables ξ^h, x^h.

This requirement (which is not in fact a restriction) will remarkably simplify our discussion. Call

$$1[\Phi] = \begin{cases} 1 & \text{for } \Phi > 0 \\ 0 & \text{for } \Phi < 0 \end{cases} \quad \text{(Heaviside step function)}$$

$$\sigma[F] \equiv \text{sgn}[F] = \begin{cases} 1 & \text{for } F > 0 \\ -1 & \text{for } F < 0 \end{cases} \quad \text{(signum)} .$$

Define

$$< F(\xi) > = \frac{1}{2^N} \sum_{(\xi'=\pm 1, \ldots, \xi^N=\pm 1)} F(\xi^1, \xi^2, \ldots, \xi^N) \quad \text{(trace)} .$$

The tensor powers of ξ have 2^N components:

$$\eta^\alpha = \begin{cases} 1 \equiv \xi^0 \\ \vdots \\ \xi^h \qquad\qquad h = 1, 2, \ldots, N \quad \text{(linear terms)} \\ \vdots \\ \xi^{h_1} \xi^{h_2} \ldots \xi^{h_r} \\ \vdots \qquad\qquad \alpha = h_1 h_2, \ldots, h_r \quad \text{(nonlinear terms)} . \\ \xi^1 \xi^2 \ldots \xi^N \end{cases}$$

$\eta \equiv \boldsymbol{\eta} \equiv \{\eta^\alpha\}$ is thus a vector in 2^N dimensions, $\eta^\alpha = \pm 1$; the α-ordering of the indices $0, 1, \ldots, 2^N - 1$ may be arranged to suit particular needs; we choose here

$$\boldsymbol{\eta} = \begin{pmatrix} 1 \\ \xi^N \end{pmatrix} \times \begin{pmatrix} 1 \\ \xi^{N-1} \end{pmatrix} \times \ldots \times \begin{pmatrix} 1 \\ \xi^1 \end{pmatrix} = \begin{pmatrix} 1 \\ \xi^1 \\ \xi^2 \\ \xi^1 \xi^2 \\ \xi^3 \\ \vdots \\ \xi^1 \ldots \xi^N \end{pmatrix}$$

Then, all the properties of Boolean functions needed here can be readily derived from the evident ones

$$(\sigma[F])^2 = +1; \quad \sigma(\sigma[F]) = \sigma[F]; \quad \sigma[FG] = \sigma[F]\sigma[G] .$$

In particular, one is the η-expansion

$$\sigma[F(\xi)] = \sum_{\alpha=0}^{2^N - 1} f_\alpha \, \eta^\alpha \equiv f^T \eta ,$$

where

$$f_\alpha = \langle \eta^\alpha \sigma[F(\xi)] \rangle = \langle \sigma[\eta^\alpha F(\xi)] \rangle .$$

If one wishes to use the $(0,1)$ Heaviside functions, posing

$$\chi \equiv \boldsymbol{\chi} \equiv \begin{pmatrix} 1 \\ x^1 \\ x^2 \\ x^1 x^2 \\ \vdots \\ x^1 x^2 \ldots x^N \end{pmatrix}$$

one easily finds,

$$\chi = C\eta ,$$

with

$$C = \begin{pmatrix} 1 & 0 \\ \frac{1}{2} & \frac{1}{2} \end{pmatrix} \times \begin{pmatrix} 1 & 0 \\ \frac{1}{2} & \frac{1}{2} \end{pmatrix} \times \ldots \times \begin{pmatrix} 1 & 0 \\ \frac{1}{2} & \frac{1}{2} \end{pmatrix} .$$
$$(N \text{ Times})$$

so that the connection between the η- and the χ-expansion is given by

$$1(\Phi[\chi]) = g^T \chi = g^T C\eta = \frac{1}{2}(1 + \sigma[\Phi]) + \frac{1}{2} f^T \eta$$

or, with

$$\tilde{f}_0 = 1 + f_0, \quad \tilde{f}_\alpha = f_\alpha \quad (\alpha > 0) \,,$$

by

$$g = \frac{1}{2}(C^T)^{-1}\tilde{f}; \quad \tilde{f} = 2C^T g \,.$$

The χ-expansion is less suited than the η-expansion for algebraic manipulations, though more directly related to logic and probabilistic considerations.

2. Linear Separable NE

2.1. Neuronic Equations

Our first work[1] considered only linear functions $\Phi(X)$ or $F(\xi)$, hence l.s. Boolean functions $1[\Phi(x)]$ and $\sigma[F(\xi)]$. It is instructive to consider this case first. The NE are (we write $u_h \equiv x_h = (0,1)$ for consistency with the notation used there)

(I Form) $$u_h(t+\tau) = 1\left[\sum_{\substack{h=1...N \\ r=1...L}} a_{hk}^{(r)} u_k(t - r\tau) - s_h\right] . \tag{1}$$

Equation (1) thus takes into account the delayed action of the neurons of the net $[a_{hk}^{(r)} > 0$ excitation, $a_{hk}^{(r)} < 0$ inhibition, $a_{hh}^{(r)}$ loop of self-excitation or inhibition, s_n threshold]. NE written as in Eq. (1), which we may call the **First Form**, describe the state of the net at time $t + \tau$; they are **state equations**. They can equivalently be written as **excitation equations**, in the Second Form

(II Form) $$w_h(t+\tau) = \sum a_{hk}^{(r)} 1[w_k(t-r)] - s_h . \tag{2}$$

Equations (1) and (2) can also be written by increasing the number of neurons from N to NL, as if no delays occured $[a_{kh}^{(r)} \rightarrow a_{hk}^{(0)} \equiv a_{hk}]$. In matrix notation, setting $(NL \rightarrow N)$

$$u_h(m\tau) = u_{h,m}; \quad \mathbf{u}_m \equiv \begin{Bmatrix} u_{1,m} \\ u_{2,m} \\ \vdots \\ u_{N,m} \end{Bmatrix}; \quad A \equiv \{a_{hk}\} .$$

we find

(I Form) $$\mathbf{u}_{m+1} = 1[A\mathbf{u}_m - \mathbf{s}] \tag{3}$$

(II Form) $$\mathbf{w}_{m+1} = A1[\mathbf{w}_m] - \mathbf{s} . \tag{4}$$

Equations (1) or (3) describes a net with binary decision elements, i.e. 2-valued logic. Equations (2) or (4) can be constructed with suitable A and $\mathbf{s}$ so as to give any wanted number k of values at each element (k need not be the same for all elements). **They therefore also describe a net working with some k-valued logic. Equations (1) or (3) show the connection with binary nets.**

Let γ_r be vectors, and form from Eq. (4) the scalar products

$$\gamma_r \cdot \mathbf{w}_{m+1} = \gamma_r \cdot A1[\mathbf{w}_m] - \gamma_r \cdot \mathbf{s} .$$

If

$$A^T \gamma = 0$$

we find

$$\gamma_r \cdot \mathbf{w}(\tau) = -\gamma_r \cdot \mathbf{s} = \text{constant} . \tag{5}$$

If A is of order N and rank R, there are $N - R$ vectors satisfying Eq. (5) and as many linear constants of motion in the net. They can be utilized as failure detectors. Nets may be computed so as to have prescribed constants of motion and no limitation on couplings within rank R. It is also possible, of course, to obtain constants quadratic and of higher orders.

We pass now to the signum representation. A particular condition appears then to simplify remarkably the form of Eqs. (3) and (4).

$$AI = 2s ;$$

This means self-duality, and if it holds Eqs. (3) and (4) become $(A \to \frac{1}{2}A)$:

$$\mathbf{u}_{m+1} = \sigma[A\mathbf{u}_m] \tag{6}$$

$$\mathbf{w}_{m+1} = A\sigma[\mathbf{w}_m] . \tag{7}$$

Equations (6) and (7) reduce immediately in turn to the form Eqs. (3) and (4) by keeping fixed the state of some neuron ($N - 1$ neurons are then free). The self-dual form of NE simplifies many computations.

2.2. Polygonal Inequalities

Take $F(\xi)$ linear

$$F(\xi) = \sum_{h=1}^{N} a_h \xi^h , \quad a_n \text{ real numbers}$$

where we assume, to begin with, that

$$a_1 \geq a_n \geq \ldots \geq a_N > 0$$

so that one readily has

$$f_1 \geq f_2 \geq \ldots \geq f_N \geq 0 .$$

These restrictions are easily removed (as shown later on). They serve only to simplify our discussion. We are concerned in particular with the relevant case

$$F(\xi) = \sum_{h=1}^{2n+1} \xi^h \tag{8}$$

where, of course, $f_1 = f_2 = \ldots = f_N$. Clearly, in Eq. 8 $F(\xi) \neq 0$ always. We also utilize, with $h \equiv (i_1, i_2, \ldots, i_h)$, the notation

$$F^h(\xi) = \sum_{\substack{i \neq i_1, \\ i_2, \ldots, i_h}} a_i \xi^i .$$

For self-dual functions the expansion $\sigma[F(\xi)]$ only has odd terms:

$$\sigma\left[\sum_{h=1}^{N} a_h \xi^h\right] = \sum_{h=1}^{N} f_h \xi^h + \ldots$$
$$+ \sum_{\substack{1,\ldots,N \\ h_1 < h_2 < h_3}} f_{h_1 h_2 h_3} \xi^{h_1} \xi^{h_2} \xi^{h_3} + \ldots + f_{1,2,\ldots,N} \xi^1 \xi^2 \ldots \xi^N . \tag{9}$$

Consider first

$$f_h = \langle \xi^h \sigma[F(\xi)] \rangle$$
$$= \frac{1}{2}(\langle \sigma[a_h + F^{(h)}(\xi)] \rangle + \langle \sigma[a_h - F^{(h)}(\xi)] \rangle)$$
$$= \langle \sigma[a_h + F^{(h)}(\xi)] \rangle$$

hence, the non-zero contributions to f_h come only from configurations $\xi^1, \xi^2, \ldots, \xi^{h-1}, \xi^{h+1}, \ldots, \xi^N$ such that

$$|F^{(h)}(\xi)| < a_h$$

so that

$$f_h = \frac{1}{2^{N-1}} \mathcal{N}_h$$

where $\mathcal{N}_h$ denotes the number of configurations for which the inequality holds. Thus, if

$$a_1 > a_2 + a_3 + \ldots + a_N , \tag{10}$$

$\mathcal{N}_1 = 2^{N-1}$, $f_1 = 1$, all other $f_h = f_\alpha = 0$ terms other than the first are irrelevant in the expansion. It is impossible to construct a polygon with sides $a_1, a_2, \ldots, a_N$. Otherwise, in general, $\mathcal{N}_h$ is twice the number of distinct triangles which one can construct by taking a_h as the base side, and aligning along the two other sides the remaining $a_i \neq a_h$ (their ordering on each of these sides is immaterial).

For linearly separable functions the f_h or the $\mathcal{N}_h$ are not independent from one another.

As an illustration (we remind that $a_1 > a_2 > a_3 > 0$):

$$\sigma[a_1\xi^1 + a_2\xi^2 + a_3\xi^3] = \frac{1}{2}\xi^1 + \frac{1}{2}\xi^2 + \frac{1}{2}\xi^3 - \frac{1}{2}\xi^1\xi^2\xi^3$$

if a_1, a_2, a_3 can be the sides of a triangle,

$$\sigma[a_1\xi^1 + a_2\xi^2 - a_3\xi^3] = \xi^1$$

if $a_1 > a_2 + a_3$.

Consider next the trilinear term

$$f_{hkl} = \langle \xi^h \xi^k \xi^l \sigma[F(\xi)] \rangle .$$

Proceeding as before and combining f_h, f_k and f_l, one finds, after due cancellations:

$$f_{hkl} + f_h + f_k + f_l = < \sigma[a_h + a_k + a_l + F^{(h,k,l)}(\xi)] = \frac{1}{2^{N-3}} \mathcal{N}_{hkl} .$$

where $\mathcal{N}_{hkl}$ is twice the number of distinct triangles which can be constructed by stretching on the base side the segments a_h, a_k, a_l and laying as before, the remaining $a_i \neq a_h, a_k, a_l$ along the other two sides.

If no such triangle is possible, then

$$a_h + a_k + a_l > \sum_{1 \neq h,k,l} a_i$$

and

$$\mathcal{N}_{hkl} = 2^{N-3}, \quad f_{hkl} + f_h + f_k + f_l = 1 .$$

The r.h.s. can also be obtained from the η-expansion of the l.s. function of $N-2$ variables

$$\sigma\left[\sum_{i \neq h,k,l} a_i\xi^i + (a_h + a_k + a_l)\xi^m \right] ,$$

where the l.h.s is now the coefficient of the linear term:

$$\xi^m \equiv \xi^h \equiv \xi^k \equiv \xi^l .$$

This property, which holds for all higher f_α, as is trivial to see, gives already a first answer to the problem of computing the nonlinear coefficients of the η-expansion of a l.s. function of N variables. They can be obtained as linear coefficients of suitable l.s. functions of $N-2, N-4, \ldots$ variables. The general formula is ($k = 2p + 1$):

$$f_{h_1 \ldots h_k} + \sum_{i_1 < i_2 \ldots < i_{k-2}}^{(h_1 \ldots h_k)} f_{i_1 i_2 \ldots i_{k-2}} + \ldots + \sum_{i_1 < i_2 < i_3}^{(h_1 \ldots h_k)} f_{i_1 i_2 i_3}$$

$$+ \sum_{i=h_1}^{h_k} f_i = \frac{1}{2^{N-k}} \mathcal{N}_{h_1 h_2 \ldots h_k} ,$$

where $N_{h_1 h_2 \ldots h_k}$ is 2^{N-k} if (10) holds with a_1 replaced by $a_{h_1} + a_{h_2} + \ldots + a_{h_k}$ and so on, or is twice the number of possible triangles of base side $a_{h1} + a_{h_2} + \ldots + a_{h_k}$.

We report here, for completeness, the general form when a constant term a_o (threshold) is added and is the only requirement $a_h \geq 0$ ($h \geq 0$). One then also has even terms in the η-expansion, such as $f_{ij} \equiv f_{oij}$, $f_{ijkl} \equiv f_{oijkl}$. Nothing changes, except that now $N_{h_1 \ldots h_k}$ counts the number of triangles (not twice that number) which can be constructed by adding to $a_1 \ldots a_N$ another segment of length a_o. If $a_o = 0$, all previous results are clearly reproduced whether or not $a_1 \geq a_2 \geq \ldots \geq a_N$. One finds

$$
\begin{cases}
f_o = \frac{1}{2^{N-1}} N_o, \\
f_1 = \frac{1}{2^{N-1}} N_1, \\
\vdots \\
f_N = \frac{1}{2^{N-1}} N_N,
\end{cases}
$$

$$
\begin{cases}
f_{12} + f_o + f_1 + f_2 = \frac{1}{2^{N-3}} N_{012}, \\
\vdots \\
f_{ij} + f_o + f_i + f_j = \frac{1}{2^{N-3}} N_{oij}, \\
\vdots \\
f_{123} + f_1 + f_2 + f_3 = \frac{1}{2^{N-3}} N_{123}, \\
\vdots \\
f_{ijk} + f_i + f_j + f_k = \frac{1}{2^{N-3}} N_{ijk}, \\
\vdots
\end{cases}
\tag{11}
$$

$$
\begin{cases}
\vdots \\
f_{h_1 h_2 h_3 h_4} + \sum_{i_1 < i_2}^{(h_1 \ldots h_4)} f_{i_1 i_2} + f_o + f_{h_1} + f_{h_2} + f_{h_3} + f_{h_4} = \frac{1}{2^{N-5}} N_{oh_1 h_2 h_3 h_4}, \\
\vdots \\
f_{h_1 h_2 h_3 h_4 h_5} + \sum_{i_1 < i_2 < i_3}^{(h_1 \ldots h_5)} f_{i_1 i_2 i_3} + f_{h_1} + f_{h_2} + f_{h_3} + f_{h_4} + f_{h_5} = \frac{1}{2^{N-5}} N_{h_1 h_2 h_3 h_4 h_5}, \\
\vdots
\end{cases}
$$

We find thus that the problem of studying l.s. Boolean functions (in particular determining the coefficients of their η-expansion) is entirely equivalent to that of computing the numbers $N_{h_1 \ldots h_{2p+1}}$. Equations (11) generalize therefore the classic triangular inequality into corresponding appropriate "polygonal inequalities" that pertain to polygons that can be constructed by assigning at will segments $a_o, a_1 \ldots a_N$ as sides. This fascinating problem of finite mathematics, group theory and topology, should be investigated on its own merits. Nothing much can be gathered regarding the values $N_{h_1 \ldots h_k}$ with the customary use of hyperplanes in the study of l.s. functions. This knowledge might possibly lead to a deeper geometrical understanding of the properties of the $\mathcal{N}_{h_1 \ldots h_k}$ and the yet unknown ways in which these numbers relate to one another.

2.3. Computation of the η-expansion of arbitrary l.s. functions

We consider again only the case $a_1 \geq a_2 \geq \ldots \geq a_N > 0$, and take a l.s. function of the form

$$\sigma[F(\xi)] = \sigma \left[\sum_{h=1}^{2n+1} \xi^h \right] ,$$

where $N = 2n + 1$ always secures that $F(\xi) \neq 0$. Clearly

$$f_{h_1 h_2 \ldots h_{2p+1}} \equiv f_{1,2,\ldots,2p+1} \equiv f_{(2p+1)} ,$$

$$\sigma[F(\xi)] = f_{(1)} \sum \xi^h + f_{(3)} \sum_{h_1 < h_2 < h_3} \xi^{h_1} \xi^{h_2} \xi^{h_3} + \ldots + f_{(2n+1)} \xi^1 \xi^2 \ldots \xi^{2n+1} .$$

The relations (11) become

$$\begin{cases} f_{(1)} = \frac{1}{2^{2n}} \binom{2n}{n} , \\[2ex] f_{(3)} + 3 f_{(1)} = \frac{1}{2^{2n-2}} \left[\binom{2n-2}{n} + \binom{2n-2}{n-1} + \binom{2n-2}{n-2} \right] , \\[2ex] f_{(5)} \binom{5}{3} f_{(3)} + 5 f_{(1)} = \frac{1}{2^{2n-4}} \sum_{p=0}^{4} \binom{2n-2}{n-p} , \\[1ex] \vdots \\[1ex] f_{(n-1)} + \binom{n-1}{n-3} f_{(n-3)} + \ldots + (n-1) f_{(1)} = \frac{1}{2^{n+2}} \sum_{p=0}^{n-2} \binom{n+2}{n-p} , \\[2ex] f_{(n+1)} + \binom{n+1}{n-1} f_{(n-1)} + \ldots + (n+1) f_{(1)} = 1 , \\[1ex] \vdots \\ \qquad\qquad\qquad\qquad\qquad\qquad\quad = 1 . \end{cases}$$

These equations are readily solved; one finds

$$f_{(1)} = \frac{1}{2^{2n}} \frac{2n}{n} = \frac{(2n-1)!!}{(2n)!!} ,$$

$$f_{(3)} = \frac{-1}{2n-1} f_{(1)} = (-1) \frac{(2n-3)!!}{(2n)!!} ,$$

$$f_{(5)} = \frac{1 \cdot 3}{(2n-1)(2n-3)} f_{(1)} = \frac{(2n-5)!!(5-2)!!}{(2n)!!!} ,$$

$$\vdots$$

$$f_{(2h+1)} = \frac{(-1)^h (2h-1)!!}{(2n-1)(2n-3)\ldots(2n-2h+1)} f_{(1)}$$

$$= (-1)^h \frac{(2n-2h-1)!!(2h-1)!!}{2n!!} .$$

It follows easily that only one half, or one half plus one, of the written coefficients need be computed; in fact

$$f_{(k)} = (-1)^{n+1} f_{(2n+2-k)} .$$

As an example, with $2n-1 = 19$, only 5 of the 2^{19} coefficients of the η-expansion differ in modulus. They are:

$$f_{(1)} = -f_{(19)} = \frac{1}{2^{18}} 48620 ,$$

$$f_{(3)} = -f_{(17)} = \frac{-1}{2^{18}} 2860 ,$$

$$f_{(5)} = -f_{(15)} = \frac{1}{2^{18}} 572 ,$$

$$f_{(7)} = -f_{(13)} = \frac{-1}{2^{18}} 220 ,$$

$$f_{(9)} = -f_{(11)} = \frac{1}{2^{18}} 140 .$$

Consider now an arbitrary l.s. function

$$\sigma \left[\sum_{h=1}^{N} a_h \xi^h \right]$$

with a_h real number and $\sum_{h=1}^{N} a_h \xi^h \neq 0$ for all ξ^h. An ε exists such that

$$0 < N_\varepsilon < \min \left[\sum_{h=1}^{N} a_h \xi^h \right]$$

and, for any a_h, two positive integers n_h, l_h such that

$$0 \leq \frac{n_h}{l_h} - a_h = \varepsilon_h \leq \varepsilon ,$$

so that

$$\left[\sum_{h=1}^{N} \varepsilon_h \xi^h \right] \leq N_\varepsilon \leq \min \left[\sum_{h=1}^{N} a_h \xi^h \right]$$

always, and

$$\sigma \left[\sum_{h=1}^{N} a_h \xi^h \right] = \sigma \left[\sum_{h=1}^{N} a_h \xi^h + \sum_{h=1}^{N} \varepsilon_h \xi^h \right]$$

$$\equiv \sigma \left[\sum_{h=1}^{N} \frac{n_h}{l_n} \xi^h \right] = \sigma \left[\sum_{h=1}^{N} m_h \xi^h \right] ,$$

where $m_h = n_h \prod_{k \neq hk}^{1...N_1}$. Hence, any l.s. function coincides with a function with integer $a_h = m_h$. Now call $F[\xi]_{\text{red}}$ any $F(\xi^1, \xi^2, \ldots, \xi^N)$ in which the following identifications (or "reductions") are made:

$$N = m_1 + m_2 + \ldots + m_N \, ,$$
$$\xi^{N+1} = \xi^{N+2} = \ldots = \xi^{N+m_1-1} = \xi^1 \, ,$$
$$\xi^{N+m_1} = \xi^{N+m_1+1} = \ldots = \xi^{N+m_1+m_2-2} = \xi^2 \, ,$$
$$\vdots$$
$$\xi^{N-m_N+2} = \xi^{N-m_N+3} = \ldots = \xi^N = \xi^N \, .$$

We find

$$\sigma \left[\sum_{h=1}^{N} m_h \xi^h \right] = \sigma \left[\sum_{i=1}^{N} \xi^i \right]_{\text{red}} \quad (N \text{ odd}) \, ,$$

and

$$\sigma \left[\sum_{h=1}^{N} m_h \xi^h \right]$$
$$= \left(f_{(1)}^N \sum_{i=1}^{N} \xi^i + f_{(3)}^N \sum_{i_1 < i_2 < i_3}^{1...N} \xi^{i_1} \xi^{i_2} \xi^{i_3} + \ldots + f_{(N)}^N \xi^1 \xi^2 \ldots \xi^N \right)_{\text{red}} \, .$$

2.4. Continuous versus discontinuous behaviour: transitions

We have considered in the text linear arguments $\sum_{h=1}^{N} a_h \xi^h$ with the condition

$$a_1 \geq a_2 \geq \ldots \geq a_N > 0 \, ,$$

that is, the canonical form of a self-dual l.s. function (threshold $= 0$). All these conditions are most easily removed, as is well known:

α) Putting in the function and in its η-expansion $\xi^h = +1$ or -1 changes a_h into a threshold.

β) A permutation of the variables $\xi^1, \ldots, \xi^N$ changes the original η-expansion into that of the permuted ones.

γ) Only positivity remains. This is removed by a change of sign of any wanted set of variables, accompanied by a change of $f_{j_1, j_2, \ldots, j_h}$ into $(-1)^{r_h} f_{j_1 \ldots j_h}$, where r_h is the number of indices among $j_1, \ldots, j_h$ corresponding to variables ξ^1 that have changed sign. Therefore every canonical l.s. function corresponds to

$$\Omega_N = \frac{2^N N!}{2^z \pi s_i!}$$

(z number of $f_i = 0$, s_i number of $f_i = f_k = \ldots = f_i$)

different l.s. functions, whose η-expansion are immediately deducible from the canonical η-expansion by the operations β) and γ) above; thresholds, and even terms, by α).

We find here the most remarkable property of such functions. We can indeed regard the operations β) and γ) as defining different sectors in N space of the same function, α) as a restriction to a semi-space in a given sector. Hence, dynamical behaviour (change of a_{hk}) will consist of jumps, or transitions, across sectors. As long as one stays within a sector, the changes are irrelevant. The continuum of real numbers reduces thus, in a perfectly defined way, to the discontinuous behaviour of l.s. functions. Since, as we have said, any Boolean function is reducible to a net of l.s. functions, and in the finite and discrete any function is expressible through Boolean functions, our initial claim that the study of l.s. NE already describes the most general nonlinear behaviour is substantiated. How to construct nets for this aim is the subject of the next Part.

3. General Boolean NE

3.1. Linearization in tensor space

We suppress here the restriction that the arguments of our NE be linear, so that the most general Boolean functions are allowed into them. It is convenient here to work directly with the tensorial signum expansion. If each neuron h of the net has as excitation function the real function $f^{(h)} = f^{(h)}(\xi^1, \xi^2, \ldots, \xi^N)$, we can write the NE for a general net as

$$\xi^h_{m+1} = \sigma[f^h(\xi^1_m, \ldots, \xi^N_m)] = \sum_\alpha f^h_\alpha \eta^\alpha_m = f^{hT} \eta_m \ .$$

3.2. Next-state matrix

The augmented equations give for each state η of the net at time t its successor at time $t + \tau$. Aside from the normalization coefficients, which is introduced to simplify the formalism, note that now the columns of the matrix Φ contain all, and only, the possible 2^N states of the nets. Joining vectors as columns of a matrix, we have then the matricial equation:

$$\Phi_{m+1} = F\Phi_m \ , \quad \text{with } \Phi_0 = \Phi \ .$$

We remark now that the effect of F on Φ_m is to permute its columns, or to suppress some column and bring one of the remaining ones in its place (degeneration). That is

$$F\Phi = \Phi P \ ,$$

where P is a $2^N \times 2^N$ permutation matrix, which may be degenerate. Of course

$$F^k = \Phi P^k \bar{\Phi} \ , \quad P^k = \bar{\Phi} F^k \Phi \ .$$

F and Φ have the same Eigenvalues (P can always be put in diagonal form). If P is degenerate, i.e. $\det(P) = 0$, P is a **projection** and the net exhibits, correspondingly, **transient** states.

3.3. Normal modes, attractors

We show next that NE exhibit **normal modes**, just as linear equations do, though more complex than the simple periodic sequence typical of linearity in N space. They intertwine into "reverberations"[1], since they stem from nonlinearity in 2^N-space.

Their interpretation is in principle the same as that expressed by Eigen and Schuster[8] for "quasi-species" in their classic discussion of hypercycles (for which NE might be an apt tool).

Let the matrix Δ, $\det(\Delta) \neq 0$, diagonalize F:

$$F\Delta = \Delta\Lambda , \quad \Lambda \text{ diagonal} ,$$

then

$$P \cdot \Phi\Delta = \Phi\Delta \cdot \Lambda ,$$

i.e. $\Phi\Delta$ diagonalizes P.

Since P is a permutation matrix, its characteristic polynomial (same as of F) is necessarily of type:

$$\lambda^a \prod_b (\lambda^b - 1)^{c_b} = 0$$

with

$$a + \sum_b b c_b = 2^N , \quad c_b \geq 0 .$$

Thus

$b = 1$ implies c_1 **invariant states**

$\lambda = 0$ implies **transients**

$b > 1$ implies c_b **cycles** of period b, corresponding to $\lambda_h = e^{2\pi h i/b}$ ($b = 1$ can of course be regarded as a cycle of period 1). If we set

$$\eta_m = \Delta \chi_m$$

the NE read

$$\Delta \chi_{m+1} = F \Delta \chi_m = \Delta \Lambda \chi_m ,$$

so that

$$\chi_{m+1} = \Lambda \chi_m$$

or

$$\chi_{\alpha,m+1} = \lambda_\alpha \chi_{\alpha,m}$$

express the wanted normal modes.

We consider now the normalized ξ-state matrix of the net $\Phi_{(N)}$; with $N = 3$, e.g., it is

$$\Phi_{(3)} = 2^{-3/2} \begin{pmatrix} 1 & -1 & 1 & 1 & -1 & -1 & 1 & -1 \\ 1 & 1 & -1 & 1 & -1 & 1 & -1 & -1 \\ 1 & 1 & 1 & -1 & 1 & -1 & -1 & -1 \end{pmatrix} .$$

We can augment the $N \times 2^N$ ξ-matrix $\Phi_{(N)}$ to the $2^N \times 2^N$ η-state matrix, from

$$\eta = \begin{pmatrix} 1 \\ \xi_N \end{pmatrix} \times \ldots \times \begin{pmatrix} 1 \\ \xi_1 \end{pmatrix} ,$$

as follows

$$\Phi_{(N)} = \begin{pmatrix} \frac{1}{\sqrt{2}} & \frac{1}{\sqrt{2}} \\ \frac{1}{\sqrt{2}} & -\frac{1}{\sqrt{2}} \end{pmatrix} \times \ldots \times \begin{pmatrix} \frac{1}{\sqrt{2}} & \frac{1}{\sqrt{2}} \\ \frac{1}{\sqrt{2}} & -\frac{1}{\sqrt{2}} \end{pmatrix} \quad \text{(N times)} .$$

$\Phi_{(N)} = \Phi$ is a Hermite matrix such that

$$\Phi = \Phi^T; \quad \Phi^2 = 1; \quad \Phi = \Phi^{-1}, \quad \det(\Phi_N) = (-1)^N .$$

We can thus also augment the N ξ-state NE to the 2^N η-state form

$$\eta_{m+1} = F\eta_m ,$$

in which F is a $2^N \times 2^N$ matrix whose first row has all elements $= 1$; N "linear" rows have the *linear* coefficients, and the remaining ones are given by tensor multiplication.

We obtain thus the **central result**, that passage from ξ- to η-space **linearizes** the NE. Thus

$$\eta_m = F\eta_{m-1} = \ldots = F^m\eta_0 .$$

That passage to functional space should linearize the NE is of course not surprising; the relevant feature is that 2^N is (of course) **finite**, and from now on standard matrix algebra can be used.

3.4. Synthesis of nets: the inverse problem

The procedure outlined thus far solves the synthesis problem if what is wanted is a net that leads each state into a prescribed successor: the N "linear" rows of $F = \Phi P \Phi$ determine the coefficients $f_\alpha^{(h)}$. We are more interested, though, in the synthesis of a net which, starting from a given state, follows in time a prescribed sequence of states. This we call the **"inverse problem"** and it is equivalent to solving linear equations with the Cramer theorem. Call the wanted sequence (assumed, for simplicity, without repetitions)

$$\eta_1 = X_1, \quad \eta_2 = X_2, \ldots, \eta_{2^N} = X_{2^N}$$

and define $\mathcal{F}$ such that

$$\mathcal{F}(\chi_1\chi_2\cdots\chi_{2^N}) = (\chi_2\chi_3\cdots\chi_{2^N}\chi_1) \,,$$

that is

$$\mathcal{F}\chi = \chi\Theta \,,$$

where the $2^N \times 2^N$ permutation matrix Θ shifts the first column to the last place or moves such shifts.

Then

$$\mathcal{F} = \chi\Theta\chi_R^{-1}$$

solves the problem.

χ_R^{-1} means right-inverse, and pseudoinverse matrices will have to be considered whenever necessary (repetitions, transients). The states X_α need not be distinct. They will actually reproduce the behaviour described by the characteristic equation if $\mathcal{F}$, χ are brought into F, Φ by an appropriate permutation. Note that χ is not normalized (unlike Φ), and that our synthesis problem requires only the knowledge of the "linear" rows of $\mathcal{F}$, which is the product of two known matrices (which, brought into the (0,1)-notation, are sparse).

3.5. Separable versus Boolean nets; connections with spin formalism

Working with permutation matrices P is far simpler than using F. If we request that the net be Boolean, it is immediately feasible. If, instead, we restrict the net to be linear separable, or of some other special type, we have to start with F:

$$P = \Phi F \Phi$$

unless the condition for a matrix P to describe a linear separable net is known (it is actually easy to see that matrices P satisfying a same equation may give rise, depending upon the phase relations one chooses for cycles, to separable or to nonseparable nets). This problem has recently been solved, and will be discussed elsewhere.[9]

Working with this matricial notation, such as with η-expansions, does not bring into light a main feature of NE: the close relation between the continuous and discontinuous aspects of the theory. This is actually an essential point for any concrete application of the formalism to adaptive or learning devices, or just for reliable design. We have already said that it is always possible, by changing the number of variables and the value of τ (in a way which we forgo but should be evident in the present formalism) to reduce any Boolean function or net to a larger linear net. The consideration of Boolean nets cannot therefore yield behaviours that one cannot also obtain from larger linear nets. Use of one or the other is therefore to be considered a matter of convenience rather than a matter of principle.

Finally, we note that analysis of the behaviour and synthesis of a net reduces thus to study of permutation matrices P. The problem is therefore reduced to the study of representations of finite groups generated by the primitive roots of 1, plus degenerations due to 0's.

It is evident to a physicist that the matrix F is the **transfer matrix** (Q) from a **spin state** to another (N neurons $\rightarrow N \frac{1}{2}$ spin elements), and that many other such connections can be established (e.g. $\Phi \rightarrow$ unitary). With neural net models we are interested primarily in finding the exact couplings that give the desired behaviour. This is not the type of information one wants when considering spin systems (taken as a paradigm for any discrete, finite, time-quantized model). One hopes however, that when the need arises to characterize — albeit schematically — the subnets of a net as individual elements of a new net, renormalization group methods may be of use, provided they do not destroy basic structures. This is just an example to reiterate that the conclusion of our task means only that new ones are posed, in a way that, pleasingly enough, becomes more and more interdisciplinary with each further step we take.

References

1. Caianiello E.R., *J. Theor. Biol.*, (1961) 1:209.
2. Caianiello E.R., *Nuovo Cimento* X, 1634 (1953).
3. Caianiello E.R., Marinaro M., *Phys. Scripts.* (1986) **34**, p.444.
4. McCulloch W.S., Pitts W. (1943) "A logical calculus of ideas immanent in nervous activity", *Bull. Math. Biophys.*, 5.
5. Caianiello E.R. (with Capocelli R.) (1971) *Kybernetik*, 8:233 and *III I.J.C.P.R.*, Nov. 1976, Coronado; (1965) Calcolo, 2:83; Coral Gables Conf. "Orbis Scientiae", Jan. 1974; (1977) *Biol. Cybern.*, 26:151; (1987) Caianello E.R., Aizerman M.A. ed.'s "Topics in the General Theory of Structures", D. Reidel, Dordrecht.
6. Caianiello E.R. (with A. de Luca) (1966) "Decision equations for binary systems-application to neuronal behaviour", *Kybernetik* 3:33; (with Lauria F.) (1970) "Il sistema nervoso centrale." Atti Conv. Med. Eur., Ist. Angelis.
7. Caianiello E.R., Simoncelli G. (1981) "Polygonal inequalities as a key to neuronic equations", *Biol. Cybern.*, 41:203.
8. Eigen M., Schuster P. (1979) "The hypercycle-principle of natural self-organization", Springer, Berlin Heidelberg, New York.
9. Caianiello E.R., Marinaro M., "The inverse problem for l.s. Boolean nets", in print.

The Dynamics of Searches Directed by Genetic Algorithms

John H. Holland
The University of Michigan

Almost all difficult problems reduce to a search for useful solutions in a large space of possibilities. Typically, the space is so large that exhaustive search is not feasible, and no nontrivial decompositions exist that allow the value of a proposed solution to be estimated by *adding up* values assigned to the factors. That is, difficult problems involve large search spaces with "nonlinear" evaluation functions. One can only *sample* such a space, searching for regularities or regions that are "enriched" in useful solutions. To have much hope of success, the search must exploit the information it acquires to bias its sampling toward "promising" regions. In short, the search procedure must *learn*.

The classic example of such a search in the machine learning literature is Samuel's [1959] checkersplayer. Samuel represents strategies for playing checkers by using a weighted linear form, $\Sigma_j w_j \theta_j(b)$, where each *parameter*, θ_j: B → *Reals*, assigns a property value (such as *pieces ahead*, *number of pieces beyond the centerline*, etc.) to each possible board configuration $b \epsilon B$, and the *weights* w_j estimate the importance of each of the parameters. Because the parameter functions θ_j are given at the outset and are fixed, the space of all n-tuples of weights $\{w_j\}$ determines the space of strategies searched. Assuming 20 parameters and 20 distinct values for each weight, this is a space with $20^{20} \simeq 10^{26}$ strategies. Because a strategy can only be evaluated through play of the game, this is certainly not a space that one would search exhaustively. Samuel controls the search (through learning) by using the n-tuples to make *predictions* about outcomes *during* the play of the game. (In effect, the n-tuple models both the strategy of the checkersplayer and that of its opponent, and is used, with a lookahead technique, to predict the value the checkersplayer expects to attain after it executes a sequence of several moves.) The predictions are checked against the actual outcome and the n-tuple of weights is revised accordingly. The details are complex, but the main point is that the information acquired biases the search to certain neighborhoods of the more successful strategies -- it is *not* a uniform random search of the space of possibilities.

A question immediately presents itself: In searching a large space with a nonlinear evaluation function, what kinds of conditions afford an advantage to *learned* biases? To give some precison to this question, we can begin by considering some large set of structures (rules, strategies, chromosomes, policies, or the like) that can be represented by k-bit strings. (There is little loss in generality in using k-bit strings -- any function can be approximated arbitrarily closely by a discrete representation of its argument space in terms of binary strings, with k large enough; and, of course, any computer model ultimately rests upon a representation in terms of binary words.) The unkown evaluation function then takes the form u: $\{1,0\}^k \to$ *Reals*. Each $x \in \{1,0\}^k$ represents a structure to tried, and $u(x)$ is the value returned when it is tried. Note that the evaluation of a single x may be a time-consuming task as, for example, when x is a strategy for playing a game. Here we are only concerned with conditions under which the information returned -- the value $u(x)$ of the structure x -- will be helpful in biasing the search of $\{1,0\}^k$.

The approach to "learned biases" presented here involves the following steps (details will follow):

1) The function u: $\{1,0\}^k \to$ *Reals* is first transformed, using a *hyperplane analysis* (somewhat reminiscent of a Fourier analysis). Under this transformation, information accumulated by sampling u's argument space $\{1,0\}^k$ is more transparently related to possibilities for further biasing the sampling process. The hyperplane analysis is accomplished by assigning a probability distribution to the argument space so that u becomes a random variable, whereupon subsets of the space become *events* having well-defined *expectations*. The expectations associated with selected sets of hyperplanes of progressively lower dimension are then used to provide a unique, transformed representation for any finite, nonlinear function u.

2) Because hyperplanes of higher dimension are larger subsets of $\{1,0\}^k$, they typically receive a larger fraction of any set of samples drawn randomly from $\{1,0\}^k$. As a consequence, estimates of the expectation $u(s)$ associated with a higher-dimensional hyperplane s will be confirmed faster than similar estimates for lower-dimensional refinements of s. A search can only reasonably be biased in terms of available information -- it makes little sense to bias the distribution with respect to hyperplanes that have received only one or two samples. Accordingly, as the number of samples increases, biases should proceed from biases based on estimates for high-dimensional hyperplanes to biases involving lower-dimensional refinements of those hyperplanes.

3) The problem, then, is to design a feasible algorithm that, as information accumulates, provides the biases suggested by the *hyperplane transform*. It is easy to see that an explicit calculation of the hyperplane transform at each step involves too much

computation. However, it can be proved that *genetic algorithms* rapidly provide the biasing implied by the hyperplane transform without explicitly carrying out the calculations involved.

[Throughout the remainder of this paper when the term *random* is used without further qualification, as in "*random*-ly generated", it implies a sample space with a *uniform* distribution.]

The Hyperplane Transformation.

The hyperplane transformation represents u in terms of its averages over certain easily defined hyperplanes in the space $\{1,0\}^k$. These hyperplanes, called *schemas* in [Holland 1975], are specified by strings from the set $\{1,0,*\}^k$. In $s \in \{1,0,*\}^k$ the "*" is interpreted as a "wildcard" or "don't care" symbol; the positions in s that are occupied by 1 or 0 (i.e., those positions *not* occupied by *'s) are called the *defining loci of s*. s specifies a subset (hyperplane) of $\{1,0\}^k$ under the rule that $x \in s$ if and only if x matches s at every defining locus of s. Thus, 1**...* designates the subset of all strings that start with a 1, and 11...1* designates the two-element subset {11...11, 11...10}. For convenience, a hyperplane defined with c defining loci will be called a hyperplane of *level c* (it is a hyperplane of *dimension* k-c).

To define averages over these hyperplanes, we must convert $\{1,0\}^k$ to a *sample space* by imposing a probability distribution, $p: \{1,0\}^k \to [0,1]$. Then u becomes a *random variable* with a well-defined *marginal expectation* ,

$$u(X) = \sum_{x \in X} p(x)u(x) \Big/ \sum_{x \in X} p(x) ,$$

over each *event* (subset) $X \subset \{1,0\}^k$. (The average of a set of samples drawn from X constitutes an estimate of u(X).) In particular, each schema s can be assigned the expectation u(s). Under an algorithm that biases p as information accumulates, p becomes a function of time, p(t); u(s,t) then designates the expected value of u on s under the current probability distribution p(t). In the development that follows, $p(s) = \sum_{x \in s} p(x)$ will designate the probability of the event s under p.

Each choice of a set of defining loci partitions the space $\{1,0\}^k$ into disjoint subsets containing equal numbers of elements. For example, schemas having positions 1 and 2 from the right as defining loci, partition $\{1,0\}^k$ into the disjoint subsets {**...*11, **...*10, **...*01, **...*00}. We will use strings from $\{d,*\}^k$ to designate the partitions specified by defining loci. Each such partition can be assigned a unique index by the simple expedient of

substituting 1's for d's and 0's for *'s throughout the string,
treating the result as a binary integer. Thus, the partition just
used as an example is named by **...*dd, and it has the index
00...011 = 3_{10}.

It will be convenient to speak of the set partitions specified by
c defining bits as being the partitions at *level c* .

Because **...** designates the whole space, u(**...**) is just the
expected value of u under p. Consider now a partition specified by
a single defining locus, say the partition **...*d. The two schemas
that are the elements of this partition, **...*1 and **...*0, have
well-defined marginal expectations under p, u(**...*1) and
u(**...*0) respectively. Using the index 1 associated with the
partition **...*d, define

$$\delta_1 = \delta(**...*d)$$
$$= p(**...*1)[u(**...*1) - u(**...**)].$$

Roughly, δ_1 measures the departure of the marginal expectation of
the elements of the partition from the overall average u(**...**).

It follows at once from the definition that
$$u(**...*1) = u(**...**) + [\delta_1/p(**...*1)].$$
It also follows that
$$u(**...*0) = u(**...**) - [\delta_1/p(**...*0)],$$
because u(**...**) = p(**...*1)u(**...*1) + p(**...*0)u(**...*0).

Clearly a δ can be assigned in the same way to each 2-element
partition specified by a single defining locus, yielding a set $\{\delta_1, \delta_2,$
$\delta_4, ..., \delta_{2k-1}\}$. Moreover, given any schema s defined on m loci,
one can partition s into two subsets by selecting one *additional*
defining locus, obtaining two schemas, call them s_1 and s_0, defined
on the m+1 loci. That is, s_1 and s_0 play much the same role with
respect to s that was played by **...*1 and **...*0 with respect to
... in the example above. With a little care, we can set up an
induction based on this analogy. It assigns a unique δ to every
indexed partition, such that for an arbitrary schema s,

$$u(s) = \delta_0 + [\sum_{\substack{s'\in s \\ s' \neq **...**}} 2^{-(m(s)-m(s'))}\sigma(s')\delta_{j(s')}]/p(s) ,$$

where $\delta_0 = u(**...**)$, the expectation of u under p,
   m(s') = the number of defining bits in s',
and $\sigma(s')$ = +1 if s' has no 0's or an even number of 0's in its
                defining bits,
      = -1 otherwise.

Because $\{1,0\}^k$ contains 2^k points and because there are 2^k selected partitions with associated δ_j's, it is easy to show that the hyperplane representation is unique for each distinct function u and distribution p. The reverse transform is given by

$$\delta_j(s) = \sum_{\substack{s' \in s \\ s' \neq **\ldots**}} (-2)^{-(m(s)-m(s'))} p(s')[u(s') - \delta_0] ,$$

for any schema for which the defining bits are all 1's.

Because the u(s) are marginal expectations, the average of any sample drawn from s under the distribution p(t) constitutes an *estimate* of u(s, t). The p(s, t) are determined by the biasing process so that the $\delta(s, t)$, as functions of the u(s, t) and the p(s, t), can also be estimated.

Using Estimates of u(s) to Bias the Search of $\{1, 0\}^k$.

The hyperplane averages take a particularly simple form when the function u is linear over individual loci and the probabilities assigned to individual alleles are independent of one another. We will consider this case first, then we'll proceed to the general case. In this discussion two useful pieces of terminology from genetics will be adopted: The positions along a string will be called *loci*, and the values that can be inserted at any position will be called *alleles*.

u is a linear function of values assigned to individual loci when
$$u(x) = \sum_h u_h(x_h),$$
where $x = x_1 x_2 \ldots x_k \in \{0,1\}^k$, and $u_h: \{0,1\}^k \rightarrow Reals^+$ assigns values to the two alleles at locus h. The probabilities of the individual alleles are independent of each other when the probability of an arbitrary string x is given by $p(x) = \Pi_h p_h(x_h)$, where $p_h(x_h)$ is the probability of allele x_h at locus h.

Consider now the δ's at level 1, that is, the $\delta_j(s)$ for which $j(s) = 2^h$. When u is linear and the probabilities of the alleles in a string are independently assigned, it is a simple exercise to show that
$$\delta_{2h} = p_h(1)p_h(0)[u_h(1) - u_h(0)].$$
Note that the formula for δ_{2h} involves only values and probabilities assigned to the alleles, 1 and 0, at locus h. It is also easy to show that the δ's for partitions of level higher than 1 are all zero. It follows that linear functions can be solved via searches involving only the level 1 hyperplanes, as intuition might suggest.

In more detail: Each hyperplane at level 1 is specified by the single allele at its defining locus. For example, the partition of index 2 has as elements the two hyperplanes **...*1* and **...*0*. For each locus, we can select the allele that corresponds to the above-average hyperplanes at that locus. The string made up of the alleles so-selected optimizes u. Stated another way, the optimizing string lies in the intersection of all the above-average level 1 hyperplanes. Thus, if the estimates of u(s) for these level 1 hyperplanes reflect the true values, an optimal string is easily determined.

In the general case, a hyperplane s supplies nontrivial information when the $\delta_{j(s)}$ associated with partition j(s) is non-zero. Drawing on the terminology of mathematical genetics, the $\delta_{j(s)}$ for which j(s) is *not* a power of 2 amount to *epistatic* effects -- departures from the expectations that would hold if u were linear. When there are epistatic effects, the search for better values of the evaluation function will be hampered until the corresponding u(s) can be estimated.

To arrive at some idea of the difficulty posed by epistatic effects note that a hyperplane at level j can be expected to receive a fraction 2^{-j} of a uniformly distributed set of samples. Thus, about half of all samples can be used to estimate u(s) for any level 1 hyperplane, while the proportion drops by half each time the level is increased by 1. Since the reliability of an estimate of u(s) depends upon the number of samples s receives, the rate at which relevant information accumulates drops by half each time the level is increased by 1. It is clear that non-zero $\delta_{j(s)}$ at higher levels greatly increase the difficulty of a search.

Matters also become more complicated when the probabilities assigned to loci are no longer independent of one another. Then even linear functions may yield non-zero δ's at deep levels. Intuitively, the information supplied by a sampling procedure depends upon the underlying distribution. A biased distribution emphasizes certain regions, and the information supplied by the δ_j's is modified accordingly.

The rate at which information accumulates about the u(s) dictates the order in which the search of the argument space $\{1,0\}^k$ should be biased. Estimates of a given level of reliability accumulate first for level 1 hyperplanes, then, at half that rate, for level 2 hyperplanes, and so on. Thus we should expect a realistic search to produce biases that, at first, are largely dependent upon estimates for u(s) associated with low-level hyperplanes. Biases based on higher-level hyperplanes enter as the number of samples increases. A search so-directed has the concomitant advantage that a more simple problem is solved more quickly.

The Genetic Algorithm as a Hyperplane-Directed Search Procedure.

(1) *Description of the genetic algorithm.*

[The algorithm acts on a set $B(t)$ of M strings $\{x_1, x_2, ..., x_M\}$ over the alphabet $\{1,0\}^k$ with observed values $u(x_j)$.]

Briefly, a genetic algorithm has the form (see Holland et al. [1986] for more details):

(1) Compute the average strength $u^(t)$ of the strings in $B(t)$, and assign the normalized value $u(x_j)/u^(t)$ to each string $x_j \epsilon B(t)$.

(2) Assign each $x_j \epsilon B(t)$ a probability $p(x_j, t)$ proportional to its normalized value. Using this probability distribution, select n pairs of strings, n << M, from B(t), and make copies of them.

(3) Apply *crossover* (and, possibly, other *genetic operators*

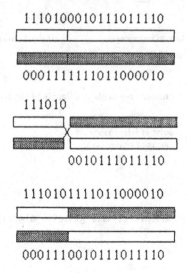

Figure 1.1 An example of the crossover operator.

118

such as *mutation*) to each copied pair, forming 2n new
strings. *Crossover* is applied to a pair of strings as follows:
Select at random a position i, $1 \le i \le k$, and then exchange the
segments to the left of position i in the two strings (see Figure
1.1).
(4) Replace the 2n lowest strength strings in B(t) with the
2n strings newly generated in step (3).
(5) Set t to t+1 in preparation for the next use of the
algorithm and return to (1).

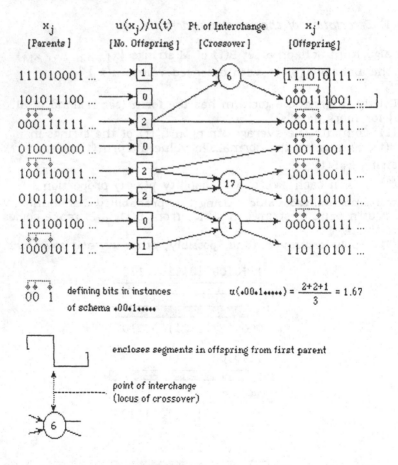

Figure 1.2 An example of the genetic algorithm acting on
    schemas.

[Step (4) in the algorithm may be modified to prevent one kind of string from "overcrowding" B(t) (see Bethke [1981] and DeJong [1975] for details). Contiguity of constituents, and the building blocks constructed from them, is significant under the crossover operator. Close constituents tend to be exchanged together. Operators for rearranging the loci, such as the genetic operator *inversion* , can bias the rule-generation process toward placements in which interacting loci are in close contiguity. Other genetic operators, such as *mutation* , have lesser roles in this use of the algorithm, mainly providing "insurance" (see Holland [1975], Chapter 6, sections 2, 3, 4 for details).]

The fundamental theorem for genetic algorithms (see Holland [1975]) can be rewritten as a procedure for progressively biasing a probability distribution over the space $\{1,0\}^k$:

$$p(s,t+1) \geq [1-\lambda(s,t)][1-P_{mut}]^{d(s)}[u(s,t)/u(t)]p(s,t).$$

Here, $p(s,t+1)$ is the expected fraction of the population that will be occupied by the instances of s at time t+1 under the genetic algorithm, given that $p(s,t)$ is the fraction occupied by s at t. The factors entering into the right hand side are:

$[u(s,t)/u(t)]$, the ratio of the observed average value of the schema s compared to the overall population average, determines the rate of change of $p(s,t)$, subject to the "error" terms $[1-\lambda(s,t)][1-P_{mut}]^{d(s)}$. If $u(s,t)$ is above-average the proportion of schema s increases, and vice-versa. The "error" terms are the result of breakup of instances of s because of crossover and mutation, respectively. Specifically,

$\lambda(s,t) = P_{cross}(l(s)/k)p(s,t)$ is an upper bound on the loss of instances of s resulting from crosses that fall within the interval of length $l(s)$ determined by the outermost defining loci of the schema, and

$[1-P_{mut}]^{d(s)}$ gives the proportion of instances of s that escape a mutation at one of the $d(s)$ defining loci of s.

(The underlying algorithm is stochastic so this equation only provides a bound on expectations at each time-step; using the terminology of mathematical genetics, this equation supplies a *deterministic model* of the algorithm under the assumption that the expectations are the values actually achieved on each time-step.)

In any population that is not too small, distinct schemas will almost always have distinct subsets of instances if the number of instances is relatively small. For example, in a randomly generated population of 2500 instances, the largest subset of any given schema with 8 defining loci can be expected to have about 10 instances. There are

$$\left|\begin{array}{c} 2500 \\ 10 \end{array}\right| \simeq 3\times10^{27}$$

ways of choosing this subset, so that it is extremely unlikely that the subsets of instances for two such schema will be identical. (Looked at another way, the chance that two schemas have even *one* instance in common is less than $10\times2^{-8} \simeq 1/25$ if they are defined on disjoint subsets of loci). Because the sets of instances are overwhelmingly likely to be distinct, the observed averages $u^(s,t)$, will have little cross-correlation. As a consequence, the rate of increase (or decrease) of a schema s under a genetic algorithm is largely uncontaminated by the rates associated with other such schemas. Loosely, the rate is uninfluenced by "crosstalk" from the other schemas. To gain some idea of how many schemas are so processed consider the following:

Theorem. Select some bound e on the transcription error under reproduction and crossover, and pick l such that $l/k \leq e/2$. Then in a population of size $M = c_1 2^{l/k}$, obtained as a uniform random sample from $\{1,0\}^k$, the number of schemas propagated with an error less than e greatly exceeds M^3.

proof:

1) Consider a "window" of 2l contiguous loci in a string of length k such that $2l/k = e$. Clearly any schema having all its defining loci within this window will be subject to a transcription error less than e under crossover.
2) There are

$$\left|\begin{array}{c} 2l \\ l \end{array}\right| \simeq 2^{2l}/[\pi l]^{-1/2}$$

ways of selecting l defining positions in the window, and there are 2^l different schemas that can be defined using any given set of l of defining loci. Therefore, there are approximately $2^{3l}/[\pi l]^{-1/2}$ distinct schemas with l defining positions that can be defined in the window.
3) A population of size $M = c_1 2^l$, for c_1 a small integer, obtained by a uniform random sampling of $\{1,0\}^k$ can be expected to have c_1 instances of *every* schema defined on l defining positions.

Therefore, for the given window, there will be approximately $M^3/(c_1)^3[\pi l]^{-1/2}$ schemas having instances in the population and defined on some set of l loci in the window.

4) The same argument can be given for schemas of length l-1, l-2, ..., and for l+1, l+2, ..., with values of

$$\left| \frac{2l}{l \pm j} \right|$$

decreasing in accord with the binomial distribution. There are also k-l-1 distinct positionings of the window on strings of length k. It follows that many more than M^3 schemas, with instances in the population of size M, increase or decrease at a rate given by their observed marginal averages with a transcription error less than e.

QED

A genetic algorithm's ability to meaningfully bias the sampling rate of a large number of schemas while processing a relatively small set of instances is called *implicit parallelism* (ne' *intrinsic parallelism*, Holland [1975]).

From the point of view of sampling theory, 20 or 30 instances of a schema s constitutes a sample large enough to give some confidence to the corresponding estimate of u(s). Thus, for such schemas, the biases p(s,t) produced by a genetic algorithm over a succession of generations are neither much distorted by sampling error nor smothered by "crosstalk".

(2) *Effects of the δ's on the search generated by a genetic algorithm.*

The fundamental theorem makes it clear that the biases p(s,t+1), produced by a genetic algorithm at time t+1, depend directly upon the observations u^(s,t) and biases p(s,t) at time t. The hyperplane transform H, applied to the sample space defined by the new distribution p(t+1), determines a new set of $\delta_i(t+1)$. Any partition i with a large associated $\delta_i(t+1)$ has at least one element (schema) s with a value u(s,t+1) considerably above the average $u_0(t+1)$. Such a schema s, accordingly, offers an opportunity for improvement. However, the opportunity will be exploited only if the genetic operators, applied to the current population, are likely to form instances of s. That is, the level of the partition must not be so deep that it cannot be reached, via a few recombinations and mutations, from currently exploited hyperplanes.

By looking at the levels in which the non-zero δ's are distributed, one can attain a qualitative understanding of the trajectory induced by a given nonlinear function u. Consider again an initial population $B(0)$ that has been generated using a uniform random distribution over $\{1,0\}^k$. If the size M of that population is 2^m, then we can expect multiple copies of all schemas with fewer than m defining loci. If, as earlier, we set a bound e on the error produced by the genetic operators, requiring $[1-\lambda(s,t)][1-P_{mut}]^{d(s)}$ < e, then in excess of M^3 of these schemas will be processed with an error less than e (for an appropriately chosen size M). Because the above-average schemas increase their instances exponentially (with exponent $[1-e][u(s,t)/u(t)]$) they soon come to occupy a substantial fraction of the population. Sampling then is concentrated on these hyperplanes and their intersections (though not exclusively).

While the above-average schemas with instances in the initial population are being exploited, new schemas exploiting deeper δ's are discovered in two ways: 1) mutation of loci that are near the loci defining one of the initially exploited schema, 2) recombination, under crossover, of fragments of the initially exploited schemas. It is worth looking at each of these discovery processes in more detail:

Mutations of loci continguous to the defining loci of an above-average schema s can provide instances of schemas not previously present in the population. Each such schema, s', is a refinement of s (i.e., s contains s') and hence is an element of a deeper level partition. Consider, then, a partition of s comprised of 2^h hyperplanes obtained by specifying the values for some set of h loci contiguous to s. Because the mutations are allocated randomly, the number of samples $n(s')$ of s' will be approximately $2^{-h}n(s)$, where $n(s)$ is the number of samples allocated to s. The *central limit theorem* assures that the averages of different sets of samples of any s' will be distributed approximately as a Gaussian distribution, whatever the probabilities assigned to the elements x in s'. The sampling process (mutation operator) thus produces an estimate of $u(s')$ with a variance that decreases as $\sqrt{[2^{-h}n(s)]}$. Very roughly, then, one would expect to reliably discover s' for which $u(s') > u(s)$ at a rate on the order of $\sqrt{[2^{-h}n(s)]}$. In other words, mutation will discover improvements in the vicinity of s at a rate that falls off as the square root of the number of samples allocated to s. In biological terms, this would correspond to an *adaptive radiation* wherein variants of the prototype s provide incremental improvements.

This process of "exploring the neighborhood" via mutations contrasts sharply with the jumps produced by crossover. To develop the contrast, consider a randomly generated population with instances of two above-average schemas s_1 and s_2, where $d(s_1) \geq d(s_2)$. Let the defining bits of s_1 and s_2 be such that there are no instances of the schema s designating the intersection of s_1 and s_2, and let $u(s) > \max\{u(s_1), u(s_2)\}$. Under these conditions, on the order of $d(s_2)/2$ mutations must accumulate in some instance of s_1 before an instance of s appears in the population. The mutation rate can of course be increased to make the accumulation more rapid, but only at the cost of making it increasingly *unlikely* that s will be "copied" into successive generations (see the example just below). Mutations tend to explore in a linear way -- the depth of the exploration is a linear function of the number of generations elapsed.

On the other hand, a single crossover between parents that are instances of s_1 and s_2, respectively, can yield an instance of s. That is, an above-average schema s at depth 2b that falls in the intersection of established schemas s_1 and s_2 at depth b can be discovered in a single generation. This doubling of depth in successive generations comes about whenever established schemas can be combined as "building blocks" to yield improved schemas. Under these conditions, crossover explores with directed exponential increases in depth.

(3) *An Example*.

The following example gives a quantitative measure of the difference between mutation and crossover:

Let schema s be defined on contiguous loci, i.e. $l(s) = d(s)-1$. Divide the defining loci of s into disjoint subsets of contiguous loci, forming schemas s' and s" that have s as their intersection. Let $e(s)$ be an upper bound on the allowable rate of transcription errors for s. (That is, if the transcription error rate exceeds $e(s)$, then even an above-average schema s with an instance in the population $B(t)$ is unlikely to have instances that persist through successive generations.)

Let s* be any schema properly contained in s' and properly containing s. Assume that for all such s*, $u(s') > u(s^*)$, while $u(s) > u(s')$. That is, the schema s' is "surrounded" by a "desert" of lesser valued schemas $\{s^*\}$, at the edge of which is a "higher peak" s. For ease of calculation, assume that s' has a single persistent instance in the population.

Consider, first, the time expected to elapse, under mutation alone, before the "peak" s is attained from s'. Note that the bound on transcription error sets an upper bound on the mutation rate because $(1-P_{mut})^{d(s)} > 1-e(s)$ and this yields the inequality

$$P_{mut} < e(s)/d(s).$$

For an instance of s' to be an instance of s, the alleles at all the defining loci $D(s")$ of s" must match the corresponding alleles of s. Schemas s* that are refinements of s' (but not equal to s) cannot persist because $u(s*) < u(s')$. As a consequence, there is no way for instances of s' to gradually accumulate the mutations that "lie on the way" to s. To attain s all the required mutations must occur simultaneously in some single instance of s'. In a randomly generated population, we can expect that half of the alleles at the loci $D(s")$ match the corresponding alleles of s. Thus $d(s")/2$ mutations must occur simultaneously in some instance of s'. This will occur with probability

$$(P_{mut})^{d(s")/2} < [e(s)/d(s)]^{d(s")/2},$$

with a corresponding search time

$$T_{mut} > [d(s)/e(s)]^{d(s")/2}.$$

If we look to the corresponding calculations for crossover we first find that the bound on transcription error sets a bound $1 - P_{cross}(l(s)/k) > 1 - e(s)$, or

$$P_{cross}(l(s)/k) < e(s)$$

which implies that

$$k > P_{cross}[l(s)/e(s)] = P_{cross}[(d(s)-1)/e(s)]$$

using the fact that $l(s) = d(s)-1$ for schema s. Under crossover, an instance of s' will have an offspring that is an instance of s if two things happen: (1) The other parent in the cross is an instance of s", and (2) the cross occurs exactly at the juncture of s' and s". In a randomly generated population, a randomly selected parent will be an instance of s" with a probability $2^{-d(s")}$ and the crossover will occur at the required juncture with probability $1/k$. Thus the overall event will occur with probability

$$2^{-d(s")}/k = 2^{-d(s")}/P_{cross}[(d(s)-1)/e(s)]$$

with a corresponding search time

$$T_{cross} < [d(s)/e(s)]2^{d(s")}$$

using the fact that $P_{cross} \leq 1$.

Even for relatively small "deserts" these waiting times are enormously different. For example, with $d(s') = 16$, $d(s") = 8$ and $e(s) = 0.1$,

$$T_{mut} > 3.4 \cdot 10^9$$

and

$$T_{cross} < 6.2 \cdot 10^4.$$

Comments.

Most searches of complex spaces conducted by genetic algorithms should exhibit the same broad characteristics:

At the outset, there should be a rapid biasing of sampling probabilities toward hyperplanes s that
(1) have instances in the initial population, and
(2) have an average value $u^\wedge(s,0)$ sufficiently above the population average $u^\wedge(0)$ that it overbalances the transcription error under the genetic operators $\lambda(s,0)$.
In particular, if the mutation rate is low, this is a requirement that the "length" $l(s)$ of the schema be short enough that $l(s)/k < [u^\wedge(s,0)/u^\wedge(0)]-1$. If the observed $u^\wedge(s,0)$ is a good estimate of the actual marginal expection $u(s)$, then the number of instances of s will increase exponentially in subsequent generations.

As the exponential increase continues, the increasing number of instances of above-average schemas can be expected to drive the population average $u^\wedge(t)$ upward. For any given s this forces $u^\wedge(s,t)/u^\wedge(t)$ ever closer to 1. The exponential increase of s then tapers off. Eventually, a decrease ensues unless some refinement of s offers further improvement, in which case the sampling is concentrated in that refinement, or else s represents the best that can be achieved

During the time that a schema has a large number of instances in the population, the genetic algorithm acts to provide many *new* instances of it. The mutation operator treats the schema as a focal point, generating a variety of new instances in its "neighborhood". Crossover, and other forms of recombination, generate new instances by treating the schema as a "building block" that can be used in combination with other building blocks.

The fate of a newly discovered instance of a deep, above-average schema s depends upon the manner of its discovery. Consider a schema s with a length $l(s)$ large enough to indicate a large transcription error. Under normal circumstances, this transcription error would be enough to quickly destroy instances of s. However, if s has been discovered by recombination of well-established building blocks, things happen differently. The well-established building blocks occupy large fractions of the population, so the parents are likely to hold several building blocks in common. As a result, crosses that normally would break up instances of s now just recreate s because they exchange pieces of identical building blocks. Thus, s increases its representation despite the large transcription error.

Generally, crossover is the operator that can be expected to yield substantial improvements based on deeper δ's. Crossover implements the heuristic that "good" structures are constucted of "good" building blocks (cf. Simon's [1981] discussion of the architecture of complexity). This amounts to a conjecture that new non-zero δ's are associated with the *intersections* of hyperplanes already known to be associated with non-zero δ's. Of course, the conjecture may prove untrue for many intersections, but it need only be true upon occasion for improvements to be made. It should be recalled that improvement is the object of the search, not discovery of the global optimum, which may involve δ's so deep that they will never be uncovered in feasible times. Implicit parallelism, by assuring that the genetic algorithm usefully searches large numbers of schema combinations in each successive generation, makes it likely that *some* useful intersections will be uncovered. It *is* possible to design a function u that often "guides" the genetic algorithm away from good regions, but it is hard to design a function that keeps the algorithm away from improvements over long intervals.

Overall, and in qualitative terms, the search exhibits continual small improvements, punctuated by saltations to schemas involving deeper δ's. This behavior is a direct consequence of the manner in which genetic operators exploit sparse deeper δ's. From a biological perspective, it is interesting that this succession of "punctuated equilibria" occurs without the intervention of higher order selection principles.

References.

Bethke, A. D. 1980. *Genetic Algorithms as Function Optimizers*. University of Michigan Ph. D. Dissertation.

DeJong, K. A. 1980. Adaptive system design -- a genetic approach. *IEEE Transactions: Systems, Man, and Cybernetics*. *10*, 9.

Holland, J. H. 1975. *Adaptation in Natural and Artificial Systems*. University of Michigan Press.

Holland, J. H., Holyoak, K. J., Nisbett, R. E., and Thagard, P. R.

1986. MIT Press.

Samuel, A. L. 1959. Some studies of machine learning using the game of checkers. *IBM Journal of Research and Development. 3.*

Simon, H. A. 1981. *The Sciences of the Artificial.* MIT Press.

PROBABILISTIC NEURAL NETWORKS

John W. Clark

McDonnell Center for the Space Sciences
and Department of Physics
Washington University, St. Louis, MO 63130 USA

1. INTRODUCTION

We are now witnessing a powerful resurgence of the Neurobiological Paradigm, which embodies the premise that the structure and principles underlying natural intelligence may be imitated to design machines displaying useful cognitive properties. In following this paradigm, the operation of the brain as a network of interconnected and interacting neurons is to be recapitulated in algorithms and computational devices for solving difficult problems. The first wave of excitement about neural networks and connectionist principles began in the early 40s with the pioneering ideas of McCulloch and Pitts[1] on binary threshold automata and culminated in the late 50s with Rosenblatt's perceptron.[2] However, progress was hindered by the primitive state of existing electronic computers and of existing theories of nonlinear dynamical systems. With the recognition of the intrinsic limitations of the elementary, single-layer perceptron,[3] interest in this "naturalistic" approach to artificial intelligence waned, and the discipline evolved along different lines, with sequential symbolic processing as the dominant theme. Although there were bursts of significant work during the 60s and 70s,[4-8] it was not until this decade that neural networks again became a growth industry. Pivotal developments included Hopfield's formulation[9] of the problem of collective computation in terms of modern nonlinear dynamical theory and statistical physics, exploiting ideas from the theory of spin glasses, and the path-breaking work of the PDP research group[10] (Hinton, McClelland, Rumelhart, Sejnowski, Williams, *et al.*). By analytic arguments and extensive computer simulations, the PDP group demonstrated the power of multilayer perceptrons containing hidden units, introducing the Boltzmann machine and the back-propagation algorithm for credit assignment.

In implementing the Neurobiological Paradigm, it is necessary first to achieve a decent understanding of the biological structures and processes which are responsible for adaptive or intelligent behavior in animals. Such behavior is understood to stem from the organized activity of assemblies of neurons, which are identified as the basic information-processing units. This activity is governed by the complex of interactions among the neurons, in conjunction with external stimuli. So it is necessary to understand the anatomy and physiology of the varied types of nerve cells and of the synaptic interactions between them. But at what level, and how precisely? The neuron has (with some frustration) been called an infernal machine. It exhibits complexity on many different levels -- from the variety of gross shapes and sizes of cell body and axon, to the intricate geometry of the dendritic tree, to the bewildering array of synaptic junctions, down to the molecular scale of neurotransmitters, receptor sites, and ionic channels. This wealth of detail may or may not be irrelevant to the ultimate description of the collective neural phenomena associated with information processing and cognitive behavior. Taking the prudent course, one would model the neuron and its interactions at successively deeper levels. Having devised a neurobiological model which captures some reasonable level of detail, the next step is to explore its dynamical behavior with the aim of determining its adaptive or cognitive capabilities and comparing them with what is observed in nature. In this way one hopes to gain further insights for the culminating step of the paradigm, namely the design of machines which mirror biology in essential structure and function and demonstrate sophistication in cognitive tasks, yet are free from various impediments imposed by evolution. The sophistication of the constructs should increase as one proceeds to deeper levels of the neurobiological description, but eventually a point of diminishing return would be reached.

Apart from its practical, technological implications within the domains of machine intelligence and computer science, this process is beneficial in another important way: Its hierarchical and recursive character leads to to an enrichment of all the other disciplines involved, including neurobiology, dynamical systems theory, statistical physics, applied mathematics, and cognitive psychology.

It is the purpose of this contribution to illuminate some selected aspects of the implementation of the neural-net paradigm, based on a view of neuron and synapse which is somewhat more refined than that adopted in most of the connectionist literature. This view focuses on the noisy nature of real neural processing, and especially on the uncertainties associated with the synaptic transmission mechanism. Neural network models incorporating such stochastic effects are formulated in Section 2. The statistical dynamics of these models is explored in Sections 3 and 4, with the aid of methods and ideas appropriated from theoretical physics, the theory of stochastic processes, and nonlinear dynamics. Dynamical response characteristics are discussed in terms of possible use of the model systems as content-addressable memories. Some final remarks, in Section 5, summarize the general features derived for probabilistic neural networks and address the role of plasticity.

2. MODELING THE NOISY NEURON

The story begins with two different approaches to the modeling of uncertainties of neuronal behavior associated with the quantal nature of communication across synaptic junctions. These approaches will lead to two rather different pictures of the statistical time development of a neural network, the one involving a linear, Markovian, Master-Equation description and the other, a polynomial mapping of the interval $[0,1]^N$ into itself. In these models, attention is focused on certain stochastic effects of presynaptic origin. In order to judge the realism of the models, some background in neuronal and synaptic anatomy and physiology is needed. What follows might be regarded as a minimalist view of a rich panorama of phenomena. At this time it is not yet clear what level of detail is required for an adequate understanding of the cognitive behavior of living nerve nets. It is possible that the description given is still too crude, or misses essential mechanisms.

2.1. Empirical Properties of Neuron and Synapse

Fig. 1 offers a schematic picture of a synaptic junction formed by the close proximity of an axon terminal of presynaptic neuron μ to a dendrite or to the cell body of postsynaptic neuron ν. Imbedded in the presynaptic structure are vesicles or sacks containing one or another neurotransmitter substance, while the postsynaptic membrane is dotted with receptor sites for transmitter molecules. A receptor site is actually a large protein molecule which penetrates through the membrane to the fluid interior of the cell. When transmitter molecules bind to a receptor, a channel is opened in the protein molecule. This allows ions (Na^+, K^+, Cl^-) to move into or out of the cell, under the influence of concentration and potential gradients.[11] Such ionic flows lead to changes in the potential across the postsynaptic membrane -- i.e., they give rise to postsynaptic potentials (PSPs). These potential fluctuations may be positive or negative in sign, as the effect of the transmitter on cell ν is excitatory or inhibitory. The anatomy of a "typical" neuron is roughly as shown in Fig. 2. The induced PSPs are relayed passively up the dendritic arborization, to the cell body, and across the cell body to the initial axon segment (axon hillock) of neuron ν, decaying in time and with distance of travel from their points of origin. If the net positive potential fluctuation at the axon hillock exceeds the net negative signal by a sufficient margin (threshold of the neuron), then neuron ν will fire an action potential, a unit pulse of positive, active excitation which travels without decrement along the length of the axon, reaching all the axon terminals of μ in a time which is generally much less than a millisecond. (The inside of the neuron being negatively charged relative to the extracellular fluid, one speaks of a positive "depolarization" due to PSPs, which may be strong

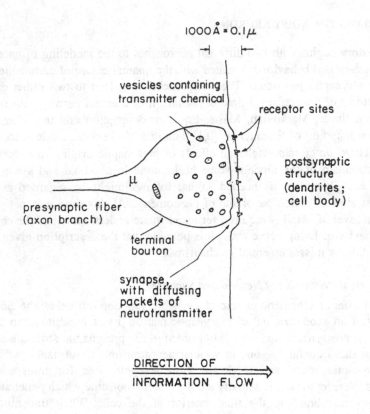

$1000 \overset{\circ}{A} = 0.1\mu$

vesicles containing
transmitter chemical

receptor sites

presynaptic fiber
(axon branch)

μ

ν

postsynaptic
structure
(dendrites;
cell body)

terminal
bouton

synapse,
with diffusing
packets of
neurotransmitter

DIRECTION OF
INFORMATION FLOW

Fig. 1. Synaptic junction of terminal of axon branch of neuron μ onto dendrite or cell body of neuron ν. Information is transferred from μ to ν via diffusion of transmitter chemical across synaptic gap. The distance scale is marked in microns and Ångstroms.

enough to produce an action potential.) If its threshold is not surpassed (or equaled) during the stimulus-response scenario, the neuron will fail to fire an action potential, and the depolarization or hyperpolarization resulting from spatio-temporal summation of PSPs will decay away within a few ms to tens of ms. Just after firing an action potential, a neuron experiences a certain dead time, or refractory period, during which its threshold for firing a second pulse is first effectively infinite (absolute refractory period) and then declines toward the resting value (relative refractory period).

Experimentally,[12] it is known that the neurotransmitter chemical is released into the synaptic cleft in packets, or "quanta," containing some 10^{4-5} molecules. Such packets are discharged upon the fusion of vesicles with the

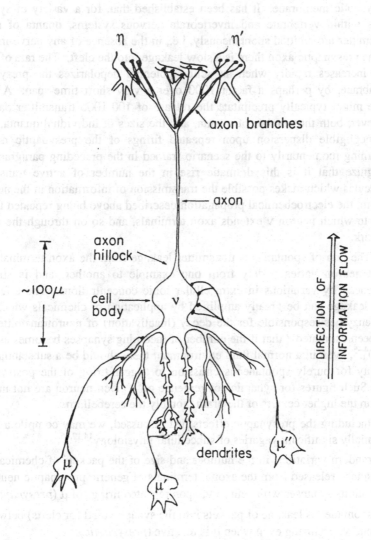

Fig. 2. Schematic neuronal anatomy, showing cell body (soma), dendritic arborization, and axon of neuron ν. Cells presynaptic (μ, μ′, μ″) and postsynaptic (η, η′) to neuron ν are indicated. The distance scale is marked in microns.

presynaptic membrane. It has been established that, for a variety of synaptic types within vertebrate and invertebrate nervous systems, quanta of neuro-transmitter are emitted spontaneously, i.e., in the absence of any nerve impulse on the presynaptic axon there is a slow leakage into the cleft. The rate of emission increases rapidly when an action potential depolarizes the presynaptic membrane, by perhaps a factor 1000 over a very short time-span. A nerve pulse might typically precipitate the release of 100-1000 transmitter packets; however, both the number of quanta, and the sizes of individual quanta, show non-negligible dispersion upon repeated firings of the presynaptic neuron. Returning momentarily to the scenario framed in the preceding paragraph, we recognize that it is this dramatic rise in the number of active transmitter molecules which makes possible the transmission of information in the nervous system, the electrochemical propagation described above being repeated for the cells to which neuron ν extends axon terminals, and so on through the neural network.

The rate of spontaneous transmitter leakage from the axon terminals of a silent neuron varies widely from one example to another, and is strongly influenced by variations in extracellular ionic concentrations. The effects of such leakage can be greatly amplified by application of chemicals which destroy enzymes responsible for the decay (inactivation) of neurotransmitters. It has been estimated[13] that if the number of incoming synapses becomes as large as 10^{4-5}, even in a normal fluid environment there should be a substantial probability for purely spontaneous emission to trigger firing of the postsynaptic cell. Such figures for synaptic convergence on a given neuron are not uncommon in the higher centers of the brain, notably the cerebellum.

Including the presynaptic effects just discussed, we may compile a list of potentially significant vagaries of subcellular physiology[14,12,15]:

[i]   random variations in the number and size of the packets of chemical substance released from the axonal terminals of generic presynaptic neuron μ, into its synapses with neuron ν, upon repeated firings of μ (*presynaptic*),

[ii] spontaneous leakage of packets into the synaptic cleft (or clefts) between μ and ν, occurring even when μ is inactive (*presynaptic*),

[iii] fluctuations in the number of transmitter molecules reaching and attaching to a given receptor site, for a specified concentration of neurotransmitter in the synaptic gap (*synaptic*),

[iv] stochastic variability in the length of time a receptor channel stays open (the channel protein-transmitter complex being randomly buffeted by molecules of the ambient medium) (*postsynaptic*), and

[v] temporal fluctuations in the membrane resting potential of neuron v (*post-synaptic*).

If a neuron is delicately poised between firing and not firing, one or another of these random elements can act to tip the balance. These stochastic phenomena are not necessarily just useless noise; rather they may have been exploited by evolution to enhance, somehow, the stability and efficiency of the brain as an information-processing system of the highest order.[16,17,13,7,14,18–22]

To get some idea of the subtleties that might be involved in unraveling the mysteries of communication and coding in the nervous system, consider the fact that, due to the dense packing of cell bodies, dendrites and axons in brain tissue, the extracellular space occupies only a small fraction (~20%) of the total volume. The activity of a given neuron and neighboring cells (e.g., but not exclusively, in action-potential traffic) can thus produce substantial perturbations on the concentrations of Na^+, K^+, Ca^{2+}, and Cl^- ions in the restricted intercellular regions. Since these ions play vital roles in the generation of nerve impulses, such variations in concentration may serve to modulate the spike activity both of the original cell and its neighbors,[23] providing a highly complex channel of interneuronal communication which has received little attention. (Referring to the above list, this mechanism may affect [i], [ii], and especially [v].) Similar fluctuations in the concentrations of neurotransmitter chemicals and neurohormones, as enhanced by the constrained geometry, further complicate the picture, providing further avenues for information transmission. While these mechanisms may be diffuse, sluggish, and noisy, their importance to neural behavior cannot easily be dismissed.

Mathematical models involving the "quantal" and stochastic aspects of synaptic communication have been devised by Taylor[13] and by Shaw and Vasudevan.[14] Working both in discrete- and continuous-time formulations, Taylor concentrates on the implications of spontaneous emission, i.e., noise source [ii], suppressing variability in the number and size of quanta secreted under presynaptic firing, i.e., noise source(s) [i]. In spite of the latter simplification, interesting complexity can arise; there is an interference of spontaneous and action-potential contributions to the transmitter concentrations affecting the postsynaptic neuron, engendering an inherently nonlinear statistical evolution. In their synthesis of the empirical information, Shaw and Vasudevan sought to justify the probabilistic neural network model of Little.[7] This alternative approach includes both kinds of presynaptic noise, i.e. both [i] and [ii] in the list, although a number of simplifications and approximations are made along the way, leading ultimately to a linear theory of the statistical time development of the system. Both approaches are interesting enough to describe in some detail.

136

2.2. Model of Shaw and Vasudevan

Let us recall first the assumptions and the principal results of Shaw and Vasudevan.[14] The microscopic dynamics is envisioned in discrete time: assuming a universal delay τ for communication of excitation from any presynaptic to any postsynaptic cell, the picture will be one of synchronous operation of the neural assembly on a time grid with spacing τ. A synapse $\mu \rightarrow \nu$ is considered for fixed postsynaptic neuron ν and generic presynaptic neuron μ. In accordance with the experiments of Katz and coworkers,[12] it is supposed that the number of quanta of neurotransmitter released at this synapse, as observed in a large sample of individual firings of neuron μ, is governed by a Poisson process with mean $\lambda_{\nu\mu}\pi_\mu$. Here $\pi_\mu = (\sigma_\mu + 1)/2$ is a variable taking the value 1 when μ is firing, and 0 when it is silent. Similarly, it is assumed that the number of quanta due to spontaneous emission is distributed according to a Poisson process with mean $\lambda_{\nu\mu}^{(s)}$. Thus, the probability that n quanta will be secreted into the $\nu\mu$th synapse due to activity of neuron μ at time $t - \tau$ is given by

$$\exp\left[-\lambda_{\nu\mu}\pi_\mu(t-\tau)\right] \frac{[\lambda_{\nu\mu}\pi_\mu(t-\tau)]^n}{n!} , \tag{2.1}$$

while a similar expression (with $\pi_\mu(t-\tau)$ factors omitted) gives the probability for obtaining n quanta by spontaneous emission. Note that the mean values $\lambda_{\nu\mu}$ and $\lambda_{\nu\mu}^{(s)}$ are allowed to be synapse dependent. To conform with Little's model, the excitation of neuron ν is reset to zero at the beginning of each step of a discrete time grid, implying decay of postsynaptic potentials in a time something less than the grid spacing τ. This assumption may be quite unrealistic, e.g. if τ is taken as a typical synaptic delay of about 1 ms, since PSP decay times can run to several ms. If τ is identified instead with the absolute refractory period,[7] it becomes less questionable; however, the courser time grid will artificially frustrate some neuronal firings. Another assumption made by Shaw and Vasudevan is that the distribution $\phi(V)$ of the ultimate contribution V of the individual quanta to the change in membrane potential at the axon hillock is independent of μ and ν. Referring to experimental results of Katz and coworkers,[12] a Gaussian function is chosen for $\phi(V)$. Such a probability distribution has a generating function of the form

$$\tilde{\phi}(K) = \exp(-v_o K + K^2\gamma^2/2) , \tag{2.2}$$

wherein v_o and γ are respectively the mean and standard deviation of the Gaussian $\phi(V)$. It is further supposed that these individual contributions to the net potential shift at the axon hillock are simply additive and that the contributions of the various afferent (incoming) synapses of neuron ν are independent of one another and likewise additive. The total change in axon-hillock membrane potential is accordingly modeled in terms of a Poisson-filtered additive process. Executing an arbitrary integral number n (≥ 0) of convolutions of the process $\phi(V)$ and summing over n with weights determined by the distribution (2.1) and its counterpart for spontaneous emission, one may derive a compact expression for the generating function $\tilde{\Phi}(K)$ for the probability density $\Phi(V)$ that neuron ν will accumulate a potential V from all its synaptic inputs, effective at time t. The result is

$$\tilde{\Phi}_\nu(K) = \exp\{-[A_\nu(t) + A_\nu^{(s)}][1 - \tilde{\phi}(K)]\} \ , \tag{2.3}$$

where $A_\nu(t) = \Sigma_\mu \lambda_{\nu\mu} \pi_\mu(t - \tau)$ and $A_\nu^{(s)} = \Sigma_\mu \lambda_{\nu\mu}^{(s)}$, with sums running only over neurons μ which are actually presynaptic to ν. At this point an approximation is made which exploits the fact that the upcoming calculation of the probability of firing of neuron ν will not involve values of V below the threshold V_ν^T of that neuron. It is assumed that V_ν^T is *large* compared to the average contribution v_0 to the axon-hillock membrane potential due to individual quanta. (This is reasonable, since the former is of order 15 millivolts, and the latter only about half a millivolt.) Thus the description is restricted to the large-V regime, implying small K. The generating function $\tilde{\phi}(K)$ of $\phi(V)$ is then expanded in a small-K Taylor series and terms $O(K^3)$ or higher are dropped. The resulting approximation to $\tilde{\Phi}_\nu(K)$ takes the form of the generating function of a Gaussian process. Some minor alterations are made to acknowledge the different effects produced by transmitter substances emitted at excitatory and inhibitory synapses. The mean miniature PSP, which we have denoted v_0, is taken positive for the former and negative for the latter, but with a common magnitude $|v_0|$. (The standard deviation γ is still considered to be synapse-independent.) With this refinement, Shaw and Vasudevan arrive at the following expression for the probability density of receiving net excitation V at the axon hillock, valid in the high-V domain, and applicable at time-step t:

$$\Phi_\nu(V) = \frac{1}{\sqrt{2\pi\gamma^2}} \exp[-(V - \bar{V}_\nu)^2/2\delta_\nu^2] \ , \tag{2.4}$$

where

$$\bar{V}_\nu = \bar{V}_\nu(t) = (A_\nu'(t) + A_\nu^{(s)'})|v_0| \ ,$$

$$\delta_\nu^2 = \delta_\nu^2(t) = (A_\nu(t) + A_\nu^{(s)})(\gamma^2 + |v_0|^2) \ . \tag{2.5}$$

The modified A's account for the distinction between excitatory ($\varepsilon_{\nu\mu} = +1$) and inhibitory ($\varepsilon_{\nu\mu} = -1$) synapses and are defined by

$$A_\nu'(t) = \sum_\mu \varepsilon_{\nu\mu} \lambda_{\nu\mu} \pi_\mu(t - \tau) \quad , \qquad A_\nu^{(s)}{}' = \sum_\mu \varepsilon_{\nu\mu} \lambda_{\nu\mu}^{(s)} \quad . \tag{2.6}$$

We observe that, within this scheme, spontaneous emission simply contributes additively to the mean and to the variance of the distribution $\Phi_\nu(V)$, in the same way as does transmitter release due to presynaptic firings.

In the final step of the Shaw-Vasudevan treatment, the approximation (2.4) is used to find the probability ρ_ν that neuron ν will fire at time t, given the firing states π_μ (or $\sigma_\nu = 2\pi_\nu - 1$) of all neurons at time $t - \tau$. The desired quantity is just the probability that the net algebraic excitation V will exceed the threshold V_ν^T at time t under the given conditions; hence

$$\rho_\nu(\sigma_\nu(t) = +1) = \int_{V_\nu^T}^{\infty} \Phi_\nu(V) dV \quad . \tag{2.7}$$

The integral over (2.4) produces an error function. Using the fact that the error function $\mathrm{erf}(x)$ is odd in x, we may compress the results for the probability $\rho_\nu(+1)$ of firing and the probability $\rho_\nu(-1)$ of not firing into the single formula

$$\rho_\nu(\sigma_\nu(t)) = \frac{1}{2} \left\{ 1 - \mathrm{erf} \left[-\frac{\sigma_\nu(t)(\bar{V}_\nu(t) - V_\nu^T)}{\delta_\nu \sqrt{2}} \right] \right\} \quad . \tag{2.8}$$

To make contact with the Little model (to be examined next), this result is approximated by a logistic (or Fermi) function,

$$\frac{1}{2}[1 - \mathrm{erf}(x)] \cong [1 + \exp(x)]^{-1} \equiv s_F(x) \quad . \tag{2.9}$$

The quality of approximation (2.9) is excellent; numerically it is correct to within 0.01 over the full range of the argument and it is exact at the limiting values $x = -\infty, 0$, and $+\infty$.

As pointed out in Ref. 14, a more realistic treatment should be pursued in terms of the first-passage problem defined by the emission of transmitter packets and the accumulation of the resulting potentials up to the firing threshold, *taking account of* the decay of PSPs (at least on the average over one time-step). It would also be of interest to include the phenomena of facilitation and accommodation.[24,11]

2.3. Model of Little

The Little model[7] of neural networks is best known nowadays as a synchronous alternative to Hopfield's discrete-time model.[9] In recent years, its equilibrium statistical properties and its capacity for memory storage have been subjected to thorough analysis,[25-28] to the extent that the ingredients and methods of the theory of spin glasses (thermodynamic limit, mean-field theory, self-averaging, replica trick, ...) are relevant and applicable (see commentary in Subsection 3.5); generally, such analysis is predicated on symmetrical couplings determined by a local learning rule.[9,29] Within the context of our discussion, the Little model may be viewed as a streamlined extension of the Shaw-Vasudevan description of quantal information transfer to a network of interconnected neurons. The Little model is specified as follows.

L1. The network consists of N (formal) neurons labeled $v = 1, \ldots N$. Each neuron v is assigned a state variable σ_v which takes on the value +1 if v is firing and -1 if it is not (representing the "all-or-none" character of the action potential).

L2. The neurons are allowed to fire only at instants belonging to a discrete set $\{0, \tau, \ldots n\tau, \ldots\}$, where τ is some elementary time interval. (The time quantum τ might be taken as the absolute refractory period, or the delay time for signal transmission from presynaptic to postsynaptic neuron.)

L3. The stimulus felt by neuron v at time $t = n\tau$, n integral, due to synaptic input(s) from neuron μ, is written $V_{v\mu}\pi_\mu(t-\tau)$, where $\pi_\mu = (\sigma_\mu + 1)/2$. The coupling matrix $(V_{v\mu})$, with $V_{v\mu}$ a real number, embodies the pattern of synaptic connections between neurons and the strengths or efficacies of these junctions. A positive [negative] value of $V_{v\mu}$ means that the synapse (or generally the set of synapses) from μ onto v is excitatory [inhibitory] in effect, and $V_{v\mu} = 0$ if there is no synapse from μ onto v. In general we must consider $V_{\mu v} \neq V_{v\mu}$ if we wish to simulate biological nerve nets; i.e., the neuronic interactions are ordinarily not symmetrical, since (for example) a synapse of μ onto v does not imply the presence of a reciprocal synapse of v onto μ. This clashes with our experience with ordinary physical systems, where Newton's third law holds.

L4. The decision of a neuron v whether or not to fire at time $t = n\tau$ is stochastic (or probabilistic). The "firing function"

$$F_v(t) = \sum_\mu V_{v\mu}\pi_\mu(t - \tau) - \theta_v \qquad (2.10)$$

of the neuron at t equals the amount by which its algebraic stimulus at that time exceeds its threshold θ_ν. As our previous discussion makes clear, various uncertainties inherent in the electrochemical synaptic transduction mechanism may be simulated by assuming that even if $F_\nu < 0$, there is a finite probability that neuron ν will fire, and even if $F_\nu > 0$, there is a finite probability that it will not. For large positive [negative] values of F_ν, the firing probability should go to unity [zero], whereas for F_ν near zero the response should be less predictable. The specific ansatz

$$\rho_\nu(\sigma_\nu(t)) = \left\{ 1 + \exp[-\beta_\nu \sigma_\nu(t) F_\nu(t)] \right\}^{-1} \equiv s_F(-\beta_\nu \sigma_\nu(t) F_\nu(t)) \quad (2.11)$$

is made for the conditional firing probability. As before, $\rho_\nu(+1)$ [resp. $\rho_\nu(-1)$] denotes the probability that ν will fire [fail to fire] at the specified time t, given the states $\sigma_\mu(t-\tau)$ or $\pi_\mu(t-\tau)$ of the neurons one time-step earlier and hence the firing function $F_\nu(t)$. The nonnegative parameter β_ν^{-1} is a (crude) measure of the noisy character of signal transmission to neuron ν. In the limiting case that $\beta_\nu^{-1} = 0$ for all ν ("zero temperature"), the dynamics becomes deterministic, if we exclude the measure-zero case that the firing function is exactly zero. Unless otherwise stated, *all* the β_ν are taken to be finite, implying a fully stochastic net.

L5. The *state of the network* as a whole is represented at time t by the set $\{\sigma_\nu(t), \nu=1,\ldots N\}$ of individual neuronal state values, i.e. by the *firing pattern* at that time. There are exactly 2^N distinct states of the model, which will be labeled $i = 0, 1, 2, \ldots 2^N - 1$. The analogy to a spin system is apparent.

L6. The conditional firing probabilities ρ_ν at time t, given by (2.11) with (2.10), are not correlated with one another, as they do not depend on the firing states σ realized by the other neurons at that time, but instead depend only on the state j occupied by the system one time-step earlier. Accordingly, the probability per unit time that the net will undergo a transition from state j at time $t - \tau$ to state i at time t may be written as the product

$$T_{ij} = \tau^{-1} \prod_{\nu=1}^{N} \rho_\nu^{(j)}(\sigma_\nu^{(i)}) = \tau^{-1} \prod_{\nu=1}^{N} \left[1 + \exp\left\{ -\beta_\nu \sigma_\nu^{(i)} \left[\sum_\mu V_{\nu\mu} \pi_\mu^{(j)} - \theta_\nu \right] \right\} \right]^{-1} . \quad (2.12)$$

L7. In *autonomous* operation, the network is isolated from external stimuli after initial excitation of a subset of neurons (or initiation of the probability distribution over states of the system). When the network is *driven*, an additional *control term* U_ν, representing stimuli from the environment, must be added to the right side of Eq. (2.10).

For extensive discussion of the properties and behavior of this model, see Refs. 7,14,16,18-22,30,25. Of more immediate concern is the connection between Little's ansatz (2.11) and the more fundamental Shaw-Vasudevan result (2.8).

If (2.9) is used to approximate the error function, the two versions of ρ_ν take the same form (at least superficially). Identifying the firing function $F_\nu(t)$ of (2.10) with $\bar{V}_\nu(t) - V_\nu^T$, we are led to identify the spontaneity parameter β_ν with $1/\delta_\nu\sqrt{2}$. The former identification is consistent with the decomposition of $\bar{V}_\nu$ given by (2.5) and the definitions (2.6) of A_ν' and $A_\nu^{(s)'}$, provided the term $A_\nu^{(s)'}|v_o|$ of $\bar{V}_\nu$ is incorporated (negatively) with V_ν^T in the definition of the threshold θ_ν. Thus, in this interpretation, one effect of purely spontaneous emission is to shift the neuronal threshold from its "deterministic" value. Such a shift could be either positive or negative, depending on the distribution and efficacy of excitatory vs. inhibitory synapses $\mu \to \nu$. The identification of β_ν with $1/\delta_\nu\sqrt{2}$ is more problematic. The variance δ_ν^2 defined in Eq. (2.5) depends, in detail, not only on the postsynaptic neuron but also on the presynaptic neurons μ, both through their Poisson mean-value parameters $\lambda_{\nu\mu}$ and $\lambda_{\nu\mu}^{(s)}$ and their activities $\pi_\mu(t-\tau)$. Indeed, since the noise sources [i] and [ii] included in the Shaw-Vasudevan model are presynaptic in origin, it can be argued that the μ dependences should play a prominent role in the ensuing description of quantal information transmission. However, no overt dependence of the variance on the presynaptic neurons survives when (2.8) is forced to conform with Little's ansatz. We must conclude that Little's model entails a significant sacrifice of precision in the description of the stochastic phenomena [i] and [ii], since for each neuron the single parameter β_ν is asked to mock up the uncertainties arising from them, some average effects of active and passive secretion of quanta having been absorbed into the definitions of the $V_{\nu\mu}$ and the θ_ν.

Taking a broader view, the path followed in Ref. 14, in essence a physiological elaboration on Little's model, conspires to suppress information about the interplay of spontaneous and induced emission, in exchange for the simplicities of a *linear* dynamical theory. The linearity of the theory -- at the level of statistical mechanics -- arises from the fact that the probability $Q_{ij} = \tau T_{ij}$ of a one-step transition from state j to state i is *independent* of the state-occupation probabilities at the time of the jump; for then the operation of updating the probability distribution over system states will indeed be a linear one. This feature will be made explicit in Section 3 and will underlie the coming analysis of the

142

approach to equilibrium in a large class of neural-network models.

2.4. Model of Taylor

We now turn to Taylor's discrete-time formulation.[13] The difference from the treatment just sketched will turn out to be more fundamental than just a different emphasis in the stochastic effects described: it penetrates to the basic description of the dynamical evolution. Whereas the deliberations of Shaw and Vasudevan culminated in a formula [viz. (2.8)] for the probability $\rho_\nu(+1)$ of firing of neuron ν at time t, given the actual firing states of all neurons μ one time-step earlier, we now seek a law for the time development of the absolute firing probability $w_\nu(t)$ of neuron ν in terms of the preceding firing probability $w_\nu(t-\tau)$, without reference to actual firing states realized by the neurons of the assembly.

The first task, which in fact does not involve a specialization to discrete time, is to evaluate the probability distribution $g^{(s)}(q)$ of the amount q of transmitter substance present in a given synaptic cleft as a result of spontaneous emission. A synapse $\mu \to \nu$ is examined, involving a presynaptic cell μ and a postsynaptic cell ν. Impulse activity is considered to be absent for the time being, and so does not contribute to the transmitter concentration. Spontaneous leakage of transmitter is assumed to be described by a Poisson process in which quanta are released with a mean frequency $f^{(s)}$ and thus a mean sojourn time $t^{(s)} = 1/f^{(s)}$. The sizes of all such packets are assumed to be the same, each containing a definite quantity $q^{(o)}$ of neurotransmitter chemical. (One recognizes a correspondence between $q^{(o)}$ and the "mean-size" parameter $|v_o|$ of the Shaw-Vasudevan model (cf. (2.2)), the analog of the variance γ^2 being zero.) Further, all quanta are supposed to have the same, quite definite lifetime t_{dec}; that is, the amount of transmitter carried by a quantum created at time 0 is governed by $q(t) = \theta(t_{dec}-t)\theta(t)$, where $\theta(x)$ is the usual step function. The mean number of packets emitted spontaneously in a time interval t_{dec}, or the mean number of spontaneous quanta present in the gap at any time, is therefore given by $\lambda^{(s)} = t_{dec}/t^{(s)}$, a parameter which corresponds to $\lambda_{\nu\mu}^{(s)}$ of the Shaw-Vasudevan description. To find $g^{(s)}(q)$ at any arbitrary time t, we only need count quanta released during $(t-t_{dec}, t)$. In terms of $\lambda^{(s)}$, the Poisson law gives $\delta(nq^{(o)} - q)(\lambda^{(s)})^n e^{-\lambda^{(s)}}/n!$ as the probability density for the release of n packets (hence an amount $nq^{(o)}$ of transmitter) during this period. Summing over all integral $n \geq 0$, one obtains a result for the probability distribution $g^{(s)}(q)$ which may be expressed as

$$g^{(s)}(q) = \frac{1}{2\pi} \int_{-\infty}^{+\infty} \exp[-iqu + \lambda^{(s)}(e^{iq^{(o)}u} - 1)]du \quad . \tag{2.13}$$

This distribution has unit normalization. For asymptotically large $\lambda^{(s)}$, it goes over to a Gaussian distribution with mean $q^{(0)}\lambda^{(s)}$ and variance $(q^{(0)})^2\lambda^{(s)}$. (The limiting distribution may be compared with the form (2.4) derived by Shaw and Vasudevan using $V >> |v_o|$. Comparison is facilitated by dropping the "active" terms A_v' and A_v from (2.5), and also omitting the dispersion parameter γ to conform with Taylor's neglect of variability in quantum size.)

Next we set up the dynamics and define the network model itself. Again time is discretized on a regular grid, the spacing τ being identified with the period which must elapse between firing of the presynaptic neuron and arrival of the resulting neurotransmitter chemical at the axon hillock of the postsynaptic cell. Every synaptic communication channel $\mu \to v$ in the system is assigned the same delay τ. Accordingly, the neural assembly is again considered to operate synchronously, with updatings of the state of the system occurring only at instants on the time grid. A nerve impulse on the axon of presynaptic neuron μ is assumed to induce the secretion into synapse $v\mu$ of a definite but synapse-dependent number $n_{v\mu}$ of quanta, all of a definite, synapse-dependent size $q_{v\mu}^{(0)}$ (positive for excitation and negative for inhibition). If neuron μ was in fact active at time $t - \tau$, the contribution from background spontaneous emission attributable to that time-step is neglected. Neuron v will fire (or not) according to a probability distribution function h_v which depends on the difference between the total amount q of transmitter released into v's incoming synapses and a "threshold" value $q_v^{(c)}$. There is an implicit supposition that any transmitter chemical present in a synapse between time $t - \tau$ and t is used at time t and that none of its effects persists to time $t + \tau$. These assumptions regarding time quantization and induced emission, together with the stochastic description of spontaneous emission outlined in the preceding paragraph, lead to the following rule for updating the (absolute) firing probability w_v of neuron v:

$$w_v(t) = \int_0^\infty dq\ h_v(q) \prod_\mu \int_{-\infty}^\infty dq_{v\mu}\ \delta(q - \sum_\mu q_{v\mu})$$

$$\times [w_\mu(t-\tau)\delta(q_{v\mu} - n_{v\mu}q_{v\mu}^{(0)}) + (1 - w_\mu(t-\tau))g_{v\mu}^{(s)}(q_{v\mu})] \ . \quad (2.14)$$

To find w_v at time-step t, we only need to know the w_μ at one time-step earlier, no information from yet earlier times being required (Markov property). The quantity $g_{v\mu}^{(s)}(q_{v\mu})$ is given by (2.13), but with synapse labels $v\mu$ attached at appropriate places and especially to the history parameter $\lambda^{(s)}$ and the quantum size $q^{(0)}$. The following interpretations may aid in understanding the result

(2.14). The first integral accounts for the fact that neuron ν fires with probability distribution $h_\nu(q)$, given that there is an amount of transmitter q available from all its incoming synapses. (In the original work of Taylor, h_ν is just a step function $\theta(q - q_\nu^{(c)})$ reflecting the assumption of a sharp as opposed to a "soft" threshold. The threshold $q_\nu^{(c)}$ evidently corresponds to the parameter V_ν^T of Ref. 14. In the formulation given here (which mainly follows Ref. 31), the distribution h_ν is introduced to simulate effects of some of the other random variations besides [ii] in the list made earlier.) The product over μ in (2.14) takes account of all neurons which are presynaptic to ν (lumping all synapses of μ onto ν into one), while the first delta function constrains the total amount of transmitter $\Sigma_\mu q_{\nu\mu}$ to the value q. The first term inside the square brackets is the contribution from induced emission and the second is the contribution from spontaneous emission. The former is proportional to the probability that μ did fire at time $t - \tau$; and the latter, to the probability that μ did not fire at that time. To this extent, the two processes are mutually exclusive. Since the second delta function in (2.14) sets the amount of transmitter in synapse $\nu\mu$ at the certain value $n_{\nu\mu} q_{\nu\mu}^{(o)}$, the former process is considered to be deterministic, whereas the latter is described by the probability distribution $g_{\nu\mu}^{(s)}$. If neuron μ is not involved in any of the incoming synapses of ν, both $g_{\nu\mu}^{(s)}$ and the coefficient $g_{\nu\mu}^{(a)}$ of $w_\nu(t - \tau)$ must vanish (e.g. by virtue of vanishing $\lambda_{\nu\mu}^{(s)}$ and vanishing $n_{\nu\mu}$).

Within this model, the state of the system (obviously, in a statistical sense) is specified in terms of the set $\{w_\nu(t)\}$ of absolute firing probabilities of the N neurons at time t; there is now a continuum of states since each w_ν has range $[0,1]$. Eq. (2.14) defines a polynomial mapping from one set of firing probabilities to the next. Thus, the operation of updating the state of the system is nonlinear, in contrast with the linear operation of updating which will characterize the time evolution of the statistical distribution in the Shaw-Vasudevan-Little model.

Given a judicious choice of the probability distribution $h_\nu(q)$, the model underlying (2.14) should be more realistic than that of Shaw and Vasudevan, on purely physiological grounds. Some additional effects and refinements, presumably of "second order," could be given attention in an improved version of Taylor's model. Certainly, one would like to remove the assumption that spontaneous emission is de-activated (or ineffectual) during action-potential stimulation of the presynaptic fiber. Neurons often operate near their firing thresholds, so even a small contribution of spontaneity during presynaptic activity could have crucial influence. To achieve this refinement, note that the delta-function form for the coefficient $g_{\nu\mu}^{(a)}$ of $w_\nu(t - \tau)$ in (2.14) may be replaced by a general probability density in $q_{\nu\mu}$ without changing the mathematical nature of the law of motion for the w_ν (cf. Ref. 31 and Section 4). With a suitable choice for this function we can take into account the spontaneous release of neurotransmitter during arrival of an impulse at the synapse $\mu \to \nu$, and more generally any stochastic variabilities in number and size of the transmitter packets which are emitted into the synapse upon the firing of μ. In contrast to the original Taylor formulation, the function $g_{\nu\mu}^{(a)}$ then describes a *non-deterministic* process or processes. Other "second-order" effects include[13] (a) modification of assumed Poisson distributions to include the "snowball effect"[32] (enhanced probability of multiple emission of quanta over short time intervals) and (b) relaxation of the assumption of a unitary quantum. Refinement (b) would involve replacement of the step-function choice of $q(t)$ by a smooth decay function such as $q^{(0)}e^{-t/t_{dec}}$. These additional features, along with stochastic variability in the size of neurotransmitter quanta released spontaneously when μ remains silent, may be incorporated in revised specifications of $g_{\nu\mu}^{(s)}$ and $g_{\nu\mu}^{(a)}$, again without affecting the essential structure of the theory.

A very important step toward greater realism, already achieved by Taylor,[13] is reformulation of the above theory in *continuous* time. Suffice it to say that the extension involves the derivation of an equation of motion, analogous to (2.14), for the probability $w_\nu(t)dt$ that neuron ν will fire in the time interval $(t, t+dt)$, in terms of firing probabilities of the neurons of the system over earlier time intervals. Aside from operation in continuous time, the basic assumptions are essentially the same as in the original model. Thus, the parameters of that model reappear; but in addition two new parameters are introduced. These are the absolute refractory period t_r (which, in one interpretation, would in fact correspond to the grid spacing τ of the discrete model) and the time delay t_d between the arrival of an action potential at the presynaptic terminal and the consequent emission of transmitter packets. Two versions of the theory arise, depending on whether t_{dec} is less than or greater than t_r. The case $t_{dec} < t_r$ is judged to be physically the more reasonable; in that case the resulting equation of motion is of polynomial form closely resembling (2.14), but of course contains extra time integrations.

Taylor measures the importance of spontaneous release of transmitter quanta in terms of a spontaneity parameter $S = m\lambda^{(s)}q^{(0)}/q^{(c)}$, where m is the number of incoming synapses of a typical neuron and the other quantities are as defined previously. If S is of order unity, spontaneous emission will significantly affect information transmission in the neural network. Insertion of reasonable values for $\lambda^{(s)}$, $q^{(0)}$, and $q^{(c)}$ leads to a critical value of about 10,000 for m, as indicated earlier. Taylor suggests that it might be advantageous for an animal to have an *adjustable* degree of spontaneity, and especially that this facility might be involved in the learning process and more broadly in adaptive behavior. He proposes a scenario in which the animal is faced with some optimization problem (e.g. involving a search for food). The value of S is first set to a high value, allowing a random search for the optimal solution. As a reasonable solution is achieved, the parameter S would decline, so as to ensure reliable behavior of the animal when the search has proven successful. This is one clear example of how the presence of noise in the nervous system could give an evolutionary edge. To our knowledge it is the first suggestion in the literature that a process akin to simulated annealing[33] may occur in animal learning, or may be useful in machine learning. By raising the level of noise in the dynamics, one can avoid getting stuck in local minima of an energy or cost surface. Simulated annealing strategies have recently seen widespread use in applications of neural-network models (or connectionist methods) to optimization problems and higher-order recognition tasks (cf. Refs. 10,34). However, there is an important additional element in Taylor's conception, namely that of control of the degree of spontaneity by mechanisms endogenous to the organism, rather than by some external agent. This control might take the form of a dynamic coupling of S with the other network parameters which are being adjusted in the learning process. In the framework of Little's model, one would allow a dependence of the spontaneity parameters β_ν on experiential stimuli and/or internal network activity, along with plasticity of the couplings $V_{\nu\mu}$ and the thresholds θ_ν. Suitable learning algorithms would then govern the coupled behavior of these "structural" variables.

3. NONEQUILIBRIUM STATISTICAL MECHANICS OF LINEAR MODELS

The long-term behavior of neural networks following exposure to external stimuli is central to attempts at modeling brain activity and to the design of physical systems imitating biological mechanisms of memory storage and recall. A set of stimuli all eliciting the same final operating condition of the network defines an equivalence class of experiences which are said to be associated with the same stored memory. A given network may exhibit one, several, or many such attractors, while a given attractor may correspond to a fixed point, a terminal cycle, multiperiodic motion, or chaos. If the dynamical law by which the system updates its state is probabilistic rather than deterministic, one is naturally led to study an ensemble, and derive information on ensemble averages, acknowledging that the stepwise motion of a particular system is unpredictable.

In the ensemble approach, which we borrow from the statistical mechanics of physical systems, a large collection of identical copies of the system is envisioned. The fraction of copies occupying microscopic state i, for all i and given time t, serves to define the instantaneous probability distribution $\{p_i(t)\}$. The initial preparation of the system attendant to the imposition of a temporary stimulus is reflected in the specification of an initial probability distribution $\{p_i(0)\}$. The response of the system is then characterized, statistically, by the behavior of the state occupation probabilities $p_i(t)$ at asymptotically large times t. To determine this behavior, we need to formulate a law of motion for the distribution $\{p_i(t)\}$, which, of course, must be rooted in the microscopic dynamics of the system. Thus we need to develop a nonequilibrium statistical mechanics of the neural network, considered as a system of interacting neurons.

Here such an approach will be outlined for a certain class of neural network models. This class includes the Little model as a prototype, and also the analogous network version of the Shaw-Vasudevan description of synaptic interactions (Shaw-Vasudevan model). As we shall see, it does not include the Taylor model.

The class of models to be entertained has the following essential features:

C1. The microscopic dynamics (and therefore the statistical evolution) takes place in discrete time, with elementary step τ.

C2. At each time-step t, each neuron makes a stochastic decision whether or not to fire, with probability depending only on the firing states $\sigma_\mu(t - \tau)$ which were actualized by all the neurons μ *one* time-step earlier, and on whatever time-independent structural and interaction parameters neural modeling dictates (e.g. the $V_{\nu\mu}$, θ_ν, and β_ν in case of the Little model without plasticity).

C3. The states i of the full system of N neurons are the firing patterns $\{\sigma_i, i = 0, 1, .. 2^N - 1\}$.

C4. Any such system state i can be reached from any other system state j with finite probability in a finite number of time-steps (true in the Little model if $\beta_\nu^{-1} > 0$ for all ν -- i.e., if the net is *fully stochastic*).

C5. If it is possible to go from state j to state i in one time-step, the reverse transition is also allowed with finite probability.

C6. The probability Q_{ij} that the system will execute a transition from state j at time $t - \tau$ to state i at time t is *independent* of the distribution $\{p_i\}$ which is applicable at the initial time $t - \tau$.

Actually, features C2 and C3 are unnecessarily specific as regards the states of the decision elements and of the system. We have used the σ firing variables and the collective firing patterns merely to aid visualization. All we really need (in place of C2 and C3) is that the network has a *finite* number of states, and

148

that, in the process of updating the state at time t, the *only* information we need about earlier states of the system is the state at time $t - \tau$ (*Markov property*). From the stated features, it follows immediately that the models in question possess a very simple mathematical characterization in terms of the theory of Markov chains, and hence their behavior is in principle well known. While the Taylor model is also Markovian, in the sense that the firing probability w_ν at time t is determined by the firing probabilities w_μ at time $t - \tau$, this model will be seen[16] to fail the linearity requirement C6. We shall return to the issue of the long-term behavior of Taylor's model in Section 4.

Having framed the problem of memory storage and recall in terms of the large-t behavior of neural nets, and in turn having cast the problem of the statistical time development of probabilistic nets in terms of the formalism of ordinary nonequilibrium statistical mechanics, we are prompted to ask the following questions: Is there a multiplicity of ordered final operating conditions of the ensemble, such that the it winds up in one or another depending on certain aspects of the initial stimuli? A many-to-few map from initial conditions to asymptotic modes of behavior is clearly desirable if the network is to function as a content-addressable memory. Or, in imitation of familiar physical systems such as a collection of gas molecules in a box, does the neural ensemble instead always approach the same equilibrium condition, independent of the initial distribution? If the latter, does the final equilibrium condition correspond to ordinary thermodynamic equilibrium, with detailed balance of forward and backward transitions? We shall find that, for the class of networks considered, the answers to these questions (with some qualifications!) are respectively *no*, *yes*, and *no*.

3.1. Statistical Law of Motion -- Markov Chain and Master Equation

For the class of models delineated above, the time development of the probability distribution over system states, i.e. the dynamics of the state occupation probabilities $p_i(t)$, $i = 0, 1, \ldots 2^N - 1$, is given rigorously by

$$p_i(t) = \sum_j Q_{ij} p_j(t - \tau) \quad , \tag{3.1}$$

in terms of the transition matrix (Q_{ij}) (which is given explicitly by Eq. (2.12) in the Little model). This dynamics defines a Markov chain with stochastic matrix (Q_{ij}), indeed, a finite, homogeneous, irreducible, aperiodic Markov chain.[35] Forming a 2^N-dimensional vector p from the occupation probabilities p_i, the law (3.1) may be written simply as

$$p(t) = Q\, p(t - \tau) \;, \tag{3.1'}$$

where Q is a *linear* operator with matrix representation (Q_{ij}). Alternatively, but approximately, we can describe the statistical development of the system by a Master Equation

$$\frac{dp_i(t)}{dt} = \sum_j [T_{ij}\, p_j(t) - T_{ji}\, p_i(t)] = \sum_j W_{ij}\, p_j(t) \;, \tag{3.2}$$

where $W_{ij} = T_{ij} - \delta_{ij} \sum_k T_{ki}$ and $T_{ij} = Q_{ij}/\tau$ is the transition rate referred to one time-step. This approximate equation of motion becomes accurate as the time delay τ becomes much smaller than the time scale of appreciable change in the $p_i(t)$. Hence the Master Equation will provide an acceptable description in the asymptotic regime where the solution $p(t) = \{p_i(t)\}$ of either (3.1) or (3.2) approaches a constant. The steady-state solutions $\overset{*}{p}$ of (3.1) and (3.2) are obviously identical. We may further remark that $\sum_i W_{ij} = 0$ by the definition of the matrix (W_{ij}) and hence $\det(W_{ij}) = 0$. The latter property implies $\operatorname{rank}(W_{ij}) \le 2^N - 1$, so that at least one steady-state solution of (3.2) *does* exist. It also implies that $\sum_i p_i(t)$ is a constant of motion, as may be verified by summing the Master Equation on i. By virtue of the linearity of (3.2), this constant of motion may be normalized to unity provided the p_i are bounded and nonnegative.

The analysis of Eq. (3.2) is facilitated by a graphical representation. Each state of the system is represented by a point or vertex (thus, 2^N points for an N-neuron system), and each nonzero transition rate T_{ij}, from state j to state i, is represented by a line or edge bearing an arrow which points from j to i. If an edge carries arrows pointing in both directions, they are omitted to reduce clutter. Drawing all state vertices and all allowed edges, we arrive at the *basic graph* G of the system. Fig. 3 shows two simple examples. We shall carry through the analysis with the aid of a systematic theory of finite, linear, Master-Equation systems due to Schnakenberg.[36] This theory is directly applicable to the class of neural networks under consideration. One key prerequisite of the formalism (certainly met by the Little model and its extension to Shaw-Vasudevan neurons) is that if there is a finite transition probability in one direction, then there is also a finite transition probability in the reverse sense, i.e., $T_{ij} > 0$ implies $T_{ji} > 0$ (property C5). Together with property C6, this feature implies that the basic graph G will be *connected*, in the sense that for each pair of vertices (or states) there exists at least one sequence edges (or allowed transitions) connecting them in both directions.

150

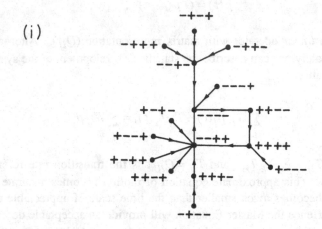

(i)

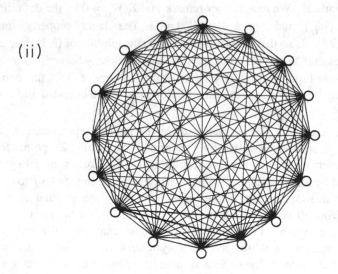

(ii)

Fig. 3. Examples of basic graphs for Little networks of $N = 4$ neurons, obeying (i) a deterministic synchronous updating rule (all β_v infinite) and (ii) fully probabilistic synchronous dynamics (all β_v finite).

We are concerned with the existence of a steady-state solution $\overset{\circ}{p}$ of Eq. (3.2) *within the physical region* circumscribed by $\Sigma_i \, p_i = 1$ and $0 \le p_i \le 1$, all i. The latter conditions must hold for the p_i to be probabilities. One of Schnakenberg's theorems establishes that if the basic graph G is connected, then such a solution is guaranteed to exist, with $0 < p_i < 1$, all i. This steady solution is known as the *Kirchhoff* solution. An elegant graph-theoretic expression for it is provided in Refs. 36,16. Another important result is that this solution is *unique* in the physical region: there is only one steady solution of (3.2) such that the p_i may be interpreted as state-occupation probabilities.

Turning to the nature of time-dependent solutions of the Master Equation, two further theorems of Schnakenberg are relevant. The first asserts that if the initial set of p_i is in the physical region, the solution of (3.2) *remains* in the physical domain, provided that the basic graph G is connected. The second result concerns approach to the Kirchhoff steady state: Again assuming a connected G, the matrix (W_{ij}) has a *nondegenerate* maximum eigenvalue 0, and all other eigenvalues have *positive* real parts. The nondegeneracy of the 0 eigenvalue implies that in fact $\mathrm{rank}\,(W_{ij}) = 2^N - 1$. The eigenvalue 0 corresponds to the nondegenerate maximum eigenvalue 1 of the stochastic matrix $(Q_{ij}) = \tau(W_{ij}) + (\delta_{ij})$. The nature of the other eigenvalues of (W_{ij}) implies that any time-dependent solution initiated at any point in the physical domain *eventually relaxes exponentially to the Kirchhoff steady state.*

Thus we know that the Kirchhoff steady state is unique in the physical region, and that, as an equilibrium point, it is absolutely stable with respect to all physical solutions of the Master Equation. When we think in terms of ordinary thermodynamics, the question naturally arises: does the Kirchhoff solution correspond to *thermodynamic* equilibrium? In order to address this issue, we must decide what thermodynamic equilibrium is to mean in this rather abstract context, where the brain (as represented by our neural network model) is not just regarded as "a piece of meat," but rather as an information-processing system with its own special brand of dynamics, operating on a plane distinct from that of mundane physical systems. Evidently, we want to identify thermodynamic equilibrium with a condition of *detailed balance* in the Master Equation, where each term in the sum over j on the right in (3.2) vanishes independently of the others,

$$T_{ij}\,\overset{\circ}{p}_j - T_{ji}\,\overset{\circ}{p}_i = 0 \quad , \quad \text{for all } ij \quad . \tag{3.3}$$

In principle, there are of course many ways in which the right-hand side of (3.2) could be zero, due to cancellations between terms with different j values. One therefore suspects that rather special choices of model parameters (couplings, thresholds, . . .) will be required, if the Kirchhoff solution is to represent thermodynamic equilibrium, and indeed for some models the condition of detailed balance may not be attainable at all. It will be shown below that this suspicion is correct.

3.2. Entropy Production in the Neural Net

Pursuing an analogy with an open but homogeneous system of 2^N reacting chemical species, the rate of entropy production in the neural network takes the bilinear form

$$\mathbb{P} = \frac{1}{2}\sum_{ij} J_{ij}A_{ij} \tag{3.4}$$

in terms of *generalized thermodynamic fluxes* $J_{ij} = T_{ij}\,p_j - T_{ji}\,p_i$ and *generalized thermodynamic forces* (or affinities) $A_{ij} = \ln\,[T_{ij}\,p_j/T_{ji}p_i]$. (For detailed argumentation, see Refs. 36,16.) The entropy of the system itself is given by

$$S = -\sum_i p_i \ln p_i \quad, \tag{3.5}$$

i.e., by the information entropy. The entropy production rate (3.4) splits into two components, $\mathbb{P} = \mathbb{P}_1 + \mathbb{P}_2$, where $\mathbb{P}_1$ is just the rate of change dS/dt of the entropy assigned to the system, and the remainder,

$$\mathbb{P}_2 = \frac{1}{2}\sum_{ij} J_{ij}\ln\,[T_{ij}/T_{ji}] \quad, \tag{3.6}$$

is ascribed to the action on the system of external thermodynamic forces which (potentially) keep it from reaching thermodynamic equilibrium. In the case of the neural network, which we suppose to be in autonomous operation, these external forces are a convenient fiction of the formalism; in fact they are inherent in the internal microscopic dynamics of the model and are ultimately expressible in terms of the transition probabilities T_{ij}. We note that in general neither $\mathbb{P}_1$ nor $\mathbb{P}_2$ is required to be nonnegative; however, $\mathbb{P} = \mathbb{P}_1 + \mathbb{P}_2 \geq 0$ applies rigorously. Obviously, $\mathbb{P}$ vanishes in a steady state corresponding to thermodynamic equilibrium.

It must be recognized that three levels of description are involved in our analysis. At the bottom level we have the *microscopic dynamics*, where attention is paid to the changing firing states σ_v of the individual neurons. The next level up is that of *statistical mechanics*, where we follow the time development of the state occupation probabilities p_i. The third or highest level is the analog of thermodynamics, generally *nonequilibrium thermodynamics*; at this level the description is framed in terms of certain *macroscopic forces and fluxes* (analogous, say, to temperature gradients and concomitant heat flows). A bridge from the first to the second level has been provided by the Master Equation (3.2) (or by the dynamics (3.1) of the Markov chain defined by the network model). To link the second and third levels, we shall equate the relevant expressions for the rate of entropy production of the network in the asymptotic steady state, denoted $\hat{P}$. We already have an expression for $\hat{P}$ at the level of statistical mechanics: just substitute the Kirchhoff solution $\hat{p}$ into (3.4).

3.3. Macroscopic Forces and Fluxes

At the macroscopic level of nonequilibrium thermodynamics, the steady-state entropy production assumes the bilinear form

$$\hat{P} = \sum_{f=1}^{F} J(\mathbf{C}_f) A(\mathbf{C}_f) \ , \tag{3.7}$$

in a set of $2F$ *macroscopic forces and fluxes* (for motivation and substantiation, see Ref. 37). A force $A(\mathbf{C}_f)$ and a flux $J(\mathbf{C}_f)$ is present for each member of a *fundamental set of cycles* of the basic graph G. A fundamental set of cycles is characterized by the property that an arbitrary cycle (i.e., closed loop of oriented edges) occurring in G may be expanded in such a set, according to rules similar to those for expanding an arbitrary vector in a basis for the linear vector space in which it resides. The graph-theoretic aspects of expression (3.7) are developed at length in Refs. 36,16 (see also Ref. 38). For our purposes it is sufficient to realize that upon setting this macroscopic version of $\hat{P}$ equal to the statistical version constructed above, one may determine the macroscopic variables $A(\mathbf{C}_f)$, $J(\mathbf{C}_f)$ in terms of the microscopic transition rates T_{ij} and the steady-state values $\hat{p}_i$ of the statistical variables p_i. The construction of a fundamental set of cycles is illustrated in a very simple case (2-neuron system) in Fig. 4. First, chose a maximal tree T(G) of the basic graph G, i.e., some connected, covering, circuit-free subgraph of G -- any will do. An edge of G which is absent from T(G) is called a chord of T(G). If G has E edges, there will be $F = E - 2^N + 1$ such chords. Now, each in its turn, add these chords to T(G) and discard any dangling edges, forming a set of subgraphs each containing exactly one circuit C. This set constitutes a fundamental set of circuits $\{C_1, \ldots C_F\}$. Upon assigning orientations to the $C_1, \ldots C_F$ (arbitrarily), we arrive at a fundamental set of cycles $\{\mathbf{C}_1, \ldots \mathbf{C}_F\}$.

154

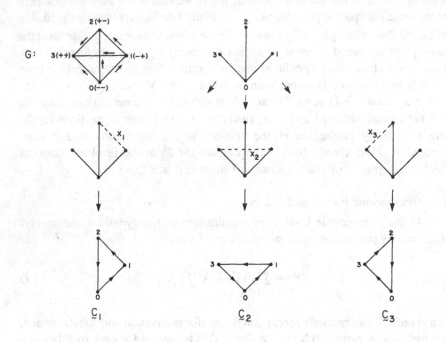

Fig. 4. Generation of a fundamental set of cycles for a fully stochastic neural network with $N = 2$. Each state point of G carries a trivial T_{ii} loop, omitted for simplicity. The arrows adjacent to the edges of the basic graph G indicate chosen reference orientations.

It is sobering to note that -- within, say, the Little model, where the basic graph G is fully connected -- the decomposition (3.7) would entail $2 \times [2^N (2^N - 1)/2 - 2^N + 1]$ thermodynamic variables, a huge number for any sizeable N. (In the human brain, N is of order 10^{11} and hence $2F$ would be of order $4^{10^{11}}$.) The thermodynamic description looks anything but economical!

We quote three beautiful results from the graph-theoretic analysis:

(A) The condition $A(\mathbf{C}_f) = 0$ for all f holds if and only if $J(\mathbf{C}_f) = 0$ for all f; either is a necessary and sufficient condition for detailed balance (3.3) and accordingly for the Kirchhoff solution to correspond to thermodynamic equilibrium.

(B) The forces $A(C_f)$ (and hence the forces $A(C)$ associated with *arbitrary* cycles C) are independent of the state-occupation probabilities $\hat{p}_i$.

(C) The force $A(C)$ around an arbitrary cycle C may be computed as the log of the ratio of the product of transition rates going in the forward direction around C to the product of transition rates going in the reversed direction around C. (Thus, in thermodynamic equilibrium, "forward" and "backward" pressures balance and $A(C) = 0$.) For a cycle $j \to k \to i \to j$, one has in particular

$$A(C) = \ln \frac{T_{ji} T_{ik} T_{kj}}{T_{jk} T_{ki} T_{ij}} \quad . \tag{3.8}$$

The findings (A)-(C) will permit an efficient explication of the conditions on neuronic couplings (and, as relevant, other network parameters) which permit the attainment of thermodynamic equilibrium.

3.4. Conditions for Thermodynamic Equilibrium

In this subsection some specific and general conclusions on the issue of thermodynamic equilibrium are stated and proved.

Theorem I. Within the Little model, the symmetry properties

$$\beta_v V_{v\mu} = \beta_\mu V_{\mu v} \ , \qquad \mu, v = 1, \ldots N \ , \tag{3.9}$$

provide necessary and sufficient conditions for the steady state of an N-neuron network to correspond to thermodynamic equilibrium, i.e. satisfy Eq. (3.3), independently of neuronal thresholds and diagonal couplings.

Subject to the assumption that all the β_v coincide, this criterion has already been established by Peretto.[25] The proof outlined below rests on the fact [Result (A) above] that thermodynamic equilibrium is equivalent to the vanishing of $A(C_f)$ for all members $f = 1, \ldots F$ of a fundamental set of cycles (selected at will). In exploiting this result, Property (B) is obviously crucial. Rule (C) is invoked to express the balance condition as an equality of the product of transition rates going forward around an arbitrary fundamental cycle (left-hand side) and the product of transition rates going backward around the cycle (right-hand side). An important feature of this expression, in the case of Little's model (or its refined version à la Shaw and Vasudevan), is that for every firing probability ρ appearing as a factor on the left, there is a ρ factor on the right with exactly the same argument (apart from the sign of the argument), and vice versa; this property is assured by the fact that the same *initial* states appear, each once and only once, on the left and on the right. In the detailed manipulations, extensive use is made of the identity

$$s_F(\pm x) = (1 + e^{\pm x})^{-1} = \frac{1}{2} e^{\mp x/2} \operatorname{sech}(x/2) \tag{3.10}$$

satisfied by the Little choice for the functional form of ρ, and of the additivity of exponents.

Before offering a proof for arbitrary N, it will be useful to demonstrate the necessity of conditions (3.9) for the simplest nontrivial case, viz. $N = 2$. We choose the fundamental set of cycles shown in Fig. 4 and consider the balance condition for C_1:

$$T_{02} T_{21} T_{10} = T_{01} T_{12} T_{20} . \tag{3.11}$$

Specializing to the Little model, and referring to the Appendix, this condition becomes

$$s\,[\beta_1(V_{11}-\theta_1)]s\,[\beta_2(V_{21}-\theta_2)]s\,[-\beta_1(V_{12}-\theta_1)]s\,[\beta_2(V_{22}-\theta_2)]s\,[-\beta_1\theta_1]s\,[\beta_2\theta_2] \tag{3.12}$$

$$= s\,[\beta_1(V_{12}-\theta_1)]s\,[\beta_2(V_{22}-\theta_2)]s\,[\beta_1(V_{11}-\theta_1)]s\,[-\beta_2(V_{21}-\theta_2)]s\,[\beta_1\theta_1]s\,[-\beta_2\theta_2] .$$

The first thing to notice is that when we insert identity (3.10) into (3.12), the sech factors will all cancel out, due to their even parity and to the coincidence of arguments on left and right (apart from sign). The second simplification concerns the thresholds θ_v and diagonal couplings V_{vv}. We observe that each neuronal threshold appears with the same number of plus [minus] signs on the right of (3.12) as it does on the left, and similarly for each diagonal coupling that enters. Thus the exponential factors $e^{\pm\theta_v}$ and $e^{\pm V_{vv}}$ all cancel out of the relation. The threshold cancellation automatically removes the transition-rate factors T_{20} and T_{10} having 0 as initial state. We are left with a condition on exponents which reads

$$-\beta_2 V_{21}/2 + \beta_1 V_{12}/2 = -\beta_1 V_{12}/2 + \beta_2 V_{21}/2 . \tag{3.13}$$

To interpret this result within a broader context, note that neuron 1 is "on" in state 2 but "off" in state 1, whereas neuron 2 is "on" in state 1 but "off" in state 2. Thus there is no neuron which has the *same* firing state in both of the system states 1 and 2; we say there are no "common" neurons. The dead state

0 does not play a role in this nomenclature. Now, Eq. (3.13) is telling us that the *only* exponents which survive to produce a nontrivial condition come from s_F factors referring to (i.e., representing the firing probability of) a non-common neuron and involving excitation of a *different* non-common neuron in the initial state. The result (3.13) of course implies $\beta_1 V_{12} = \beta_2 V_{21}$. The same condition is reached by reduction of the balance conditions for cycles C_2 and C_3, thus validating the theorem for $N = 2$. However, the situation for these other two cycles differs somewhat from that for C_1. For C_2, neuron 2 is "common" and neuron 1 is "non-common," being excited in state 3 but not in state 1; for C_3 there is again one common and one non-common neuron. Accordingly, for C_2 there is a cancellation of the s_F factor for neuron 1 in T_{13} against the s_F factor for that neuron in T_{03} because the initial state is the same and neuron 1 is "off" in both final states -- and similarly for C_3. (Alternatively, with diagonal couplings removed, one may cancel the s_F factor for non-common neuron 1 in T_{01} against the corresponding s_F factor in T_{03}, since the final states coincide and only the common neuron 2 is active in both initial states -- and again similarly for C_3.) In these cases, the only surviving exponents come from s_F factors referring to a common neuron and involving excitation of a non-common neuron in the initial state, or referring to a non-common neuron and involving an excited common neuron in the initial state; there is the obvious additional restriction that each reference neuron is in different firing states on the left and right.

This simple example was treated in such painful detail to prefigure a general proof, which we now present.

(i) That the relevant conditions on the network parameters must be independent of the thresholds θ_v may be traced to the fact that the same system states appear as *final* states on the left and right in the generic balance equation. For then a given θ_v enters with the same number of minus signs on the left and right, and the same number of plus signs. Hence, in applying (3.10) and combining exponents, the θ_v's all cancel out.

(ii) That the conditions we seek must be independent of the diagonal couplings may be seen as follows. Suppose neuron v remains excited for n successive state points around an arbitrary cycle, just before which and just after which it is not excited. Then in using (3.10) and adding exponents, $\beta_v V_{vv}$ will enter $n - 1$ times with a minus sign and once with a plus sign when going *either* forward *or* backward around the cycle. Transitions from states in which v is not excited patently contribute no terms in $\beta_v V_{vv}$. Extension of the argument to cycles in which v is alternately excited for n_1 successive vertices, de-excited for the next n_2, then excited for the next n_3, etc., is trivial.

158

(iii) To affirm the necessity and sufficiency of the symmetry conditions (3.9), it is convenient (as well as permissible) to choose a very simple set of fundamental cycles, each member of which contains the dead state 0 (i.e., the state with all neurons "off") and only three edges. (That such a set can always be formed is easily seen in terms of elementary graph theory[38,36,16]: just take a maximal tree with the $2^N - 1$ edges $0i$, $i \neq 0$.) Thus we may restrict our considerations to the generic cycle $0ji$, a triangle with vertices 0, j, and i, where i and j are any states that (directly) communicate with one another, 0, j, and i being distinct. Accordingly, thermodynamic equilibrium hinges on the balance condition

$$T_{0i} T_{ij} T_{j0} = T_{0j} T_{ji} T_{i0} . \tag{3.14}$$

(This corresponds to Peretto's triangle equality.[25]) Appealing to results (i) and (ii) above, we may make the further simplifications of setting all thresholds θ_ν and diagonal couplings $V_{\nu\nu}$ equal to zero, without sacrifice of generality. With zero thresholds, the transition rates T_{j0} and T_{i0} with 0 as initial state both reduce to 1.

Let $\kappa_1 \ldots \kappa_k$ label the neurons which are *on* in state i but *off* in state j; and let $\lambda_1 \ldots \lambda_l$ denote the neurons which are *on* in j but *off* in i. The remaining neurons will again be called "common"; they are denoted generically by ζ. It is easy to see that the s_F factors involving *only* the "common" neurons, which have by definition the same firing states in i and j, are the same on the left and right of (3.14) and hence may be removed. There is also a cancellation of all s_F factors bearing *only* the indices $\kappa_1 \ldots \kappa_k$ or *only* the indices $\lambda_1 \ldots \lambda_l$. This happens because we are comparing factors which arise from the same initial state (viz. i in the pure κ-index case, j in the pure λ-index case), the non-common neurons all being "off" in the respective final states (viz. 0 or j in the κ case, 0 or i in the the λ case). Finally, there is still another sort of cancellation, of factors in T_{0i} against factors in T_{0j}. The final state being the same, the s_F factors referring to non-common neurons and involving excitation only of common neurons match exactly. The ensuing general balance condition is again best reduced -- having invoked the identity (3.10), along with the even parity of sech(x) and the property that the same arguments (apart from sign) appear on the left and right -- to a condition on exponents of the form $\pm\beta_\xi V_{\xi\omega}$. In a rather schematic notation, this relation reads

$$2\sum_{\kappa}^{k}\sum_{\zeta^+}\beta_\kappa V_{\kappa\zeta^+}+\sum_{\zeta}\text{sgn}_\zeta\sum_{\lambda}^{l}\beta_\zeta V_{\zeta\lambda}+\sum_{\zeta}^{l}\sum_{\lambda}\beta_\zeta V_{\zeta\lambda}+2\sum_{\kappa}^{k}\sum_{\lambda}^{l}\beta_\kappa V_{\kappa\lambda} \tag{3.15}$$

$$=2\sum_{\lambda}^{l}\sum_{\zeta^+}\beta_\lambda V_{\lambda\zeta^+}+\sum_{\zeta}\text{sgn}_\zeta\sum_{\kappa}^{k}\beta_\zeta V_{\zeta\kappa}+\sum_{\zeta}^{k}\sum_{\kappa}\beta_\zeta V_{\zeta\kappa}+2\sum_{\lambda}^{l}\sum_{\kappa}^{k}\beta_\lambda V_{\lambda\kappa} \ .$$

Integral subscripts have been dropped from the κ and λ neuron labels. The sign written sgn_ζ is plus if the common neuron ζ is "on", minus if it is "off"; active common neurons are denoted specifically by ζ^+.

Consider now the special case $k = 1$, $l = 1$, with all common neurons silent. (We note that this effectively reduces the situation to that of the preliminary example for $N = 2$ and cycle C_1.) Appealing to the fact that neurons κ_1 and λ_1 are completely arbitrary and may be relabeled at will, the generic condition (3.15) is readily seen to collapse to the symmetry condition (3.9) stated in the theorem, establishing its *necessity*. It is a straightforward matter of induction to show that (3.9) is also *sufficient* to guarantee the generic relation (3.15).

The proof just given supersedes that of Ref. 16. The condition (8.17) appearing in the latter proof rests on the unstated assumption that all common neurons are silent, while the sum over η in condition (8.17') should include only active common neurons.

Remark 1. It must be emphasized that in establishing Theorem I we have relied on certain special properties of the conditional firing probability. This probability may be conveniently expressed as

$$\rho_v(\sigma_v) = s(-\beta_v\sigma_v F_v) \ , \tag{3.16}$$

wherein Little's special choice (2.11) has been made for the function s. Other functions s of sigmoid character could have been made in setting up the network model, without affecting the formal development which preceded Theorem I. (For example, we could have adopted $s(x) = 1 + (1/\pi)\tan^{-1}(\pi x/2)$, or the result (2.8) of Shaw and Vasudevan, for that matter.) However, our proof of Theorem I will only go through if s has the form

$$s(x) = e^{cx} y(x) \ , \tag{3.17}$$

where c is an arbitrary constant and $y(x)$ some *even* function of x. This *does not* mean that our strongly negative conclusions regarding the occurrence of thermodynamic equilibrium are tied to (2.11). Quite the contrary: with Result

(A) imposing F relations (not necessarily independent) among the $N^2 + N$ hardware parameters $\beta_\nu V_{\nu\mu}$, $\beta_\nu \theta_\nu$ of the network, *highly restrictive* conditions for detailed balance will still prevail. But now there is the added complication that, in general, these conditions involve not only the $\beta_\nu V_{\nu\mu}$ with $\nu \neq \mu$, but also the quantities $\beta_\nu V_{\nu\nu}$ and $\beta_\nu \theta_\nu$. Moreover, the form of these conditions will depend on N, in contrast to the universality of relations (3.9) within the context of Theorem I. Similar statements can be made for other realizations of microscopic network structure, interactions, and dynamics consistent with the specification C1-C6 made at the beginning of this section.

Remark 2. From the mathematical point of view, we may redefine the couplings via $\beta \tilde{V}_{\nu\mu} \equiv \beta_\nu V_{\nu\mu}$ (as implicit in Ref. 25), β being a neuron-independent noise parameter. Condition (3.9) then states that the redefined interactions $\tilde{V}_{\nu\mu}$, $\tilde{V}_{\mu\nu}$ between neurons ν and μ are symmetrical or reciprocal ("Newton's Third Law for neurons").

Higher-order synapses are known to exist in the nerve networks of invertebrates[39] and in the peripheral nervous systems of vertebrates. Presumably they also play a significant role in the central nervous systems of vertebrates, considering the observed density and complexity of synaptic junctions.[40,41] A nth-order synapse is one at which neuroanatomical processes (axon terminals, dendrites) from n presynaptic neurons $\mu_1, \mu_2 \ldots, \mu_n$ come together to influence the action of postsynaptic cell ν; that is, the change in membrane potential of neuron ν attributable to such a synapse depends on the constellation of neurons of the set $\{\mu_1, \mu_2, \ldots \mu_n\}$ which were simultaneously active during a relevant interval of the recent past (cf. Ref. 42). In the framework of synchronous discrete modeling, there will be a nontrivial dependence of the input from this synaptic complex at time t on the correlated activity of the presynaptic neurons at time $t - \tau$. Using the language of many-particle theory, the neurons experience interactions of up to $(n+1)$-body $[((n+1)$-neuron] character. This increased realism and increased richness may be accommodated in the Little model by a transparent generalization of the firing function F_ν of Eq. (2.10):

$$F_\nu(t) = \sum_{m=1}^{n} \sum_{\mu_1 \cdots \mu_m} V_{\nu\mu_1 \cdots \mu_m} \pi_{\mu_1}(t - \tau) \cdots \pi_{\mu_m}(t - \tau) - \theta_\nu \ , \quad (3.18)$$

where the $m+1$-body interactions are described by the quantities $V_{\nu\mu_1 \cdots \mu_m}$, which define tensors of order $m+1$. We are interested in determining the constraints on these tensors (in particular, the symmetries) which are imposed by thermodynamic equilibrium. For the simple case of binary interactions, $n = 1$, the answer is given by Theorem I. We may elaborate the same reasoning to extend this theorem to $n = 2$ (and higher order, as desired).

Theorem II. Within the Little model extended to include three-neuron couplings according to (3.18) with $n = 2$, the following relations among the neuronic interactions are necessary and sufficient for the ultimate attainment of thermodynamic equilibrium:

$$\bar{V}_{\nu\mu} = \bar{V}_{\mu\nu} , \tag{3.19}$$

$$\bar{V}_{\nu\mu_1\mu_2} + \bar{V}_{\nu\mu_2\mu_1} = 0 , \tag{3.20}$$

Here ν and μ are distinct, as are μ_1 and μ_2, these labels being otherwise generic; and

$$\bar{V}_{\nu\mu} \equiv \beta_\nu(V_{\nu\mu} + V_{\nu\mu\mu}) . \tag{3.21}$$

The proof follows the same pattern as that of Theorem I. Argument (i) applies without alteration and argument (ii) is trivially generalized; thus the salient conditions are again independent of the thresholds θ_ν and diagonal couplings $V_{\nu\nu}$, $V_{\nu\nu\nu}$, etc. Argument (iii) proceeds as before, up to the explication of the generic balance condition, which -- relative to (3.15) -- acquires ten new terms on each side, involving three-neuron couplings. To derive *necessary* conditions from this generic relation among exponents, we consider three cases: (a) $k = 1$, $l = 1$, with all common neurons silent, (b) $k = 1$, $l = 0$, with all common neurons silent except ζ_1, and (c) $k = 2$, $l = 1$, with all common neurons silent. In case (a), balance implies (3.19). Imposing this necessary condition in case (b), we generate, as a second necessary condition, the version of (3.20) in which one of μ_1, μ_2 coincides with ν. Invoking these results, case (c) yields the version of (3.20) in which all three indices are distinct. An inductive process based on the generic condition then establishes the sufficiency of (3.19)-(3.20).

Remark 3. We observe that while the barred two-body interaction must be *symmetric*, the barred three-body interaction tensor must be *antisymmetric* in its second and third indices. Extending the arguments to higher orders n, (3.20) is supplemented by further conditions requiring the vanishing of sums of coupling components $\bar{V}_{\nu\mu_1\cdots\mu_m}$ over permutations of the indices $\mu_1, \ldots \mu_m$, while

$\overline{V}_{\nu\mu_1\cdots\mu_m}/\beta_\nu$ becomes, generally, $V_{\nu\mu_1\cdots\mu_m}$ plus a suitably normalized sum of all partially diagonal couplings bearing the same set of indices.[43]

3.5. Implications for Memory Storage: How Dire?

We are now well equipped to deal with the questions posed at the beginning of this section, for the specified class of finite, linear, Markovian neural networks. Consider again that an ensemble of such nets is exposed momentarily to stimuli which produce a given initial distribution over system states i. It is a rigorous mathematical result that if the subsequent statistical evolution is allowed to proceed autonomously for a sufficiently long time, the ensemble *will* assume a steady-state (equilibrium) distribution. In the Master Equation formulation, this equilibrium distribution is, in fact, *absolutely stable* in the 2^N-dimensional space of the probability distributions $\{p_i\}$ over firing patterns i; in the Markov chain formulation, which is exact, it is the *unique* stationary distribution. On the other hand, the condition which is reached is generally *not* one of *thermodynamic* equilibrium, wherein the probability flow $T_{ij}p_j$ from state j to state i exactly balances the flow from state i to state j, for all pairs ij. Instead, the final stationarity is generally achieved only by cyclic processes involving triples, quadruples, ... of states. In this sense we may speak of the neural system as operating away from (thermodynamic) equilibrium.

The relations among synaptic interactions under which thermodynamic equilibrium does prevail, laid out in detail in the preceding subsection, are highly restrictive and unlikely to be met by biological systems, except approximately in sharply defined contexts.[44] Thus, to the extent that the class of models studied here is relevant, it must be concluded that biological nerve nets typically do not obey the principle of detailed balance. Hence they cannot be described by a Hamiltonian (Gibbsian) formalism. The description initiated in Subsections 3.2-3.3, leading to the definition of macroscopic forces and fluxes, offers an alternative which deserves further development and exploration.

Since neural network modeling is largely motivated by the hope of gaining new insights into human memory and learning on the one hand, and the goal of building efficient content-addressable memory devices and adaptive automata on the other, the memory storage capability of proposed models is a prime concern. Based on the mathematical results obtained here, the storage capacity of probabilistic nets of the Little type, or of the broader class C1-C6, would appear to be limited in the extreme. As we have seen, for a network with given parametric specifications (e.g., given $\beta_\nu V_{\nu\mu}$, $\beta_\nu\theta_\nu$) there is only one final condition in $\{p_i\}$ space (the Kirchhoff steady state), *independent of initial conditions*. In this sense such a network can only store one memory, since all stimuli elicit the same ultimate response. However, information can also be carried by the *transient* statistical response of the system, which may subsist for a very long time if there happens to be more than one eigenvalue of the stochastic matrix $Q = (Q_{ij})$ with modulus near unity.

Of course, when N and all the β_ν are finite, we know from the Master-Equation analysis, or from the theory of Markov chains, or -- basically -- from the Perron-Frobenius theorem,[45] that there can be only one eigenvalue q of Q with modulus exactly unity, all the others satisfying $|q| < 1$. However, consider the noiseless limit of the Little model, with the number of neurons N and the coupling and threshold parameters held fixed. For simplicity we take all β_ν equal to a common value β, and suppose that F_ν is never exactly zero. In the limit $\beta \to \infty$, the dynamics becomes deterministic, and one is left with the McCulloch-Pitts hard-threshold model. The basic graph G of this deterministic model may be disconnected,[16] in which case the associated Markov chain will be reducible. Corresponding to each disconnected subgraph of G, i.e., corresponding to each recurrent class of the Markov chain,[35] there will exist a separate unit eigenvalue of the stochastic matrix $Q_\infty \equiv \lim_{\beta \to \infty} Q(\beta)$. (A *recurrent state* is one to which the system is certain to return. States in a *recurrent class* communicate with each other through an appropriate sequence of allowed transitions, and no communication is possible between the states of different recurrent classes.) If the Markov chain has a recurrent class containing d states, all of the dth roots of unity are eigenvalues of Q_∞, and conversely. The states of such a class are repeated with period d. Graphically, this is reflected in the presence of a d-cycle. (By a d-cycle we mean a terminal cycle or cyclic mode, *including* the trivial case $d = 1$ corresponding to a steady state.) Algebraically, this repetitive behavior is captured in the propagation relation $p(t + n\tau) = Q^n p(t)$, which follows by iteration of (3.1'). The repertoire of cyclic modes which may be excited in the deterministic limit depends in detail on the choices made for the network couplings $V_{\nu\mu}$ and thresholds θ_ν. These and related matters are discussed extensively in the papers of Little, Shaw, and coworkers[7,14,18-21] and Clark,[16] and, especially, by Thompson and Gibson[22] and Gibbs[46].

It becomes clear, then, that for finite, but suitably large, values of β, and for suitably chosen couplings and thresholds, the moduli of a certain subset of the eigenvalues of Q will be close enough to unity that the network may be considered to display *ordered* behavior, even if such order does not last forever. Here, *order* is considered to be present if a correlation exists between network states which may be many time-steps apart. A more precise definition in terms of the eigenvalues and eigenvectors of Q is provided in Ref. 22. At finite N, the transition from disorder to order upon increase of β can never be as sharp, say, as it is in the phase transition between paramagnetism and ferromagnetism in a bulk spin system, where the thermodynamic limit is implied. However, there can indeed be *order* in the neural model to the extent that the system manifests useful content-addressable memory properties. For instance, choosing β large enough that the two largest eigenvalues of Q are degenerate to within 1%, allows temporal correlations to extend over $\sim 10^2 \tau$ (Ref. 14). With judicious choice of the couplings and thresholds, the system will be characterized by a number of different "persistent states" (Refs. 7,14,18), which may be elicited by appropriate stimuli. Thus the *effective* memory capacity of networks of the sort considered here is potentially of reasonable size.

Thompson and Gibson[22] have offered numerous examples which demonstrate this potential as well as the rich behavior available to the Little model. Of special interest are two extreme cases, the one in which the eigenvalues of Q_∞ having modulus unity consist only of the dth roots of unity, with $d > 1$; the other in which the eigenvalues of Q_∞ with modulus unity are all equal to 1. We are concerned with the behavior when β is decreased to some finite but still reasonably large value, disregarding the initial transient phase. In the former example, a particular stochastic net may be expected (generally) to perform a nontrivial terminal cycle, visiting in turn each of a certain set of d microscopic states with period d; eventually a random out-of-sequence transition will disrupt the basic rhythm, but this merely resets the cycle. In the latter case, a sample net is expected to occupy some microscopic state for an extended period, then abruptly change its firing pattern to that of another state, which is occupied for another lengthy period and then suddenly exited, and so on. Thus, transitions from one "persistent state" to another would be observed, if by persistent state we mean a steady firing pattern (cyclic mode of period 1). Any single neuron would display a sequence of burst discharges, a phenomenon widely observed in biological nerve nets. More generally, a stochastic network with sizeable β will behave in *much the same way* as its deterministic (noiseless) counterpart: An instantaneous initial stimulus leads to a periodic mode (either a persistent state or a nontrivial cycle) which may endure for some time; *however*, because of the possibility of random firings or misfirings, the same mode need not always be excited, and spontaneous transitions between various modes can occur. In all these behaviors there are interesting manifestations of order, or collective phenomena, arising out of the microscopic dynamics of the network model. It is worth reiterating[22] that the existence of order does not necessarily imply a preference for some "persistent state" in the sense of Little. Indeed, non-trivial terminal cycles must be expected to play a prominent role, if the couplings $V_{\nu\mu}$ are supposed to be asymmetric.

Turning from mathematics to biology, we see that in fact our disappointment with the long-term behavior of the Little model and its relatives is quite misplaced. In real neural systems there will in general be significant dispersion in the signal-transmission times between directly communicating neurons or in the absolute refractory periods of individual neurons. There are also appreciable secular variations of neuronal and synaptic properties. Consequently, the strict synchronism of these models becomes untenable after some moderate number of time-steps which will depend on (among other things) the size of the system. For example, Thompson and Gibson[22] regard 100 time-steps as a "typical" upper limit. Thus in biological applications we should only be concerned about the "large"-t behavior of the model on a modest time scale. The above considerations ensure that interesting and possibly useful memory capabilities can easily be realized within such a restricted time frame.

En passant we should note that the asynchronous dynamics of the Hopfield model[9] (wherein at most one neuron changes its state at each updating event in discrete time) is at least as unrealistic as the synchronous dynamics of the Little model. In the Hopfield model, synaptic and other delays are suppressed, as is the fact that firings actually take a finite time and may overlap. Indeed, it has been argued[25] that synchronism is a reasonable assumption for small systems. If the system is so small that every action potential is transmitted to its destination in a time less than the relevant absolute refractory period, and if the absolute refractory periods of the neurons all have nearly the same value, this irreducible delay tends to synchronize the activity. Obviously the argument breaks down as the system grows larger.

In seeking a practical and informative description of complex N-body systems within the general arena of macroscopic, equilibrium thermodynamics, the physicist is pressed to take two limits, namely $t \to \infty$ (to achieve a stationary, equilibrium situation, if it exists) and $N \to \infty$ (to smooth out fluctuations due to the small number of particles, or units, in the system). Here we have only paid attention to the former limit, keeping N finite, though potentially large. In any such discussion, the proper order of limits becomes an issue. The conventional theory of phase transitions would dictate that the thermodynamic limit $N \to \infty$ be taken first. We might then expect to see truly ordered phases in the Little model and sharp phase transitions under variation of the "temperature" $T = \beta^{-1}$ and other thermodynamic variables, including the ratio α of the number of stored patterns to the number of decision units. In certain domains of these thermodynamic variables, we might expect the system to function reliably as a content-addressable memory, with substantial storage capacity. This is indeed what is found analytically in mean-field-theoretic treatments. Explicit analytic results can only be obtained for special choices of the interneuronal couplings; the ones which have been studied derive from ideas about learning mechanisms and/or from the theory of spin systems (for representative work, see Refs. 25-28). In particular, phase diagrams for the Little model are presented in Ref. 28, for a fully-connected net with symmetrical couplings determined by a prescription[9] which has received much attention. If the vectors $\{\sigma_\nu^{(p)}, \nu = 1, \ldots N\}$, $p = 1, \ldots s$, represent s firing patterns to be stored as memories, the off-diagonal couplings are taken as

$$V_{\nu\mu} = \frac{1}{2N} \sum_{p=1}^{s} \sigma_\nu^{(p)} \sigma_\mu^{(p)} . \qquad (3.22)$$

An adjustable constant $J_0/2$ (rather than zero[9]) is inserted for the diagonal couplings, and normal thresholds $\theta_\nu = (1/2)\Sigma_\mu V_{\nu\mu}$ are assumed.

While such efforts -- which have been vigorously pursued both for Hopfield and Little models -- are of considerable theoretical importance, their relevance to biological neural networks may be questioned, in the same spirit as we have questioned the relevance of the large-time behavior of the Little model. As to the pertinence of the thermodynamic limit of the Little model, we need only recall that the basic assumption of synchronous updating becomes more unrealistic as the size of the system increases. But a more general (and more interesting) consideration is the following: Compared with models of connectivity and interaction drawn from physics, brain tissue is heterogeneous /anisotropic in the extreme, and there is strong evidence of hierarchical organization in both structure and function (e.g. into minicolumns, cortical columns, etc.). Moreover, there are many different types of neurons, which interact with other neurons in many different ways. Thus, the temporal nonuniformities which confound the ansatz of synchronism are complemented by spatial heterogeneity and diversity. Accordingly, real neural systems may not be at all amenable to such theoretical prescriptions as the thermodynamic limit, or to mean-field theory. The same reservation applies for synthetic-intelligence devices built on the principles of neural organization. While the results obtained in this subsection, based on finite N, may be regarded as quite limited, they have the virtue of being free from restrictive assumptions about the pattern of synaptic connections in the network and about the laws of interneuronal interactions.

4. DYNAMICAL PROPERTIES OF NONLINEAR MODELS

We now return to the discrete-time model of Taylor,[13] which, on the face of it, should display very different behavior -- and possess very different memory capabilities -- than the Little model and its offshoots. Taylor's model may serve as the prototype of another class probabilistic network models, for which a nonlinear description of statistical dynamics is unavoidable. This contrasts sharply with the situation we found for the "Little-class" of models, characterized by C1-C6 in Section 3. Although the microscopic dynamics of models so specified is undeniably nonlinear, we were able to construct a linear Master Equation for the state-occupation probabilities $p_i(t)$, $i = 0, \ldots 2^N - 1$. More precisely, we were able to exploit the fact that a model of this kind functions as a very simple kind of Markov chain, to describe the statistical time development by the *linear* law of motion (3.1)-(3.1'), the transition matrix elements Q_{ij} being constants independent of the state-occupation probabilities and Q being a linear operator.

4.1. Views of Statistical Dynamics

Suppose we try for a similar Markov-chain description of the dynamics of the Taylor model. Again, the underlying microscopic states are taken to be the firing patterns $i = \{\sigma_v^{(i)}\}$. The state-occupation probabilities are constructed from the absolute firing probabilities w_v of (2.14) according to[16,47]

$$p_i = \prod_{v=1}^{N} w_v^{\pi_v^{(i)}} (1 - w_v)^{1 - \pi_v^{(i)}} , \tag{4.1}$$

with (as usual) $\pi_\mu = (\sigma_\mu + 1)/2$. The relations (4.1) together with (2.14) allow us to form an evolution equation of the form (3.1'). However, the transition operator Q leading from $p(t - \tau)$ to $p(t)$ will then turn out to be *nonlinear*, except under very special choices of the available parameters; the right-hand side of (3.1') will generally involve polynomials of degree N.

This statement is easily checked, upon recognition that (2.14) may be recast in the general form[31]

$$w_v(t) = \sum_i \alpha_i^{(v)} p_i(t - \tau) , \tag{4.2}$$

wherein $p_i(t - \tau)$ is given by (4.1), with v replaced by a dummy neuron label. For example, at $N = 2$ we have

$$w_1(t) = [\alpha_0^{(1)} \bar{w}_1 \bar{w}_2 + \alpha_1^{(1)} \bar{w}_1 w_2 + \alpha_2^{(1)} w_1 \bar{w}_2 + \alpha_3^{(1)} w_1 w_2](t - \tau) ,$$

$$w_2(t) = [\alpha_0^{(2)} \bar{w}_1 \bar{w}_2 + \alpha_1^{(2)} \bar{w}_1 w_2 + \alpha_2^{(2)} w_1 \bar{w}_2 + \alpha_3^{(2)} w_1 w_2](t - \tau) , \tag{4.3}$$

where the notation $\bar{y} = 1 - y$ has been adopted. From (2.14), the explicit expressions for the $\alpha^{(1)}$ coefficients are

$$\alpha_0^{(1)} = \int_0^\infty dq\, h_v(q) \int_0^\infty dq_1\, g_{11}^{(s)}(q_1) g_{12}^{(s)}(q - q_1) \;,$$

$$\alpha_1^{(1)} = \int_0^\infty dq\, h_v(q) \int_0^\infty dq_1\, g_{11}^{(s)}(q_1) g_{12}^{(a)}(q - q_1) \;,$$

$$\alpha_2^{(1)} = \int_0^\infty dq\, h_v(q) \int_0^\infty dq_1\, g_{11}^{(a)}(q_1) g_{12}^{(s)}(q - q_1) \;,$$

$$\alpha_3^{(1)} = \int_0^\infty dq\, h_v(q) \int_0^\infty dq_1\, g_{11}^{(a)}(q_1) g_{12}^{(a)}(q - q_1) \;, \tag{4.4}$$

with corresponding relations for the $\alpha^{(2)}$ parameters. In Subsection 2.4, $g_{v\mu}^{(s)}(q)$ is determined by (2.13), while $g_{v\mu}^{(a)}(q)$ is given by $\delta(q - n_{v\mu} q_v^{(0)})$. In the present context, these functions can assume more elaborate forms. Returning to the task at hand, the advertized polynomial mapping of $\{p_i(t - \tau)\}$ into $\{p_i(t)\}$ results upon insertion of (4.2) (with i replaced by a dummy state index) into the right-hand side of (4.1). We see that, in general, this mapping is characterized by $N \times 2^N$ parameters $\alpha_i^{(v)}$.

It is important to realize that once the connection (4.1) between the p_i variables and the w_i variables is imposed, the occupation-probability description is beset with a problem of redundancy. There are 2^N of the p_i but only N of the w_i. Connection (4.1) says that at most N of the p_i can be independent; hence there must exist a set of $2^N - N - 1$ identities among them (besides the trivial linear normalization condition $\Sigma_i\, p_i = 1$). Applying (4.1) systematically for particular choices of i, one finds ${}^N C_s$ nontrivial identities of degree s, for integers s in 'the range $1 < s \le N$. For $N = 2$, there is a single nontrivial identity $p_{10} p_{01} = p_{00} p_{11}$; for $N = 3$ there are four, viz. $p_{110} p_{101} = p_{100} p_{111}$, $p_{101} p_{011} = p_{001} p_{111}$, $p_{110} p_{011} = p_{010} p_{111}$, and $p_{110} p_{101} p_{011} = p_{000} p_{111}^2$ (Ref. 47). (Here the notation $\pi_1^{(i)} \ldots \pi_N^{(i)}$ has been used to explicate the i-state subscripts on the p_i.) Within the Taylor model, these identities are patently inconsistent with the linearity of (3.1)-(3.1') when $N > 1$, unless highly restrictive conditions are met by the parameters of the theory.

If we wish to maintain, through (4.1), an unambiguous relationship between p_i and w_i variables, the indicated set of identities must obviously be preserved by the equations of motion of the $p_i(t)$. This feature would be guaranteed by construction within the framework of Taylor's theory,[16,47,31] at the sacrifice of linearity. But what about Little's model? If the identities are satisfied at time $t - \tau$, are they satisfied at t? The answer appears to be no, except perhaps under unacceptably restrictive conditions on the $Q_{ij} = \tau T_{ij}$ of (2.12), and in turn on the $V_{\nu\mu}$ and θ_ν. Taylor[47] has examined $N = 2$ in detail for identical thresholds, with pessimistic results. For large N, the situation looks hopeless, since the number of identities grows as 2^N, while the model has only $N^2 + N$ effective adjustable parameters. Be that as it may, there is nothing within the Little-Shaw-Vasudevan description *per se* to prevent one from chosing the p_i independently, in contradiction to the identities. The initial probability distribution over system states may be set up without constraint. One can only conclude that this description is incompatible with a view of statistical dynamics based on absolute firing probabilities $w_\nu(t)$ -- i.e., that these quantities are not uniquely determined by the available information.

Further discussion of the two alternative views of statistical time development may be found in Refs. 16,47,31.

4.2. Multineuron Interactions, Revisited

A brief digression is prompted by inspection of relation (4.2) with (4.1) inserted. The linear combination of products vividly reveals the intricacy of the theory: with ν as the postsynaptic neuron, it involves synapse-synapse interactions to all orders. The Little model, containing only simple pairwise neuronic couplings, is devoid of synapse-synapse interactions. Although we did consider the stepwise generalization of the Little model to 3-, 4-, ... neuron couplings (and thereby 2-, 3-, ... synapse interactions), the evolution of the statistical variables p_i remained linear at every stage. The dynamics of the RAM nets to be defined in Subsection 4.4 will embody the same multi-node correlation structure as the Taylor model. A similar structure underlies the sigma-pi units investigated by the PDP Research Group.[10]

4.3. Cognitive Aspects of the Taylor Model

Since we are most interested in cognitive features of the Taylor model, and especially its memory storage capacity, let us consider its asymptotic behavior in time. In this model, we have the choice of following the time development of the state-occupation probabilities p_i or the time development of the firing probabilities w_ν. The law of motion of the former involves, in general, a polynomial mapping of degree N of the hyperinterval $[0,1]^N$ into itself. On the other hand, if each neuron is "wired" to receive inputs from K neurons of the pool, where $2 \le K \le N$, the law of motion (2.14) for the w_ν involves a polynomial mapping of degree K of $[0,1]^N$ into itself. (For the extreme case $K = 1$ the mapping is linear.) Thus if the network is not fully connected (as will ordinarily be the case), it would appear to be simpler to work with the $w_\nu(t)$. The nonlinearity of the problem gives hope of a richer memory map than we found for the Little model: one anticipates a more elaborate menu of final conditions, which might include a multitude of fixed points, as well as nontrivial terminal cycles. Ideally, there would be a substantial number of stable fixed points, attractors for the dynamics which may represent stored memories. Although perhaps not a desirable feature if the goal is to simulate cognitive functions, one would not be surprised to encounter strange attractors, corresponding to chaotic activity,[48] in certain ranges of the parameters of the model.

However, the available evidence[47,31] -- analytic and numerical -- sketches instead a picture of asymptotia which is qualitatively the same as that for the linear Markov models! *Away from singular surfaces in the parameter - initial value space, the dynamical evolution converges to a unique, stable fixed point.* This conjectured *Unique Fixed Point Theorem* has been checked analytically for solvable 2-neuron nets, and by extensive computer simulation for 3- and 4-neuron nets with $K = 2$ and 20- and 50-neuron nets with K running from 1 to 10. (We note that the Brouwer fixed point theorem[49] ensures that the map (2.14) always possesses *at least one* fixed point.) An additional finding which emerges from the computer experiments is summarized in the *Rapid Convergence Theorem* (also conjectured): *Away from singular boundaries, the convergence to a stable fixed point is very rapid, occurring within four iterations for* $N \gg K \geq 10$. It would appear that the Taylor model is afflicted with the same disease as the models explored in Section 3, in that, strictly, it can only store one memory. The cure for the disease may be the same as was administered there. We suggest that the desired flexibility might be realized by working close enough to the singular boundaries, where there may arise a variety of long-lived modes of steady or periodic activity. These modes, though transient, could effectively represent stored memories. In other words, we suggest that there is a parameter domain which corresponds to the large-β regime of the Little model. Again such a resolution would be in accord with the recognition that the asymptotic behavior of synchronous models is irrelevant to biological behavior, in view of the difficulty of maintaining synchronism over long periods. Further numerical studies are needed to test the practicality of this proposal for enhancing the effective memory storage capacity of the model. The rapid convergence observed so far in the vast majority of computer experiments does not give cause for optimism.

In "on-boundary" cases, the existence of nontrivial terminal cycles has been demonstrated both analytically and in numerical simulations. On the other hand, all indications with regard to the occurrence of strange attractors in the dynamics of the firing probabilities w_v are *negative*. In fact there is analytic support[31] for the contention that there are no strange attractors for *any* allowed values of the parameters. This may be considered a salutary feature of the model, if chaotic activity is viewed as pathological. Superficially, one might have expected strange attractors to arise, since the Henon map[50] is contained in the class of polynomial maps under consideration. However, the parameter values leading to chaos in the Henon map violate constraints arising from the interpretation of the w_v as probabilities. It may be noted that chaotic activity has been found to occur in some other neural models involving continuous dynamical variables.[51–53] (Note that Refs. 52,53 are concerned with continuous state variables in continuous time, whereas Ref. 51 focuses on the Little model and establishes that the ensemble-averaged activity $<(\sigma_v(t)+1)/2>$, a continuous variable, can display chaotic activity in the discrete time-frame.)

Taylor's model -- and obvious refinements and extensions[13,16,31] -- affords a more realistic description of real neurons and their synaptic interactions than the models which are most widely used in the connectionist approach to artificial intelligence. Past successes of the Neurobiological Paradigm justify a deeper investigation of the dynamical and equilibrium properties of this model. In particular, studies in the thermodynamic limit should be pursued, with the goal of constructing phase diagrams analogous to those which have been obtained for the Little and Hopfield models. However, the conventional techniques[25–28] may not be adequate for this task.

4.4 Noisy RAMS and Noisy Nets

Interest in the Taylor model is heightened by a recent finding[31] that establishes an intimate formal connection with random-access memory units (RAMs), which are widely used in the computer industry. A K-RAM is defined as a logic element with exactly K binary inputs, hence 2^K addresses a. Activation of any such address produces a binary output. This means that a given K-RAM performs one of 2^{2^K} possible Boolean functions. (For binary threshold neurons, with linear summation of inputs, the repertoire of logic functions is reduced dramatically.[54]) The output of a given RAM can be directed to inputs of other RAMs (and possibly to itself). In this way networks of RAM units may be assembled. Each unit has a fixed number of inputs, but its output can branch any number of times (without change of the binary message). Some inputs may be external, in that information is provided to them from the environment rather than internally, i.e., from RAMs of the net. Similarly, some RAMs may send their outputs to the environment, affecting it in some way. RAM nets are classified as homogeneous or inhomogeneous, depending on whether the units have the same or differing numbers of inputs. As a matter of convenience, RAM networks are assumed to operate synchronously, the output of a given RAM unit computed at time t being effective as input at time

$t + \tau$, i.e., after a universal delay τ. Such networks of Boolean units, with various prescriptions for their assembly, have been extensively studied, particularly in the context of theoretical genetics (cf. Kauffman,[55] and references cited therein).

It seems natural (and may prove advantageous in practical applications[56]) to introduce noise in the output function of the RAM, thus defining a probabilistic RAM (PRAM). In the generalization to be described,[31] the output of a K-PRAM is unity with a probability α_a which depends on which of the 2^K addresses a is activated. Since nets of PRAMs will be formed, it becomes necessary also to consider noisy inputs. Let x_κ be the probability that there is a "one" on the κth input, $\kappa = 1, \ldots K$. Thus in the noisy case the input is characterized by $x = (x_1, \ldots x_K)$ instead of some particular sequence of binary digits indicating certain activity of some subset of the input lines. To determine the probability w of obtaining "one" for the output of the RAM unit we need to compound the probability that a particular set of *binary* input values will occur with the probability of obtaining "one" from this specific input. The first probability factor is the same as the probability p_a that address $a = (a_1, \ldots a_K)$ will be activated, where the a_κ are particular binary digits. Appending the second probability factor and summing over all addresses, we arrive at

$$w = \sum_a \alpha_a \, p_a \quad , \tag{4.5}$$

wherein p_a has a composition analogous to (4.1):

$$p_a = \prod_{\kappa=1}^{K} x_\kappa^{a_\kappa} \, (1 - x_\kappa)^{1 - a_\kappa} \quad . \tag{4.6}$$

The wiring diagram of a network of N RAMs is characterized by the connectivity matrices $C^{(\nu)}$ of the individual RAM units $\nu = 1, \ldots N$. If ν is a K-RAM, $C^{(\nu)}$ is a $K \times N$ matrix in which the $\kappa\mu$th element is 1 if the κth input line of element ν comes from unit μ and is zero otherwise. Accordingly, the input to the κth line is

$$x_\kappa = [C^{(\nu)} w]_\kappa \quad , \tag{4.7}$$

given that the current output of the PRAM net is $w = (w_1, \ldots w_N)$.

Construction of the full equations of motion of the PRAM net is now straightforward. Using (4.6)-(4.7) and the discrete, synchronous updating rule,

(4.5) becomes

$$w_\nu(t) = \sum_a \alpha_a^{(\nu)} p_a(t-\tau) \qquad (4.8)$$

and we obtain

$$w_\nu(t) = \sum_a \alpha_a^{(\nu)} \prod_{\kappa=1}^{K} [C^{(\nu)}w(t-\tau)]_\kappa^{a_\kappa} (1 - [C^{(\nu)}w(t-\tau)]_\kappa)^{(1-a_\kappa)} . \qquad (4.9)$$

We have adopted a notation which overlaps the notation used in discussing Taylor's model. Possible confusion is obviated by the fact that, in their dynamics, the noisy nets of Taylor and the noisy RAM nets just defined are formally identical!

This basic equivalence, discovered by Gorse and Taylor,[31] is stated more precisely in the *Noisy Net Identity Theorem*: *The dynamical behavior of any Taylor net of noisy neurons can be mirrored by that of some net of noisy RAMs, and conversely.* Such an identification is made plausible by reviewing (4.5)-(4.9) in the light of the neural-net relations (4.1)-(4.2) and the attendant discussion. Considering the case of full connectivity ($K = N$ for all nodal units) for simplicity, the validity of the theorem is cemented by the observations that (i) for a given neural network, the parameters $\alpha_a^{(\nu)} = \alpha_i^{(\nu)}$ of the equivalent PRAM net are determined, for each ν, by the 2^N moments of the distribution function $h_\nu(q)$ convoluted with the N probability densities $g_{\nu\mu}^{(s)}$ or $g_{\nu\mu}^{(a)}$, and (ii) any possible choice of the $2^N \times N$ output probability parameters $\alpha_a^{(\nu)} \in [0,1]$, and thus any fully connected PRAM net of the specified class, can be realized by suitably chosen $h_\nu(q)$, $g_{\nu\mu}^{(s)}$, and $g_{\nu\mu}^{(a)}$ (indeed, in many different ways). The arguments may be extended in a straightforward manner to cases of partial and of inhomogeneous connectivity.

If it is agreed that the probabilistic neural network model of Taylor potentially embodies much of the neurophysiology required for a realistic nodal description of certain brain processes, exploration of the general dynamical behavior implied by (4.2) [or more specifically (2.14) with or without better choices of the densities $g_{\nu\mu}^{(s,a)}$] may help us to understand how cognitive functions are carried out by neural networks in the brain. Insights gained from such studies could in turn be exploited in the design of intelligent machines, containing circuits of PRAMs, to perform the same or similar functions. The power of this implementation of the Neurobiological Paradigm would be greatly enhanced by the identity theorem established by Gorse and Taylor. Reversing the coin, specially constructed nets of PRAMs may prove valuable for realistic simulation studies of biological nerve nets.

5. THE END OF THE BEGINNING

A continuing pursuit of the Neurobiological Paradigm will draw upon neural models which capture ever more subtle properties of neurons and their interactions. Here we have examined and compared two models which incorporate -- in more or less realistic fashion -- important stochastic features of quantal synaptic transmission. Both models are formulated in discrete time, with synchronous updating of the state variables. The statistical dynamics of the *Shaw-Vasudevan model* and of the approximate form of this model proposed earlier by *Little*, is most naturally described in terms of the linear, Markovian evolution of the occupation probabilities of the microscopic system states. The statistical dynamics of the *Taylor model* is intrinsically nonlinear, and is most naturally described in terms of the absolute firing probabilities of the individual neurons. Thus the theories arising out of these models are inherently different -- even incompatible. Nevertheless there is a remarkable unanimity in qualitative behavior: In both cases there exists, at finite N, a unique stable fixed point for the statistical dynamics, implying that only one memory can be stored as a stable attractor. In both cases, the possibility of a nontrivial memory map arises only upon close approach to boundary values of certain model parameters, and even then the modes which are to represent additional memories must decay in the long term. The absence of infinite-range temporal correlations in these models may be viewed with some dissatisfaction: interesting cognitive behavior must be forced by working near the "edges" of the theory. This feature -- which has been discussed at some length in Subsections 3.5 and 4.3 -- may or may not be fundamental. The conditions under which it survives when rigid synchronism is relaxed need further investigation, although it is certainly maintained in simple generalizations of the Shaw-Vasudevan or Little models,[25] and preliminary studies[31] indicate the same for Taylor's model. If the unique fixed-point property is indeed robust under asynchronous updating, and if the asynchronous (or synchronous) versions of the models provide sufficiently accurate pictures of real neural processes, an intriguing conclusion is suggested: True long-range temporal order in neural systems, analogous to the long-range spatial order of some phases of matter, is an *emergent property*, attainable only in the limit of a very large number of neuronal elements. Whether such a thermodynamic limit is actually relevant to subsystems of the brain is an open question.

Our consideration of the cognitive aspects of finite noisy neural networks, natural and artificial, has been quite limited. Among other things, we have only peripherally addressed the dynamical phase of the memory storage process, or the dynamics of learning more generally. In the absence of concrete results, we offer some qualitative remarks on this aspect of the larger problem.

The notion that the underlying mechanism of learning is plasticity of the nervous system -- and, especially, the modification of synaptic properties and organization -- is very old, going back to Ramón y Cajal and the beginnings of neuroscience as a modern discipline. This idea took shape and gained impetus with Hebb's proposal[57] that *synaptic facilitation* leads to the formation of "cell assemblies." In particular, it was postulated that the effectiveness of an excitatory synapse in causing a postsynaptic cell to fire increases when firing of the presynaptic neuron is followed immediately by firing of the postsynaptic neuron. Hebb's arguments were so compelling that synaptic plasticity is almost universally regarded not only as a physiological correlate of learning and early cognitive development, but indeed as the primary cellular or subcellular basis of these phenomena. In fact, most of the specific empirical evidence from neurophysiology is of recent vintage. Results for *excitatory synapses* include the following: (i) Experiments on rat hippocampus[58] indicate that temporal correlations between pre- and postsynaptic firings strongly enhance the synaptic efficiency, and that if the presynaptic neuron is inactive and the postsynaptic neuron is active, the synaptic strength diminishes substantially. (ii) Electrophysiological conditioning experiments on the visual cortex of cat[59] strongly support the thesis that synaptic modifications depend critically on the occurrence of postsynaptic activity, and in detail on the extent of temporal correlation between pre- and postsynaptic firings. Moreover, it is found that the synaptic strength declines slowly with time, whatever the presynaptic activity, if the postsynaptic neuron remains inactive. There is little empirical information on the modification of *inhibitory synapses* due to neural activity; they may be less plastic than excitatory synapses. The available experimental results may be interpreted in terms of *local learning rules*, i.e., plasticity algorithms in which the synaptic change depends *only* on the activities of the presynaptic neuron μ and the postsynaptic neuron ν (thus local in space) and refers only to the activity state of μ *just prior* to that of ν (thus local in time). The reader should consult Refs. 60,30,29 for systematic classification and theoretical analysis of such plasticity algorithms.

Taking a broader view of plasticity, it is evident that the properties of individual neurons and individual synapses may carry two kinds of time dependence, *explicit* and *implicit*. The former is supposed to describe time variations which are not influenced by on-going neural activity, and might correspond to much of the embryological, fetal, and neonatal development of the nervous system, being in large part genetically programmed. The latter stems from a dependence of the salient properties on current and recent neuronal firing states; it is believed to be responsible for learning in the mature animal, but it is presumably also involved in aspects of development which are often referred to as self-organization. Our present discussion is concerned exclusively with *activity-induced* plasticity.

Within the framework of the Shaw-Vasudevan model, activity-induced synaptic plasticity may be conveniently incorporated through an appropriate local algorithm for updating the mean numbers of quanta released in stimulated and spontaneous emission, $\lambda_{\mu\nu}$ and $\lambda_{\mu\nu}^{(s)}$. (It is usually regarded as inappropriate to change the signs of synapses, which are given by the $\varepsilon_{\nu\mu}$.) Specializing to the Little model, one or another local plasticity algorithm would be imposed on the interneuronal couplings $V_{\nu\mu}$. In Taylor's model as delineated in Subsection 2.4, the synaptic parameters $n_{\nu\mu}$, $q_{\mu\nu}^{(o)}$, and $\lambda_{\mu\nu}^{(s)}$ which enter (2.14) are available for modification. Alternatively, the parameters $\alpha_i^{(\nu)}$ of (4.2) could be subjected to some learning algorithm,[31] although the ties to neurophysiological quantities then become somewhat blurred.

Evidently, plasticity brings into play another level in the hierarchy of dynamical processes and dynamical time scales associated with neural phenomena. We generally expect the time scale for appreciable plastic change to be much longer than the time interval between successive updatings of the individual neuronal state variables σ_ν or w_ν (e.g, seconds compared to milliseconds). It might then become reasonable to invoke a sort of neural Born-Oppenheimer approximation -- called by Caianiello the *adiabatic learning hypothesis*[4] -- which permits one to solve the problem of the time evolution of the neural state, or of the statistical distribution over firing patterns, with the synaptic parameters frozen at time-averaged values. Only over the long term would the solutions for given initial conditions and given inputs, i.e., the response of the system, actual or statistical as appropriate, show appreciable change corresponding to the secular variation of plastic parameters. If such a decoupling of neuronal and synaptic dynamics is justified, it becomes easier to follow the changing memory store implied by the altered neuronic interactions. (Of course, in machine-learning applications this decoupling can always be arranged by fiat.) On the other hand, the nonlinearity of the full problem offers the prospect of novel dynamical phenomena which may have cognitive parallels. As mentioned at the end of Section 3, it will also be of great interest to explore the consequences of extending plastic dynamics to spontaneity parameters such as the $\lambda_{\mu\nu}^{(s)}$, or the parameter S defined by Taylor, or the β_ν of Little's model.

ACKNOWLEDGMENTS

Some of the research described here was supported by the the Condensed Matter Theory Program of the Division of Materials Research of the U. S. National Science Foundation under Grant No. DMR-8519077. I thank J. G.

Taylor for communication of recent results.

APPENDIX. TRANSITION PROBABILITIES IN 2-NEURON NETWORKS

The state-transition probabilities of the general $N = 2$ net of the Little model are listed for reference. We consider transitions from an arbitrary state $j = \{\sigma_1^{(j)} \sigma_2^{(j)}\}$ to an arbitrary state $i = \{\sigma_1^{(i)} \sigma_2^{(i)}\}$ in one time-step τ. The probability $Q_{ij} = \tau T_{ij}$ of such a transition is given by

$$Q_{ij} = \rho_1^{(j)}(\sigma_1^{(i)}) \rho_2^{(j)}(\sigma_2^{(i)}) \tag{A.1}$$

with

$$\rho_v^{(j)}(\sigma_v^{(i)}) = s\left[-\beta_v \sigma_v^{(i)}\left[\sum_\mu V_{v\mu}(\sigma_\mu^{(j)} + 1)/2 - \theta_v\right]\right] . \tag{A.2}$$

Adopting a base-10 notation for the system states, i.e., $0 = \{-1,-1\}$, $1 = \{-1,1\}$, $2 = \{1,-1\}$, and $3 = \{1,1\}$, the results are as follows:

$$Q = (Q_{ij}) = \begin{bmatrix} Q_{00} & Q_{01} & Q_{02} & Q_{03} \\ Q_{10} & Q_{11} & Q_{12} & Q_{13} \\ Q_{20} & Q_{21} & Q_{22} & Q_{23} \\ Q_{30} & Q_{31} & Q_{32} & Q_{33} \end{bmatrix}$$

$$= \begin{bmatrix} a_+b_+ & c_+d_+ & e_+f_+ & g_+h_+ \\ a_+b_- & c_+d_- & e_+f_- & g_+h_- \\ a_-b_+ & c_-d_+ & e_-f_+ & g_-h_+ \\ a_-b_- & c_-d_- & e_-f_- & g_-h_- \end{bmatrix} , \tag{A.3}$$

where

$$a_\pm = s\left[\mp\beta_1\theta_1\right] , \quad b_\pm = s\left[\mp\beta_2\theta_2\right] ,$$
$$c_\pm = s\left[\pm\beta_1(V_{12}-\theta_1)\right] , \quad d_\pm = s\left[\pm\beta_2(V_{22}-\theta_2)\right] ,$$
$$e_\pm = s\left[\pm\beta_1(V_{11}-\theta_1)\right] , \quad f_\pm = s\left[\pm\beta_2(V_{21}-\theta_2)\right] ,$$
$$g_\pm = s\left[\pm\beta_1(V_{11}+V_{12}-\theta_1)\right] , \quad h_\pm = s\left[\pm\beta_2(V_{21}+V_{22}-\theta_2)\right] . \tag{A.4}$$

REFERENCES

1. W. S. McCulloch and W. H. Pitts, A logical calculus of the ideas immanent in nervous activity, Bull. Math. Biophys. **5**, 115 (1943).
2. F. Rosenblatt, *Principles of Neurodynamics: Perceptrons and the Theory of Brain Mechanisms* (Cornell Aeronautical Laboratory, Buffalo, New York, 1961).
3. M. L. Minsky and S. Papert, *Perceptrons* (MIT Press, Cambridge, MA, 1969).
4. E. R. Caianiello, Outline of a theory of thought processes and thinking machines, J. Theoret. Biol. **2**, 204 (1961).
5. H. R. Wilson and J. D. Cowan, Excitatory and inhibitory interactions in localized populations of model neurons, Biophys. J. **12**, 1 (1972).
6. S. Grossberg, *Studies of Mind and Brain* (Reidel, Hingham, MA, 1982).
7. W. A. Little, The existence of persistent states in the brain, Math. Biosci. **19**, 101 (1974).
8. T. Kohonen, *Associative Memory: A System-Theoretic Approach* (Springer-Verlag, Berlin, 1977).
9. J. J. Hopfield, Neural networks and physical systems with emergent collective computational abilities, Proc. National Academy of Sciences **79**, 2554 (1982).
10. D. E. Rumelhart, J. L. McClelland, and the PDP Research Group, eds., *Parallel Distributed Processing: Explorations in the Microstructure of Cognition, Vols. 1 and 2 (MIT Press, Cambridge, MA, 1986).*
11. S. W. Kuffler, J. G. Nicholls, and A. R. Martin, *From Neuron to Brain*, 2nd Edition (Sinauer Associates, Sunderland, MA, 1984).
12. B. Katz, *Nerve, Muscle, and Synapse* (McGraw-Hill, New York, 1966); B. Katz, *The Release of Neural Transmitter Substances* (Thomas, Springfield, 1969).
13. J. G. Taylor, Spontaneous behavior in neural networks, J. Theoret. Biol. **36**, 513 (1972).
14. G. L. Shaw and R. Vasudevan, Persistent states of neural networks and the random nature of synaptic transmission, Math. Biosci. **21**, 207 (1974).
15. C. F. Stevens, The neuron, in *The Brain*, A Scientific American Book (W. H. Freeman, San Francisco, 1979), p. 15.
16. J. W. Clark, Statistical mechanics of neural networks, Physics Reports **158**, 91 (1988).
17. J. S. Griffith, *Mathematical Neurobiology* (Academic Press, New York, 1971).
18. W. A. Little and G. L. Shaw, A statistical theory of short and long term memory, Behav. Biol. **14**, 115 (1975).
19. W. A. Little and G. L. Shaw, Analytic study of the memory storage capacity of a neural network, Math. Biosci. **39**, 281 (1978).
20. G. L. Shaw, Space-time correlations of neuronal firing related to memory storage capacity, Brain Research Bulletin **3**, 107 (1978).
21. G. L. Shaw and K. J. Roney, Analytic solution of a neural network theory based on an Ising spin system analogy, Phys. Lett. **74A**, 146 (1979).
22. R. S. Thompson and W. G. Gibson, Neural model with probabilistic firing behavior. I. General considerations, Math. Biosci. **56**, 239 (1981); Neural model with probabilistic firing behavior. I. One- and Two-Neuron Networks, Math. Biosci. **56**, 255 (1981).
23. C. Nicholson and J. M. Phillips, Diffusion in the brain cell microenvironment, in *Lectures on Mathematics in the Life Sciences*, Vol. 15 (American Mathematical

Society, Providence, RI, 1982), p. 103.

24. C. F. Stevens, *Neurophysiology: A Primer* (Wiley, New York, 1966).

25. P. Peretto, Collective properties of neural networks: A statistical physics approach, Biol. Cybern. **50**, 51 (1984).

26. D. J. Amit, H. Gutfreund, and H. Sompolinsky, Spin-glass models of neural networks, Phys. Rev. A **32**, 1007 (1985); Statistical mechanics of neural networks near saturation, Ann. Phys. (NY), **173**, 30 (1987).

27. J. L. van Hemmen, Spin-glass models of a neural network, Phys. Rev. A **34**, 3435 (1986).

28. J. F. Fontanari and R. Köberle, Information storage and retrieval in synchronous neural networks, Phys. Rev. A **36**, 2475 (1987); Information processing in synchronous neural networks, J. de Physique **49**, 13 (1988).

29. P. Peretto, On learning rules and memory storage abilities of asymmetrical neural networks, J. de Physique **49**, 711 (1988).

30. J. W. Clark, J. Rafelski, and J. V. Winston, Brain without mind: Computer simulation of neural networks with modifiable neuronal interactions, Physics Reports **123**, 215 (1985).

31. D. Gorse and J. G. Taylor, On the equivalence and properties of noisy neural and probabilistic RAM nets, King's College preprint (1988); An analysis of noisy RAM and neural nets, King's College preprint (1988).

32. A. W. Liley, Investigation of spontaneous activity at neuromuscular junction of rat, J. Physiol. (London) **132**, 650 (1956).

33. S. Kirkpatrick, C. D. Gelatt, and M. P. Vecchi, Optimization by simulated annealing, Science **220**, 671 (1983).

34. D. H. Ackley, G. E. Hinton, and T. J. Sejnowski, A learning algorithm for Boltzmann machines, Cognitive Science **9**, 147 (1985); T. J. Sejnowski, P. K. Kienker, and G. E. Hinton, Learning symmetry groups with hidden units: Beyond the perceptron, Physica **22D**, 260 (1986); R. W. Prager, T. D. Harrison, and F. Fallside, Boltzmann machines for speech recognition, Computer Speech and Language **1**, 3 (1986).

35. N. T. J. Bailey, *Elements of Stochastic Processes with Applications to the Natural Sciences* (Wiley, New York, 1964), Chapters 3 and 5; D. R. Cox and H. D. Miller, *Theory of Stochastic Processes* (Chapman and Hall, London, 1965); S. Karlin, *A First Course in Stochastic Processes* (Academic Press, New York, 1966).

36. J. Schnakenberg, Network theory of microscopic and macroscopic behavior of master equation systems, Rev. Mod. Phys. **48**, 571 (1976).

37. I. Prigogine, *Introduction to the Thermodynamics of Irreversible Processes,* Third Edition (Wiley, New York, 1967); P. Glansdorff and I. Prigogine, *Thermodynamic Theory of Structure, Stability, and Fluctuations* (Wiley, New York, 1971).

38. W.-K. Chen, *Applied Graph Theory* (North-Holland, Amsterdam, 1971).

39. E. R. Kandel, *Cellular Basis of Behavior* (Freeman, San Francisco, 1976); E. R. Kandel and L. Tauc, Heterosynaptic facilitation in neuron of the abdominal ganglion of Aplysia depilans, J. Physiol. (London) **181**, 1 (1965).

40. G. H. Shepherd, *The Synaptic Organization of the Brain* (Oxford University Press, Oxford, 1979).

41. K. J. Roney, A. B. Scheibel, and G. L. Shaw, Dendritic bundles: survey of anatomical experiments and physiological theories, Brain Res. Rev. **1**, 225 (1979);

G. L. Shaw, E. Harth, and A. B. Scheibel, Cooperativity in brain function: assemblies of approximately 30 neurons, Exp. Neurol. **77**, 324 (1982).

42. P. Peretto and J. J. Niez, Long term memory storage capacity of multiconnected neural networks, Biol. Cybern. **54**, 43 (1986).

43. J. W. Clark, Probabilistic neural networks: In or out of equilibrium?, in *Condensed Matter Theories*, Vol. 3, ed. J. Arponen, R. F. Bishop, and M. Manninen (Plenum, New York, 1988).

44. J. J. Hopfield and D. W. Tank, Computing with neural circuits: A model, Science **233**, 625 (1986).

45. E. Seneta, *Non-negative Matrices* (George Allen and Unwin, London, 1973).

46. W. R. Gibbs, The eigenvalues of a deterministic neural net, Math. Biosci. **57**, 19 (1981).

47. J. G. Taylor, Noisy neural net states and their time evolution, King's College preprint (1988).

48. R. M. May, Simple mathematical models with very complicated dynamics, Nature **261**, 459 (1976); E. Ott, Strange attractors and chaotic motions of dynamical systems, Rev. Mod. Phys. **53**, 655 (1981).

49. J. Cronin, *Fixed Points and Topological Degree in Nonlinear Analysis* (American Mathematical Society, Providence, RI, 1964).

50. M. Henon, A two-dimensional mapping with a strange attractor, Comm. Math. Phys. **50**, 69 (1976).

51. M. Y. Choi and B. A. Huberman, Dynamic behavior of nonlinear networks, Phys. Rev. A **28**, 1204 (1983).

52. K. E. Kürten and J. W. Clark, Chaos in neural systems, Phys. Lett. **114A**, 413 (1986).

53. K. L. Babcock and R. M. Westervelt, Stability and dynamics of simple electronic neural networks with added inertia, Physica **23D**, 464 (1986).

54. S. Yajima, T. Ibaraki, and I. Kawano, On autonomous logic nets of threshold elements, IEEE Trans. Computers **C-17**, 385 (1968).

55. S. A. Kauffman, Metabolic stability and epigenesis in randomly constructed genetic nets, J. Theoret. Biol. **22**, 437 (1969); Emergent properties in random complex automata, Physica **10D**, 145 (1984). See also: K. E. Kürten, Correspondence between neural threshold networks and Kauffman Boolean cellular automata, J. Phys. A **21**, L615 (1988).

56. I. Aleksander, A probabilistic logic neuron network for associative learning, IEEE Proceedings of the First International Conference on Neural Networks, June, 1987; The logic of connectionist systems, IEEE Trans. Computers special issue on neural networks and Imperial College preprint (1987).

57. D. O. Hebb, *The Organization of Behavior: A Neuropsychological Theory* (Wiley, New York, 1949).

58. W. B. Levy, Associative changes at the synapse: LTP in the hippocampus, in *Synaptic Modification, Neuron Selectivity and Nervous System Organization*, W. B. Levy, J. A. Anderson, and S. Lehmkuhle, eds. (Lawrence Erlbaum Associates, London, 1985), p. 5.

59. J. P. Rauschecker and W. Singer, The effects of early visual experience on the cat's visual cortex and their possible explanation by Hebb synapses, J. Physiol. (London) **310**, 215 (1981).

60. G. Palm, *Neural Assemblies* (Springer-Verlag, Berlin, 1982).

Part Two
ARCHITECTURAL DESIGN

Some Quantitative Issues in the Theory of Perception

A. ZEE

Institute for Theoretical Physics
University of California
Santa Barbara, California 93106

ABSTRACT: We study various quantitative issues in the theory of visual perception centering around two fundamental questions. How perceptive are we and how are we as perceptive as we are?

The problem of understanding visual perception[1,2,3] surely ranks as one of the outstanding scientific problems of our time. Of all the problems that bear upon the ultimate mystery of understanding how the brain works, the problem of visual perception is perhaps the one most amenable to rigorous experimental study and quantitative theory making. In this brief review, I would like to discuss some work I have done with my collaborators, W. Bialek and R. Scalettar.

The study of visual perception represents a vast subject, of course, to which an enormous amount of scientific efforts have been devoted. Approaches to the problem range from the neurophysiologist's painstaking mapping of the visual area of the cerebral cortex to the psychologist's careful elucidation of various perceptual representations.[4] What I can bring to all this is rather limited in scope. My collaborators and I have attempted to quantify some of the issues involved, with the hope of sharpening these issues and of bringing some ideas into direct confrontation with experiments. When I first started to look at the literature on visual perception, I found it to be disconcertingly vague. Perhaps a physicist's quantitative perspective is not appropriate for this subject, but whether or not that is the case surely deserves to be explored.

The questions and issues we addressed can be divided into three areas.

(1) We humans in general, and physicists in particular, tend to think that our brains function extremely well, especially at processing visual information. In truth, however, a quantitative and even semi-quantitative measure of the ability of the brain to process information sent to it by the sensory system is lacking. We simply do not know how well the brain tackles computational problems of varying degrees of complexity.

How perceptive are we? Suppose that it could be established that the visual system performs optimally. Then we can go on to ask what sort of computation is necessary in order to achieve this level of performance, and to explore in detail the types of design that may be required.

(2) In the perception literature, a number of models of visual processing have been proposed. We raise the question of how such models can be falsified by experiment. Obviously, it is of prime importance to progress from qualitative discussions to quantitative tests, in order to determine whether a given model is in fact viable. We try to compute the performance the visual system is capable of according to each of these models. If this performance falls significantly below the experimentally measured performance, then clearly the model can be ruled out. A large class of perceptual models corresponds to the steepest descent or mean field approximation in our formulation. An important issue is whether this approximation is adequate in explaining human performance.

(3) The study of neural networks offers the exciting prospects of elucidating some aspects of mental function. We have studied how feed forward networks can study various visual tasks. In particular, we have addressed issues such as generalization. We have also studied how real neurons, as in contrast to the model neurons used in the neural network literature, code and process information with spike trains.

I. PERFORMANCE

Optimal Performance

To discuss how well the visual system performs, we must immediately raise an obvious question: What are we to compare the performance of the visual system with? The only natural standard, it appears to us, is the optimal performance allowed by information theory, that is, the performance attainable if every bit of information received by the visual system is used.

First, we must choose a "naturalistic" task well suited to the visual system but which also allows a precise mathematical formulation amenable to rigorous analysis. Since we suspect that the visual system can in fact perform at or near optimum, we also want the task to be computationally difficult. We chose the discrimination between patterns with noise and distortion added.[5,6,7]

More precisely, we propose an experiment in which the subject is first acquainted with two patterns, described by $\phi_0(x)$ and $\phi_1(x)$. Here x denotes the coordinates of the two dimensional visual field. A black and white pattern is described by a scalar field $\phi(x)$ where $\phi(x)$ is equal to the contrast, that is, the logarithm of the intensity of the pattern at the point x.

For each trial, the experimenter chooses either ϕ_0 or ϕ_1, with equal probability say. Suppose ϕ_0 is chosen. Then the pattern is distorted and obscured with noise so that the subject actually sees $\phi(x) = \phi_0(y(x)) + \psi(x)$. Here $x \to y(x)$ defines an arbitrary one-to-one mapping of the plane onto itself. The noise $\psi(x)$ is taken for simplicity to be Gaussian and white. Thus, the conditional probability of seeing $\phi(x)$ were ϕ_0 chosen is given by $P(\phi|\phi_0) = \frac{1}{Z} \int Dy \; e^{-W(y) - \beta \int d^2 x [\phi(x) - \phi_0(y(x))]^2}$. The functional $W(y)$ should be such as to favor gentle distortions, for which $y(x) \sim x$. More on W later. Here Z is a normalization factor required by $\int D\phi \; P(\phi|\phi_0) = 1$. Henceforth, we will often neglect to write the normalization factor. Evidently, a probability $P(\phi|\phi_1)$ can also be defined by substitution. The subject is to decide whether the pattern seen corresponds to ϕ_0 or ϕ_1.

Discriminability

The information-theoretic optimal performance can then be computed according to standard signal detection theory. A particularly relevant quantity is the discriminability. Define the discriminant as $\lambda(\phi; \phi_0, \phi_1) = \log \frac{P(\phi|\phi_0)}{P(\phi|\phi_1)}$. With this definition, the discriminant is positive when the probability $P(\phi|\phi_0)$ is larger than $P(\phi|\phi_1)$ and negative when the opposite holds. (The logarithmic form for the discriminant is chosen for convenience. Some other monotonic function of the ratio of the two probabilities $P(\phi|\phi_0)$ and $P(\phi|\phi_1)$ may serve equally well.) It can be shown that optimal discrimination is accomplished by maximum likelihood. In plain English, the optimal strategy is to identify the pattern as ϕ_0 if $\lambda(\phi;\phi_0,\phi_1)$ is positive, and as ϕ_1 if $\lambda(\phi;\phi_0,\phi_1)$ is negative. This is of course precisely the strategy that any sensible person capable of knowing λ will adopt. Having seen the image $\phi(x)$, we have to decide whether it is more likely that the image "came" from $\phi_0(x)$ or from $\phi_1(x)$. (Thus, the experiment implies a "learning phase" in which the subject tries to "figure out" the relevant probability distributions. We are interested in the performance reached after learning. This of course accounts for our insistence on "naturalistic" tasks, for which the necessary learning has already been accomplished through eons of evolution.)

The probability distribution of λ if ϕ_0 is chosen is defined by $P(\lambda|\phi_0;\phi_0 \text{ vs } \phi_1) = \int D\phi \ \delta(\lambda(\phi;\phi_0,\phi_1) - \lambda)P(\phi|\phi_0)$. Similarly, $P(\lambda|\phi_1;\phi_0 \text{ vs } \phi_1)$ can be defined. The discriminability, conventionally called $(d')^2$, is defined as $(d')^2 = \frac{(\langle\lambda\rangle_0 - \langle\lambda\rangle_1)^2}{\frac{1}{2}[\langle(\delta\lambda)^2\rangle_0 + \langle(\delta\lambda)^2\rangle_1]}$ where the subscript $i = 0, 1$ indicates that the corresponding expectation value should be taken in the distribution $P(\lambda|\phi_0;\phi_0 \text{ vs } \phi_1)$ and $P(\lambda|\phi_1;\phi_0 \text{ vs } \phi_1)$ respectively. The meaning of $(d')^2$ is obvious: it measures the overlap between the two probability distributions when the two distributions are bell-shaped. As the name suggests, the discriminability $(d')^2$ limits the extent to which one can discriminate between ϕ_0 and ϕ_1. We are generally interested in the regime $(d')^2 \sim 0$ when the visual discrimination task is highly "confusing." (When the distributions are bell-shaped, the discriminant can obviously be related to the percentage of correct guesses. Incidentally, the discriminant $(d')^2$, rather than some other more-or-less equivalent quantity, is used because, being formed of "naturally occuring" expectation values, it can be readily computed for certain simple problems and because experimentalists in this field typically quote their observations in terms of $(d')^2$. The discriminant provides a convenient summary of the information contained in the two $P(\lambda)$'s. The interested reader is referred to a text by Green and Swets on the use of signal detection theory in psychophysics.)

Field Theory and Statistical Mechanics

As is well-known, quantum field theory and statistical mechanics can both be described by functional integrals. Thus, the considerable body of knowledge accumulated about two-dimensional field theories and statistical mechanical systems may be brought to bear on the theoretical problem of determining the various probability distributions and $(d')^2$.

In the last two decades or so, studies in quantum field theory and statistical mechanics have revealed that apparently fairly simple systems can exhibit exceedingly intricate collective behavior. In particular, phase transitions are possible. As the parameters (β and parameters in W, in our case) appearing in a functional integral vary, the behavior of the functional integral may also change discontinuously or at least drastically. The computational complexity involved in evaluating the functional integral may also change correspondingly.

Thus, in the actual experiment, it may be interesting to see to what extent the actual performance tracks the optimal performance. It may happen that for a region of the parameter space the actual performance would agree with the optimal performance, but as the experimenter vary the parameters, the actual performance may abruptly deviate from the optimal performance, or it may drop drastically even as it tracks the optimal performance.

Likely and Unlikely Distortions

To choose a reasonable W, we appeal to the observational fact that, to first approximation, we have little difficulty in recognizing a pattern that has been rigidly translated, dilated, and rotated. In other words, $W(y)$ should be zero for y equal to one of these rigid transformations. We chose $W = \int d^2 x \left[\frac{1}{g_1^2} (\partial_\mu \partial_\nu A_\lambda)^2 + \frac{1}{g_2^2} (\partial_\mu \partial A)^2 \right]$ where $\mu, \nu, \lambda = 1, 2$ denote the two Cartesian directions. We have also written $y^\mu(x) = x^\mu + A^\mu(x)$. (Unfortunately, this simple form also allows, at no cost, anamorphic transformations,[6] whereby the pattern is stretched or compressed by two different factors in two orthogonal directions. To make these transformations cost, we can add terms quadratic in ∂A.)

A potential point of confusion is that, while we disavow any theorizing about how the visual system works, the functional integral *can* be regarded as such a theory: we summon from memory storage either the prototype pattern ϕ_0 or ϕ_1, apply distortion, add noise, and try to find particular forms of the distortion and noise so as to match the seen pattern ϕ, all the while weighing the likelihood of the particular distortion and noise. This "theory", while simple, is not implausible. Subjectively, we feel that when varying a pattern we gauge the likelihood of various distortions. In other words, we carry in our heads a functional W. We can easily imagine designing a machine along these lines.

Suppose experiments show the actual performance to be substantially below optimal performance. What would that mean? It might mean that the visual system is capable of only a crude approximation in evaluating the functional integral involved. It would then be interesting to determine what approximation the visual system uses. This is certainly possible in principle. Alternatively, it might mean that the visual system can evaluate the relevant functional integral fairly accurately but that the W used by the experimenter does not correspond to the W we "carry in our heads."

In principle, the experimenter can carry out a series of experiments, each with a different W, all corresponding to "reasonable" choices. Suppose the optimal perrformance can be determined for each W. It could happen that the actual performance does not come close to the optimal performance for any of these W's. Perhaps more interestingly, it could also happen that the actual performance reaches or comes close to the optimal performance for some W's.

The correspondence with statistical mechanics also suggests the question of whether some sort of universality might play an essential role in visual perception. We can also ask whether the corresponding statistical mechanical system exhibits short ranged or long ranged correlation. In this connection, we may perhaps emphasize that two logically distinct issues surface in our program. First, we have the question of whether the visual system can attain the optimal performance theoretically attainable. Next, given that this optimal performance is in fact attained, we can ask what are the computations necessary to attain this performance.

We would like to conjecture that actual performance does in fact come close to optimal performance for some reasonable W. If experiments verify our conjecture, then we are confronted by the interesting issue of the type of circuitry and algorithm capable of effectively evaluating the functional integral involved.

Local versus Non-local Computations

In our work, we construct tasks in which arbitrarily long-ranged and multi-point correlations must be computed if optimal performance is to be reached, at least in certain limits which are controllable as the different image ensembles are generated. It is known that,[7] in discrimination among simpler image ensembles, human observers can approach optimal performance in the sense defined here. This suggests experiments in which the performance of humans is measured as a function of the parameters which control the relevant correlation lengths. If the visual system can only compute local functionals, as with feature detectors, performance should follow the optimum only for a restricted range of correlation lengths and then fall away dramatically. If on the other hand the system can adapt to compute strongly non-local functionals of image intensity, no such abrupt drop will be observed. These experiments will be difficult, but they have the potential of providing serious challenges to our understanding of computation in the nervous system.

Our suspicion is that the system *can* solve non-local problems, and that there are interesting theoretical questions to be answered about the algorithms and hardware responsible for such contributions. Suspicions aside, the approach described here[5,6] provides the tools for asking very definite questions about the computational abilities of the brain.

Unfortunately, it is well-nigh impossible to evaluate functional integrals exactly. After all, the exact evaluation of a functional integral amounts to the exact solution of a statistical mechanical system or of a quantum field theory. The history of statistical mechanics and quantum field theory testifies amply to the difficulty of the task. Thus, in our work we are reduced to trying various approximations to the functional integrals, often reaching only qualitative conclusions. (Of course, the functional integrals can also be evaluated numerically.)

We could, of course, choose the functional integral corresponding to an exactly soluble statistical mechanical system or quantum field theory. The trouble is that the corresponding discrimination task may not be particularly "naturalistic." We thus have to search for a compromise. An interesting possibility which we have studied involves a colored picture with constant intensity and saturation. In the crudest possible approximation, the pattern is described by a vector field $\vec{\phi}(x)$ of constant length instead of the scalar field $\phi(x)$. The angle of $\vec{\phi}(x)$ measures the hue according to some color wheel. In this case, the relevant functional integral may be seen to describe a driven sine-Gordon system solvable in certain limits.

Amusingly, the functional integral defining $P(\phi|\phi_0)$ is reminiscent of that occuring in string theory to describe the motion of a string moving in a background metric, with the correspondences $x \rightarrow$ spacetime coordinates of a two dimensional space, $y \rightarrow$ string coordinates, $W(y) \rightarrow$ string action, and $\phi_0 \rightarrow$ the background metric.

We describe briefly our efforts to evaluate $P(\phi|\phi_0)$. Details can be found elsewhere.[6] For instance, in the "smooth picture" approximations in which we take ϕ_0 to vary slowly, we can show that the correlation length of the vector field $A^\mu(x)$ increases as the noise increases $(\beta \rightarrow 0)$. (The appearance of a finite correlation length, that is, a mass in field theory terms, for $A^\mu(x)$ is reminiscent of the Higgs phenomenon.)

We have also considered the mean field approximation. In other words, we evaluate $P(\phi|\phi_0) \equiv \int Dy\, e^{-E(y)}$ by steepest descent. In particular, we take the picture to consist of N circular "blobs" with N large: $\phi_0(x) = \sum_n B(x - x_n)$ where $B(x)$ is a rapidly decreasing function centered at $x = 0$. The "energy landscape" $E(y)$ is the sum of a deterministic and a noise piece. The deterministic piece contains a true, that is, global minimum of depth of $O(N)$ and $O(N^2)$ false, that is, local minima of depth of $O(1)$. The noise piece contains fluctuations of amplitude of $O(\sqrt{N})$

It would appear at first sight that noise of $O(\sqrt{N})$ can not obscure the true minimum of $O(N)$. But in fluctuating noise we may get a deep well of depth $\gg O(\sqrt{N})$ once in a while. The question is how often. To answer this question, we use a generalization of Rice's method[8,9] to compute the probability $P(E)$ of encountering a well with depth E in the noise. We find that $P(E) \sim E^{2N} e^{-E^2/\alpha N}$ where α is a measure of the noise. We see that with the "Boltzmann factor" alone the root mean square value of E would be of order $\sqrt{N}$. The "entropic factor" of E^{2N}, however, negates this naive conclusion so that in fact $P(E)$ peaks at $E^* \sim \sqrt{\alpha}N$. (Amusingly, the calculation of $P(E)$ can be formally regarded as describing fermions in a background field. Also, for obvious reasons, similar calculational techniques are relevant to the theory of biased galaxy formation.) Thus, we conclude that performance should drop drastically as noise is increased past a certain value of $O(1)$.

It is worth remarking that much of the work in the literature associated with neural networks is what we would call the steepest descent approximation. An energy or cost function is invented to describe some perceptual problem such as stereopsis[10] and a neural network is then designed to search out the minima in the "energy landscape." In other words, the visual system is thought of as searching for a "best match." In particular, in our problem, minimizing the energy $E(y)$ amounts to searching for a distortion $y^*(x)$ which would match the seen image $\phi(x)$ best with the learned image $\phi_0(x)$ essentially a biased (by W) template matching. As Hopfield and Tank[11] have shown, neural networks are well suited to perform optimization tasks. Whether or not the steepest descent approximation is adequate, however, is open to question. More on this potentially important issue later.

Instead of trying to evaluate $P(\phi|\phi_0)$, we have also tried to extract the general features. In particular, we have considered using renormalization group to study the properties of $P(\phi|\phi_0)$ in an attempt to discover a strategy for "universal computation" in processing visual information.[12]

It is the interplay between noise and distortion that makes the evaluation of the field theory defined by $P(\phi|\phi_0)$ so difficult. If either noise or distortion is omitted, the task of evaluating $P(\phi|\phi_0)$ becomes considerably simpler. (In particular, with no distortion, the problem becomes Gaussian and trivial.) Why do we make life miserable for ourselves? Because we want to appreciate the difficulty of a task that the visual system performs extremely well (at least according to the subjective evidences). Indeed, our work represents largely a record of our awakening to how difficult the computations involved are. The difficulty of this task is also reflected in the fact that machines with artificial vision have not mastered this task of "invariant perception." Indeed, as far as we know, current machines have difficulty recognizing images if the image can be arbitrarily rigidly translated, rotated, and dilated. Of course, it may also turn out that our visual system does not perform as well as we think it does.

Some Questions

In summary, the research program outlined here consists of asking the following questions.

(1) What is the optimal performance allowable for various perceptual tasks?

(2) What are the computations needed to reach this optimal performance? Can we identify the issues involved (for instance, local versus non-local computation)?

(3) Is the visual system actually capable of this optimal performance? How close does it come? (These questions can be answered only be experiments of course.)

(4) If the performance of the visual system approximates optimal performance, how does it perform the computations identified in (2)? What neural circuitry and algorithm can carry out these computations?

(5) Are there universal features and properties in the sense of statistical physics?

As is evident by the preceeding discussion, we have touched only on the beginnings of this program and have reached only qualitative conclusions. Many challenging problems remain.

Performance of Neural Nets

Meanwhile, while waiting for the relevant experiments to be done and while torturing ourselves trying to evaluate various functional integrals, we can try out this program on neural networks instead of the human visual system. At the least, it provides an instructive example of how we may unravel the underlying architecture of a neural system.[13]

We take a feedforward network with binary units. The input is a configuration of a one-dimensional chain of Ising spins. The configuration belongs to the equilibrium ensemble at temperature β^{-1} of either a ferromagnet or an anti-ferromagnet. The network is to decide whether the configuration is ferromagnetic or anti-ferromagnetic. Another task is to discriminate between ferromagnetic configurations at two different temperatures. In particular, the task of discriminating between an infinitesimal $\beta > 0$ and $\beta = 0$ corresponds to the detection of incipient order.

To determine the optimal performance, we now merely have to evaluate the partition function of a one-dimensional Ising system. The problem chosen is sufficiently simple that we can see explicitly the computations necessary and the correlation lengths involved and so on.

The network learns the task by the standard back propagation algorithm. We are interested in the performance reached asymptotically after learning. Different neural architectures can be tried out. Sure enough, for certain tasks, there are architectures which allow the network to reach optimal performance and others which do not.

II. MODELS

Feature Detectors

In I, our discussion does not make any reference to the visual system. Rather, it is an information theoretic analysis. In the second part of our work, we attempt to capture, in quantitative models, the essence of some leading theories of perception. (This is clearly not the place to review these theories. Nor do we have the ability to do so. For an introduction to the field, see Refs. 1, 2, and 3.) We then compute the predicted performance at a "naturalistic" perceptual task, with the aim of ultimately comparing whatever results we may obtain with actual experiments on the human visual system.

As in I, the experiment consists of showing the subject the image $\phi(x)$ when a prototype image $\phi_0(x)$ is chosen, with $\phi(x)$ generated probabilistically according to the distribution $P(\phi|\phi_0)$. The visual system processes the perceived image ϕ in some unknown way and it is at this point that theorizing about perception enters.

Consider the feature detector theory which originated in the neurophysiological experiments of the 1950s. (For a brief review of the history of feature detectors, see the article by Barlow in Ref. 2.) Neurons in the visual system are assumed to compute nonlinear functionals of the image intensity and thus signal the presence of features in the image. Thus the continuous pattern $\phi(x)$ is converted into a set of discrete "feature tokens" to be processed by subsequent layers of neurons. We attempt to capture the essence of this theory by taking the simplest possibility for the feature tokens: they are Ising spins σ_μ located at $x_\mu, \mu = 1, 2, \ldots N$, with σ_μ taking on values ± 1. The image is sampled at x_μ to give $\phi_\mu = \int d^2 x f(x - x_\mu)\phi(x)$ where $f(x - x_\mu)$ represents the response function of a feature detector neuron located at x_μ. The response function $f(x)$, with its excitatory center and inhibitory surround, is well-known to neurophysiologists.[1] It is often modelled as the Laplacian of a Gaussian $\nabla^2 G$ or as the difference of two Gaussians.[14] Our model[15] is that σ_μ tends to be $+1$ when ϕ_μ is positive and -1 when ϕ_μ is negative, as described by some probability distribution $P(\sigma|\phi)$. Putting it together, we have the conditional probability $P(\sigma|\phi_0) = \int D\phi\, P(\sigma|\phi)P(\phi|\phi_0)$. In other words, the experimenters (or the natural environment we live in) turns the known image ϕ_0 into ϕ. The seen image ϕ is then processed into the "feature tokens" σ_μ.

We believe that this "Ising" model is prototypical of a large family of models which replace the continuous image $\phi(x)$ by discrete and local feature tokens. It contains one of the classic feature detector ideas concerning the extraction of edges, a concept formalized by Marr, Poggio, and others as the location of contours where some appropriately filtered version of the image vanishes. Here the "domain walls" between spin up and down regions mark the zero crossing contours, so in fact this spin representation has a bit more information than a "sketch" based on zero crossing contours alone.

As described in I, various quantities measuring performance, such as the discriminability $(d')^2$, can be computed using $P(\sigma|\phi_0)$. The distribution $P(\sigma|\phi_0) = e^{-H(\sigma;\phi_0)}$ may be thought of as describing the statistical mechanics of an Ising spin system and these measures of performance can be related to physical properties of the spin system. In contrast, the optimal performance can be determined directly from $P(\phi|\phi_0)$. The quantity of interest is the efficiency defined by $\epsilon = (d')^2/(d')^2_{optimal}$.

To proceed, we have to take the simplest $P(\phi|\phi_0)$ possible. We do not include distortion and write $P(\phi|\phi_0) = e^{-\frac{\gamma}{2} \int d^2x (\phi(x)-\phi_0(x))^2}$. (Our emphasis here differs from that in I: here we want to test various theories of perception by their performance at various tasks. We might as well choose the simplest non-trivial task. After all, $P(\phi|\phi_0)$ is under experimental control.) We will also take $\phi_1 = 0$ so that the discrimination between ϕ_0 and ϕ_1 reduces to the detection of an image hidden by noise. The subject is to say whether or not the signal ϕ_0 is present. The optimal performance here is defined (as in I) as the performance attainable if all the information contained in $\phi(x)$ can be exploited. With these simplifying assumptions, we find easily that $(d')^2_{optimal} = \gamma \int d^2x \, \phi_0^2(x)$. (This form is of course due to our taking the noise to be simply Gaussian and white.)

Ising Spins in Random Fields

Even with these simplifying assumptions, the evaluation of $(d')^2$ is a daunting challenge. We began by choosing $P(\sigma|\phi) = \prod_\mu (e^{\beta\sigma_\mu\phi_\mu - \log 2 \, ch \, \beta\phi_\mu})$. Here β, a measure of noise in the visual system, may be thought of as an inverse temperature, and ϕ_μ as an external magnetic field. The log ch term, which makes life miserable, is required by normalization. We have thus here a system of Ising spins interacting with a randomly fluctuating field ϕ_μ.

Our first thought is to expand $P(\sigma|\phi_0)$ in the large γ limit. This limit is, however physically uninteresting. Instead, we go to the limit in which $\phi_0 \to 0$, with the signal disappearing under the noise. (By scaling, this is equivalent to $\gamma \to 0$.) The problem is then tractable, but only in the zero temperature $\beta = \infty$ deterministic limit $P(\sigma|\phi) = \prod_\mu \theta(\sigma_\mu \phi_\mu)$, in other words, in the limit of neglecting noise in the visual system. (The inclusion of noise can only lower the efficiency.)

By using the replica method, for example, we are able to obtain an expression for $(d')^2$ from which various properties of this model can be read off. For details, see Ref. 15. If the range of the response function $f(x)$ is small compared to the intercellular spacing, (which is not biologically reasonable) the maximum efficiency can be seen to be $2/\pi = 0.64$ which appears low compared to experimental reports of efficiency ranging from 0.5 to 0.95. We conclude that overlaps of the receptive fields of neighboring cells are essential for understanding the observed efficiency of visual perception. Furthermore, these overlaps must be negative to enhance $(d')^2$, which necessitates a excitatory center inhibitory surround type of organization found for real neurons. We are now considering detailed numerical study of this model.

Obviously, we can go on to consider variations of this model. For instance, the work of Hübel and Wiesel established that certain cells are selectively sensitive to directions.[1] Thus, instead of Ising spins, we can consider "Heisenberg" spins $\vec{s}_\mu$, that is, unit vectors which respond to $\vec{\phi}_\mu = \int d^2x\, f(x - x_\mu)\vec{\nabla}\phi(x)$ according to some probability $P(\vec{s}_\mu|\vec{\phi}_\mu)$. Alternatively, we can have, at each site, a group of directional sensitive Ising spins $\sigma_\mu(\theta)$ labelled by a (possibly discrete) angular variable and responding to $\phi_\mu(\theta) = \int d^2x\, f(x - x_\mu)\hat{\theta} \cdot \vec{\nabla}\phi(x)$ with $\hat{\theta}$ a unit vector in the direction defined by θ.

Linear Filters

Another class of models we have considered supposes that the detectors in the visual system act as linear filters.[17] In other words, each detector functions as a narrow-band Fourier analyzer centered at some characteristic spatial frequency. Models of this type are suggested by the work of Campbell, Robson, Lawden, and DeValois.[18]

Perception by Steepest Descent

In I, we mentioned that neural network can solve optimization problems efficiently and in parallel. Thus, it is an attractive theory that the functional integrals relevant to visual perception are evaluated in the steepest descent approximation. We feel that this is an important and urgent issue that should be settled by experiment.[19] Does the brain merely do steepest descent? Or is it considerably more sophisticated?

To answer this question, we need to have a version of the problem outlined in I but simplified to such a degree that we can solve it both exactly and in the steepest descent approximation. We settle on a simple experiment described by

$$P(\phi|\phi_0) = \frac{1}{Z'} \int dp \; e^{-\beta \int d^2 x [\phi(x) - \gamma \phi_0(x_p)]^2}$$

where p parametrizes a family of distortions. For example, p can be an angle θ and x_p is equal to x rotated through θ. We also take $\phi_1 = 0$ for simplicity. Thus, the subject is to decide on the presence or absence of the prototype pattern ϕ_0 with the pattern obscured by noise and presented with a randomly chosen orientation on each trial. We are interested in the regime in which the noise becomes overwhelming. (By simple scaling, we see that quantities such as $(d')^2$ depend only on $\beta^{\frac{1}{2}}\gamma$, so the β can be absorbed.) If $\int d^2 x \; \phi_0^2(x_p) = \int d^2 x \; \phi_0^2(x)$ as is the case for the examples we have considered, the relevant functional integral can be organized in the suggestive form

$$P(\phi|\phi_0) = \frac{1}{Z} \int dp \; e^{-\beta H_0(\phi) - \gamma H_1(\phi,p)}$$

with the bare Hamiltonian $H_0 = \int d^2x\phi^2$ and the "perturbing" Hamiltonian $H_1 = -\beta \int d^2x$ $\phi(x)\phi_0(x_p)$. Our task is to evaluate the integral over p exactly and in the steepest descent approximation and hence to obtain the efficiency $\epsilon = (d')^2_{\text{steepest descent}}/(d')^2$. Of course, in the large γ limit, ϵ tends to 1 as a mathematical statement of the efficacy of the steepest descent approximation. We are interested in the opposite noisy limit. Thus, we evaluate the small γ expansion exactly and by steepest descent.

We find to lowest order in γ that

$$(d')^2_{\text{steepest descent}} = \gamma^2 \frac{[\langle \bar{H}_1 H_1^* \rangle_0 - \langle H_1^* \rangle_0 \langle \bar{H}_1 \rangle_0]^2}{[\langle H_1^{*2} \rangle_0 - \langle H_1^* \rangle_0^2]}$$

Here $\bar{H}_1 = \int dp \; H_1(\phi,p)$ (with the integral over p normalized so that $\int dp = \int \frac{d\theta}{2\pi}$ if p is an angle) and $H_1^* \equiv H_1(\phi, p^*(\phi))$ is the minimum of $H_1(\phi,p)$ as a function of p. Notice that it has a highly non-trivial dependence on ϕ. As indicated the various expectation values are to be taken with the bare Hamiltonian $\langle \cdots \rangle_0 = \frac{1}{Z_0} \int D\phi \; e^{-\beta H_0(\phi)} (\cdots)$. In comparison, we have $(d')^2 = \gamma^2 [\langle \bar{H}_1^2 \rangle - \langle \bar{H}_1 \rangle^2]$. The obvious condition that $\epsilon \leq 1$ is satisfied by Schwarz's inequality.

The evaluation of expectation values such as $\langle \bar{H}_1 H_1^* \rangle_0$ is a rather non-trivial exercise in functional integration. Again, we have to use generalizations of Rice's method determining the distribution of minima. Here we merely summarize the result, which turns out to depend on the correlation function $\Delta(p,p') = \langle H_1(\phi,p)H_1(\phi,p') \rangle = 2\beta \int d^2x \; \phi_0(x_p)\phi_0(x_{p'})$. (In the examples we considered, $\Delta(p,p') = \Delta(p - p')$ is "translation invariant.") Let $\Delta_0 = \Delta(0), \Delta_2 = \frac{d^2\Delta(0)}{dp^2}$, and $\Delta_4 = \frac{d^4\Delta(0)}{dp^4}$. Then

$$(d')^2_{\text{steepest descent}} = \gamma^2 \left(\int dp\Delta(p) \right)^2 \frac{\Delta_4}{(\Delta_0\Delta_4 + (1 - \frac{\pi}{2})\Delta_2^2)}.$$

$$\cdots$$

To see what is actually going on we can now try out various specific prototype pictures $\phi_0(x)$. For example, we have considered a wedge or leaf shaped picture $\phi_0(x) = f(r)e^{-\frac{1}{2C(r)}\theta^2}$ where r and θ are the polar coordinates of x. The important quantity here is the width of the wedge $C^{\frac{1}{2}}(r)$ (which we take to be small). We find

$$\epsilon = \frac{\langle C^{\frac{1}{2}} \rangle}{\pi^{\frac{1}{2}}} \frac{1}{1 + \frac{1}{3}(1 - \frac{\pi}{2})\frac{\langle C^{-1} \rangle^2}{\langle C^{-2} \rangle}}$$

where $\langle \cdots \rangle$ denotes some average of $(\cdots)$ over the radial direction weighted by $f^2(r)$ and geometrical factors. If $C(r) =$ constant (so that the picture is wedge shaped) we have $\epsilon = \frac{C^{\frac{1}{2}}}{\pi^{\frac{1}{2}}}\frac{3}{4(1-\frac{\pi}{3})}$. While if the picture is very jagged so that $\langle C^{-2} \rangle \gg \langle C^{-1} \rangle^2$, then $\epsilon = \frac{\langle C^{\frac{1}{2}} \rangle}{\pi^{\frac{1}{2}}}$.

How can we use this analysis to find out if the visual system is actually an efficient device that locates minima or "best matches" (as simple neural network models would suggest)? Suppose the experiment outlined is done and the measured efficiency comes out to be equal to the value predicted by steepest descent. That would offer dramatic support for the idea of "best match." On the other hand, if the efficiency is measured to be greater than the predicted efficiency, that would rule out or at least cast grave doubt on the "best match" theory. Unfortunately, the situation is complicated by the possibility that information is lost by processing, for instance, by feature detectors. Thus, one would have to consider the steepest descent approximation in evaluating the integral over p not in $P(\phi|\phi_0)$ but in $P(\sigma|\phi_0) = \int D\phi P(\sigma|\phi)P(\phi|\phi_0)$ (with σ denoting some feature "tokens"). For the calculation outlined here to be relevant, we have to suppose that processing affects both $(d')^2_{\text{steepest descent}}$ and $(d')^2$ in the same proportion so that the effect cancels out in ϵ. Note, however, that the experiment can be repeated and the theoretical expression for ϵ can be evaluated (numerically at least) for a wide variety of prototype pictures ϕ_0.

III. NETWORKS

Feed Forward Net and Grandmother Cells

In part I we approach the visual system via information theory, trying to determine the optimal performance attainable and to understand, at least in qualitative terms, the computational complexity involved. In part II we study mathematical models in which the information reaching the visual system is processed in some way. Finally, in this part we consider neural networks as a possible realization of the information processing system.

While the study of neural networks has had a long history,[20] the interest of physicists in the subject was sparked by the remarkable work of Hopfield.[21] In particular, a fully connected symmetric network is essentially a spin glass. In our work, we have largely focussed on the feed forward network in which information is processed and passed on from one layer of neurons to the next. As with all complex systems, the highly non-linear "dynamics" of such nets can lead to fascinating behavior.

Clearly, a feed forward net can serve as a memory device by associating outputs with inputs. It can also be thought of as a model of the visual system, in which the visual image (the input) is recognized as a definite object on which some decision can be made (the output). The net "learns" to produce the correct output for each input by adjusting the weights or connections between neurons according to, say, the back propagation algorithm of Rumelhart, Hinton, Williams,[22] and others.[23] We have studied what would happen if the input is corrupted with noise, in line with the general philosophy expressed throughout this article. Perhaps the most outstanding feature of the visual system is its ability to perform in the presence of noise and distortion.

We found that under certain conditions each of the hidden neurons (that is, the neurons in the "hidden" layer between the output and input layers) becomes specialized in remembering a particular input pattern. In other words, the hidden cells assume the role of "grandmother cells." The existence of grandmother cells has long been a matter of debate in neurobiology. The notion is that a particular neuron in the brain fires when one's grandmother comes into view. Many authors believe that for each of the objects and symbols commonly known to us, a particular neuron specializes in recognizing that object or symbol. The emergence of grandmother cells in this context is perhaps suggestive.

Since our work has been described elsewhere,[24] we will limit ourselves here to a summary of our main results.

(1) By requiring tolerance to error in the input, as is the case in the real world, we find that a feed forward network evolves into a grandmother type memory when the number of memories is equal to the number of hidden neurons.

(2) When the number of hidden neurons is larger than the number of memories, we introduce a mechanism by which the weights decay (*i.e.*, "forgetfulness"). We propose that weaker connections decay faster than stronger connections. As the collection of memories to be stored grows incrementally, the additional memories become associated with previously unused hidden units in an orderly one-to-one fashion.

(3) The memory capacity exceeds that of a Hopfield type memory and the error tolerance can be comparable.

We hope that these results may be of interest to researchers constructing large capacity content addressable memory devises for practical use and possibly even to brain theorists interested in issues of concept representation and learning in biological networks.

In some sense, grandmother type memory[25] stands at the opposite extreme from the fully distributed memory discussed by Hopfield and earlier workers. A fully distributed memory is clearly robust in the sense that if some cells are ripped out the memory will continue to function to a large extent, while if a hidden cell in a grandmother type memory is ripped out, the memory which that cell is responsible for is lost. A simple and obvious way of making the grandmother type memory more robust fairly leaps to mind. Suppose that, whenever we speak of a hidden cell, that hidden cell actually corresponds to γ hidden cells, where γ is a number say of order 3. The price for this added robustness is of course a decrease in the memory capacity measured per cell. Hopfield has emphasized to us a second drawback of the grandmother type memory for practical applications, namely that a hidden cell has to "fan out" to a large number of output cells.

While any connectons between this work and neurobiology is remote at best, we are nevertheless tempted to say that the results may be vaguely suggestive. If we see a pattern repeatedly, a cell may become specialized to that pattern, but only if the pattern is presented with noise. We are aware that the notion of grandmother cells is the subject of considerable controversy among neurobiologists. Nevertheless, it appears to us not unreasonable to imagine that there are cells specialized in recognizing common symbols, such as numbers and letters. We have shown that the grandmother cell arrangement is "optimal" in the context of back propagaton. Perhaps more interestingly, we have shown that grandmother cells can be called up in an orderly fashion when new patterns are presented, provided a decay mechanism (=forgetfulness?) operates.

We find the role of forgetfulness rather intriguing[26] and have explored whether forgetfulness can increase the memory capacity of a Hopfield net. Various workers have studied how to reduce the rather (and also biologically unrealistically) large number of connections ($O(N^2)$ for a net with N neurons) of fully connected nets by thinning out the connections. Forgetfulness may provide a complementary and natural mechanism: the weak connections may simply atrophy. By sharpening the contrast between the connections, forgetfulness may increase the useful capacity of the net.

Visual Perception by Neural Nets

We have studied how well networks can solve problems in visual perception. As a trial problem, we focus on the one dimensional problem of determining whether an object is to the right or to the left of another object.

Specifically, we enter the input by allowing five of the N input units or neurons to be on; the other $(N-5)$ units are off. Of the five units that are on, four are required to be adjacent. For ease of writing, we refer to this clump of four adjacent on units as the "house" and the isolated one unit as the "tree." We require that there be at least one off unit between the house and the tree.

In undertaking this investigation,[27] our aim is not only to determine how well neural nets can learn to perform perceptual tasks but also to address and discuss various issues of interest in network modelling such as learning efficiency, ability to generalize, the construction of large networks out of smaller networks, versatility, and so on. We found that to master this task the strict feed forward architecture of the network must be modified. Of particular interest is the emergence of "feature detectors" in the hidden layer (the "visual cortex") specialized to recognizing the tree and the house respectively. As alluded to above, the notion that neurons can be specialists in detecting certain visual features has captured the attention of workers in vision ever since the celebrated work of Hubel and Wiesel. The notion is somewhat controversial, with some workers maintaining that neurons can only function as frequency detectors.[29] But it seems plausible that, at the very least, groups of frequency detectors suitably connected together might function as feature detectors.

Generalization

That neural nets can learn, while mystifying to the uninitiated, is essentially trivial. Weights are adjusted so as to move the actual output closer to the desired output. The central issue is not learning a prescribed set of input-output association so much as generalization. After the net has learned what outputs to associate with inputs taken from a certain subset (the "training" or "learning" set) from the set of all possible inputs, the net is then presented with inputs it has never seen before. Can it immediately produce the correct output? The ability to generalize is clearly a hallmark of true learning.

The difficulty, of course, lies in the very term "correct output." Who is to define what is correct? In our left-right problem, for instance, the notion of left versus right can be arbitrarily reversed for inputs outside the training set (that is, for inputs forming what is called the "test" set). The situation is reminiscent of those intelligence test questions in which one is given a sequence of numbers and asked to determine the next one. Any answer is possible. Obviously, we can test generalization only if the problem contains an "intrinsic" notion of smoothness such as in our left-right problem.

Is generalization merely interpolation and extrapolation starting with the learning set? Or is generalization significantly more than mere interpolation and extrapolation? To the extent that a neural net is simply a device to fit a given output-input relationship by sums of nested hyperbolic tangent functions, it is clearly capable of some amount of interpolation and extrapolation. In contrast, it would appear that humans are capable of considerably more. True generalization involves thinking and insight, and in tasks that appear to involve only interpolation and extrapolation human performance is impressive. For instance, I know from first hand observation that a young child can recognize as "elephant" anything ranging from a photograph of an elephant, to a rough pencil sketch of an elephant, and to a whimsical picture of a French-speaking elephant wearing a tuxedo and a crown.

We can formulate the problem along the lines discussed in part I by introducing a conditional probability $P(\alpha|\phi)$ that the seen image $\phi(x)$ represents the object α (or the "category" α). For instance, α may be "elephant." For certain $\phi(x)$, $P(\alpha|\phi)$ would be very high, while for most ϕ, $P(\alpha|\phi)$ would be infinitesimal. By Bayes' theorem, $P(\alpha|\phi) = \frac{P(\phi|\alpha)P(\alpha)}{P(\phi)}$. Thus, the problem of deciding which α a seen image represents amounts to maximizing $P(\phi|\alpha)P(\alpha)$ as a function of α.

The problem in understanding human perception is of course that we do not know how $P(\phi|\alpha)$ is computed (estimated) or learned from examples. The example discussed in part I represents our attempt to capture some features of this problem: α is replaced by an abstract pattern $\phi_0(x)$ and $P(\phi|\phi_0)$ is computed by estimating the likelihood of various amounts of distortion as measured by the "cost" function $W(y)$ and noise. In human perception, the true W function is certainly enormously more sophisticated. For instance, in recognizing elephants, our "W" function may tell us that the trunk is essential and that if the trunk is shortened below a certain characteristic length W increases drastically.

Putting such vague speculations aside, we have studied[29] how neural nets can generalize at a concrete task. We chose as our example the problem of recognizing left and right. First of all, the problem contains an intrinsic notion of smoothness. Secondly, to speak of interpolation and extrapolation at all, we need a notion of a "metric" in the space of all possible inputs. What does it mean to say that a given input lies between previously seen inputs or far from previously seen inputs? The left-right problem contains a natural distance between inputs, namely the relative separation between the tree and the house. Note that two inputs with the same separation between the tree and the house can be far apart in Hamming distance.

We use this problem to address several issues about generalization. We take as our learning set inputs with tree and house separated by less than S_L units. After the net has learned, we measure the error it made on inputs with separation S larger than S_L. Sure enough, the error increases slowly with S. An interesting question is how this ability to extrapolate scales with S_L. Is the error a universal function of S/S_L? How does this function grow? If the error approaches a constant or grows very slowly, we might suggest that, in some sense, the net has "understood" the concept of left and right.

While training the net, we may vary the sizes of the tree and the house (all the while defining the larger object as the house). We might expect that of all the possible minima in the energy landscape that would enable the net to master the left-right problem with fixed tree and house sizes the net would now be forced to choose the one "best" enabling it to tackle the problem of varying sizes. This is clearly suggested by human experience. If, when we teach a child the concept of left and right, we vary the objects used, then we may expect the child to realize sooner that the concept does not depend on the specific objects used. Sure enough, we find the net generalizing considerably better if trained on inputs with varying sizes for the tree and the house. This provides another example of how input noise can sharpen the net's ability.

Suppose we consider a subset A of the learning set L and train a number of nets on A until they perform equally well on A. Next we hold a tournament in which we compare the performance of the various nets on the set $(L - A)$ and determine a champion. Does the champion net generalize the best as measured by performance on the test set T? In our simulation, we take A to be all those inputs with separation between S_0 and $S_L > S_0$ and T to be all those inputs with separation greater than S_L so that $(L - A)$ and T are not contiguous. We find indeed that the net which extrapolates best "downward" also generalizes best "upward."

There are a number of open questions. For instance, how is a net's ability to generalize reflected in its weights, perhaps in some statistical properties such as the variance in the distribution of weights? Can answers to such questions be possibly independent of the task?

If we can answer such questions, then we can encourage nets to generalize by including appropriate terms in the energy function.

The issue of learning in neural nets is clearly of some importance. As currently formulated, learning amounts to an optimization problem. The energy, that is, the mismatch between the desired output and the actual output, is to be minimized as a function of the weights and the back propagation algorithm of Rumelhart *et al.* is simply minimization by gradient descent. Obviously, many other minimization algorithms are possible. How does one choose a particular algorithm? We propose[30] that the neural net itself be allowed to choose. The algorithm chosen varies not only from task to task but it also depends on what stage of the learning process the net has arrived at. We call the algorithm for choosing which algorithm to use the algorithm of algorithms. For instance, we have tried the following: an algorithm is used for a certain number of times, say five times, and if the total amount of energy lowered is less than a certain threshold value, the net uses some other algorithm instead. Thus, gradient descent may proceed rapidly at first but as the energy approaches a local minimum the net may switch over to some other algorithm. We have experimented with various versions of this approach.

The Discriminant in Neural Nets

To make contact with parts I and II, let us go back to the problem of determining the presence or absence of a signal obscured by noise. In the Ising feature detector model, the signal ϕ_0 is converted into the discrete variables σ_μ according to the probability $P(\sigma|\phi_0)$. We then computed the performance assuming that all the information contained in $\{\sigma_\mu\}$ is used fully. We do not specify how this information is actually processed in the visual cortex.

As a model of the processing of visual information by the visual cortex, we can consider a neural net whose input is the two-dimensional array of Ising spins σ_μ. The output is a binary unit which indicates the presence or absence of the signal by being on (+1) or off (−1). On the other hand, according to the signal detection theory outlined in part I, the optimal decision should be based on the sign of the discriminant $\lambda = \log P(\sigma|\phi_0)/P(\sigma|0)$. As we have seen, λ is a fearfully complicated function of σ.

A neural net learning to master this perceptual problem may be thought of as striving to approximate λ by a $\lambda_{net}(\sigma)$ formed out of sums of nested hyperbolic tangents. Thus, the question of how well a net performs at this visual task rests on how well $\lambda_{net}(\sigma)$ approximates the true λ. In particular, the historical failure of the perceptron is the statement that a λ_{net} linear in σ makes a miserable approximation.

The processing of visual information in this context can only be studied numerically. To make analytic progress, we have analyzed[15] the performance in the drastic approximation in which λ_{net} is linear in σ_μ. In spite of the general failure of the perceptron, we find that the performance in this linear processing model can be rather high under certain circumstances.

Neural Spike Trains

Incidentally, current work on neural nets begin by assuming that the output of a neuron is some continuous function of its input, typically of some hyperbolic tangent form.

In fact, while a neuron receives its input information in the form of an electrical signal $\xi(t)$ from the synaptic junctions on its dendrites, where $\xi(t)$ denotes the instantaneous electrical voltage measured in suitable units, its output consists of a train of spikes, occurring at time $\{t_i\}$, traveling rapidly down its axon. The height of the spikes is independent of $\xi(t)$. Information about $\xi(t)$ is encoded in the times $\{t_i\}$ themselves.[1] Workers on neural nets, however, simply assume that the pattern $\{t_i\}$ can be replaced by a continuous rate function which saturates at rather moderate stimulus intensities. That neurons fire discrete impulses is viewed only as contributing an effective noise level. It has been known for some time, however, that the timing of individual spikes and spike clusters can play a significant role in neural information processing, and recent evidence from the fly visual system indicates that two or three spikes from one neuron encode essentially all of the sensory information available about sudden movements across the visual field. Studies of primary neurons from the vibratory organ of the frog inner ear suggest that modulations of the firing rate are an essentially linear measure of stimulus amplitude over much of the behaviorally relevant range. In auditory neurons of the mammalian ear the firing rate exhibits relatively soft saturation at 30-40 dB above the threshold of hearing in quiet, but even this may be traced to saturation of the pre-synaptic signal. In the mammalian visual cortex, many neurons exhibit nearly linear responses to spatial patterns with contrasts of up to 30%, which is typical of many natural scenes.[31]

Taken together, these and other experiments suggest that we study models in which (a) the information carried by individual spikes is not ignored or averaged away, and (b) saturation of the neural response does not play an essential role. The lack of saturation implies that spikes are typically separated by intervals long compared to the electro-chemical time scales which determine the maximum firing rate. In the limit that the inter-spike intervals are very long the occurence of one spike cannot influence the generation of the next, and neural firing becomes a Poisson process. Indeed, evidence for the near-Poisson character of neural firing has been found in the mammalian auditory nerve, and in retinal ganglion cells firing has been modelled as a Poisson process driven by the Poissonian arrival of photons at the retina and slightly modified by "dead time."[31]

We are thus led to introduce a conditional probability $P[\{t_i\}|\xi(t)]$ of observing N spikes at time $\{t_i\}$ given the signal $\xi(t)$. We have undertaken a study of neural information processing from an information theoretic point of view. We address several rather broad questions:

[1] Given that we observe a set of spikes $\{t_i\}$, how much information have we obtained about the signal $\xi(t)$? How might one reconstruct this input signal?

[2] Let $\xi_r(t)$ denote the signal reconstructed from spikes $\{t_i\}$. Suppose that we re-encode this signal in the spike train $\{t_a\}$ of a second neuron. How much information is lost in the two stage process $\xi(t) \rightarrow \{t_i\} \rightarrow \xi_r(t) \rightarrow \{t_a\}$? Although this is a formal question, since simple decoding and re-encoding almost certainly never occurs in the brain, this estimate of information loss should give us an indication of whether signals are significantly degraded as they pass through successive layers of cells.

[3] How can we perform simple analog computations on the spike trains, such as multiplying two signals $\xi_1(t)$ and $\xi_2(t)$ encoded in the firing of two neurons $\{t_i^{(1)}\}$ and $\{t_i^{(2)}\}$?

While the problem we studied in part I can be thought of as a quantum field theory or a statistical mechanical system, the problem here, because of its one-dimensional character, can be cast as quantum mechanical system with $\xi(t)$ representing the position of a particle at time t. The particle is kicked at the time t_i.

Details of our work can be found in Ref. 32. Our conclusions point to the following broad notions: the need for time delays in optimal processing; the use of presynaptic filtering to remove context-dependence of the code and optimize the information capacity; the role of "sloppy coincidence" in analog computation. Some of these results already seem to have experimental correlates.

ACKNOWLEDGEMENTS

As indicated in the references, the work reviewed here was done in collaboration with

W. Bialek and R. Scalettar. I am indebted to them for numerous stimulating and interest-ing discussions. This research was supported in part by the National Science Foundation under Grant No. PHY82-17853, supplemented by funds from the National Aeronautics and Space Administration, at the University of California at Santa Barbara.

REFERENCES

*We make no attempt to give a complete set of relevant references to the perception literature.

1. M.W. Levine and J.M. Shefner, *Fundamentals of Sensation and Perception*, (Random House, 1981).

2. D.G. Albrecht, ed., *Recognition of Pattern and Form*, (Springer-Verlag, 1982).

3. D. Marr, *Vision*, (W.H. Freeman & Co., 1982).

4. S.E. Palmer, "The Psychology of Perceptual Organization: a Transformational Ap-proach," in J. Beck *et al.* eds. *Human and Machine Vision*, (Academic Press 1983).

5. W. Bialek and A. Zee, *Phys. Rev. Lett.* **58**, 741 (1987).

6. W. Bialek and A. Zee, "Invariant Perception: A Functional Integral and Field Theo-retic Approach", in preparation.

7. Experiments similar to (but simpler than) the ones proposed here have been done by Barrow, in Ref. 2, and *Phil. Trans. Roc. Soc. London* **B290**, 71 (1980).

8. S.O. Rice, in *Selected Papers in Noise and Stochastic Processes*, N. Wax ed., p.133 (Dover, New York 1954).

214

9. M.S. Longuet-Higgins, *Proc. R. Soc. London* **A389**, 241 (1983); *J. Opt. Soc. Am.* **50**, 851 (1960); and references therein.

10. For example, K. Schulten, to be published.

11. J.J. Hopfield and D.W. Tank, *Biol. Cybern.* **52**, 141 (1985).

12. W. Bialek and A. Zee, "Recognizing Ensembles of Images: Universality at Low Resolution," in preparation.

13. W. Bialek, R. Scalettar, and A. Zee, in preparation.

14. For example, S.W. Kuffler, "Discharge Patterns and Functional Organization of Mammalian Retina," *J. Neurophysio.* **16**, 57 (1953); A. Parker and M. Hawken, "Capabilities of monkey cortical cells in spatial resolution tasks," *J. Opt. Soc. Am.* **2**, 1101 (1985); and Ref. 1, p.117 & ff.

15. W. Bialek and A. Zee, "Understanding the Efficiency of Human Perception," ITP preprint.

16. For example, H. Voorhees and T. Poggio, "Detecting Blobs as Textons in Natural Images," MIT report; H. Voorhees, "Finding Texture Boundaries in Images," MIT report, and references therein.

17. W. Bialek and A. Zee, in preparation.

18. See the articles by R.L. DeValois and by F.W. Campbell and M. Lawden in Ref. 2 and references therein.

19. W. Bialek and A. Zee, "Inadequacy of mean field approximation in visual perception," in preparation.

20. J.S. Griffith, *Mathematical Neurobiology*, (Academic Press, 1971).

21. J.J. Hopfield, "Physics, Biological Computation and Complementarity" in J. de Boer *et al.* eds. *The Lesson of Quantum Theory*, (Elsevier, 1986); and references therein.

22. D. Rumelhart, G.E. Hinton, and R.J. Williams, "Learning Internal Representation by Error Propagation," in D. Rumelhart *et al.* eds. *Parallel Distributed Processing*, (MIT Press, 1986).

23. D.B. Parker, "Learning-Logic," Invention Report S81–64, Office of Technology Licensing, Stanford University.

24. R. Scalettar and A. Zee, "Emergence of Grandmother Memory in Feed Forward Networks: Learning with Noise and Forgetfulness," to appear in *Cogn. Sci.*

25. Various aspects of grandmother memory have been analyzed by E. Baum, J. Moody, and F. Wilczek, ITP preprint.

26. See also J.J. Hopfield, D.I. Feinstein, and R.G. Palmer, *Nature* **304**, 158 (1983).

27. R. Scalettar and A. Zee, "Perception of left and right by a feed forward net," ITP preprint, to appear in *Bio. Cybernetics*.

28. R.L. DeValois, in Ref. 2.

29. R. Scalettar and A. Zee, in preparation.

30. R. Scalettar and A. Zee, unpublished.

31. For references to the neurophysiological literature, see Ref. 32.

32. W. Bialek and A. Zee, "Coding and computation with neural spike trains," ITP preprint.

SPEECH PERCEPTION AND PRODUCTION
BY A SELF-ORGANIZING NEURAL NETWORK

Michael A. Cohen*, Stephen Grossberg*, and David G. Stork*+

Ctr. Adaptive Systems (*)   Department of Physics (+)
Boston University   and Program in Neuroscience
111 Cummington Street   Clark University
Boston, MA 02215   Worcester, MA 01610

Abstract

Considerations of the real-time self-organization of neural networks for speech recognition and production have lead to a new understanding of several key issues in such networks, most notably a definition of new processing units and functions of hierarchical levels in the auditory system. An important function of a particular neural level in the auditory system is to provide a partially-compressed code, mapped to the articulatory system, to permit imitation of novel sounds. Furthermore, top-down priming signals from the articulatory system to the auditory system help to stabilize the emerging auditory code. These structures help explain results from the motor theory, which states that speech is analyzed by how it would be produced. Higher stages of processing require chunking or unitization of the emerging language code, an example of a classical grouping problem. The partially compressed auditory codes are further compressed into item codes (e.g., phonemic segments), which are stored in a working memory representation whose short-term memory pattern is its code. A masking field level receives input from this working memory and encodes this input into list chunks, whose top-down signals organize the items in working memory into coherent groupings with invariant properties. This total architecture sheds new light on key speech issues such as coarticulation, analysis-by-synthesis, motor theory, categorical perception, invariant speech perception, word superiority, and phonemic restoration.

1. The Learning of Language Units

During a human's early years, an exquisitely subtle and sensitive speech recognition and production system develops. These two systems develop to be well-matched to each other, enabling rapid and reliable broadcast and reception of linguistic information. The development of these systems can be viewed as resulting from two fundamental processes: self-organization through *circular reaction* and through *chunking* or *unitization*. This chapter sketches some issues concerning these processes in speech and provides a summary of its key neural components, developed to address more general cognitive problems.

2. Low Stages of Processing: Circular Reactions and the Emerging Auditory and Motor Codes

The concept of circular reaction (Piaget, 1963) is illustrated in Figure 1. For our purposes, the reaction links the *motor* or *articulatory* system (mouth, tongue, velum, etc., and the neural structures controlling them) with the *auditory* system (ear and its neural

M.A.C. was supported in part by the Air Force Office of Scientific Research (AFOSR F49620-86-C-0037) and the National Science Foundation (NSF IRI-84-17756), S.G. was supported in part by the Air Force Office of Scientific Research (AFOSR F49620-86-C-0037 and AFOSR F49620-87-C-0018), and D.G.S. was supported in part by the Air Force Office of Scientific Research (AFOSR F49620-86-C-0037).

Acknowledgements: We wish to thank Cynthia Suchta and Carol Yanakakis for their valuable assistance in the preparation of the manuscript.

218

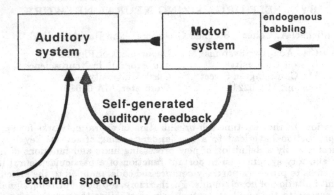

Figure 1. Circular reaction linking the motor system to the auditory system. Such a loop permits imitation of novel sounds from an external speaker.

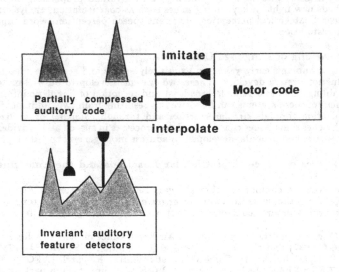

Figure 2. Neural interactions between a partially-compressed auditory code and a motor code permits the imitation of novel heard sounds.

perceptual mechanisms). In a developing infant, endogenously generated babbling signals in the motor system lead to auditory feedback, thereby allowing the auditory system to tune its evolving recognition codes. Moreover, the auditory system can compare the self-generated sounds to those from external speakers.

Figure 2 shows in slightly greater detail relevant neural interconnections in the auditory model. After processing by low-level auditory feature detectors (detecting energy in various frequency wavebands, "sweeping" frequency signals, broad-band or burst energy distributions) the auditory information is partially compressed and passed to a subsequent level, where it is represented by significant activity in a smaller number of neurons.

There is a learned auditory-to-articulatory associative map at this level, important for the following purposes. First, it permits the motor system to *interpolate* novel heard sounds. That is, if a novel sound leads to an auditory code "between" those for other, previously coded sounds, then this novel sound will be mapped to a motor code "between" those for the sounds previously heard. Second, the associative map permits the motor system to *imitate* such sounds. In this manner, a novel sound will lead to a novel, interpolated motor code. When accessed, this new motor code will lead to an utterance closer to the novel one heard. This (imitated) utterance then accesses an auditory code very similar to the interpolated one.

The auditory code at the level for this interpolation and imitation must be only *partially* compressed; a fully compressed (or *unitized*) code would map to a previously organized motor code, precluding interpolation of novel sounds. Furthermore, the auditory level for interpolation must be above stages of invariant preprocessing—only in this way can effects such as vocal tract normalization be explained (Lieberman, 1984, pp.219–223). It has been argued (Lieberman, 1984, p.222) that such normalization is due to the existence of innate mechanisms, and hence is not modifiable in the manner of the auditory-to-motor map.

3. The Vector Integration to Endpoint Model

The motor code in our network is based on the recent Vector Integration To Endpoint (VITE) model of arm movement control (Bullock and Grossberg, 1987), due to functional similarities between speech articulation and arm movement problems. Moreover, we agree with Lieberman (1984) that phylogenetically the speech system appropriated the speech articulators and their neural controlling structures from their original tasks of swallowing, chewing, and so forth—tasks more typical of standard motor control concerns. The VITE model posits three interacting neural levels: (1) a Target Position Command (TPC) level, whose spatial distribution of activity codes where the limb "wants to go," (2) a Present Position Command (PPC) level, which generates an outflow movement command, and (3) a Difference Vector (DV) level, which compares the TPC and PPC codes. Such a structure has been used to explain a range of motor control psychophysics and physiology results, in particular (for our speech system) the simultaneous contraction of several muscle groups in a synergy, even at different overall rates. The learning of a motor task, in this scheme, involves the printing (i.e., modification of synapses for long-term memory) of the motor code when the limb is at or near the target position. Put another way, learning occurs when the present position and the target position form a near match (i.e., when $DV < \epsilon$). Hence in our speech system the Difference Vector layer can act as a learning *gate*, regulating the formation of the auditory-to-articulatory map during the near match condition, as shown in Figure 3.

Speech articulators, however, do not all function as a single, unitized system; rather, there are several muscle synergies or *coordinative structures* (Fowler, 1980) working quasi-independently. For instance, one coordinative structure might link the jaw and front of the tongue for bringing the top of the tongue to the hard palate in order to utter [t], while a different coordinative structure is controlling the back of the tongue to utter a (coarticulated) [a]. Each of the coordinative structures must have its own TPC, PPC,

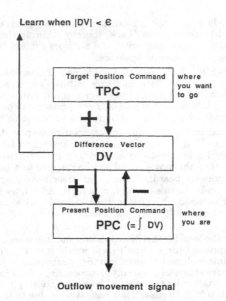

Figure 3. Basic VITE module and its learning gate, for use in encoding the TPC codes.

and DV layers, to preserve such quasi-independence. Figure 4 shows how the TPC's of different coordinative structures are chunked into distinct motor control commands. Thus the imitative map can associate different aspects of the partially compressed auditory code with different coordinative structures. Figure 4 also shows the basic structure of the circular reaction loop linking the auditory system and the motor system, incorporating the VITE circuit and its learning gate.

4. Self-Stabilization of Imitation via Motor-to-Auditory Priming

In a self-organized system, a key issue concerns the ability of the system to *self-stabilize* its learning under natural conditions (Carpenter and Grossberg, 1987a, 1987b). During speech the auditory code varies (in general) *continuously* due to its representation of a stream of varying sounds, whereas the controlling motor code varies more *discretely* due to the fact that new target position commands (TPCs) are printed by the imitative associative map only when the motor system achieves an approximate match (Figure 3), either at an initial TPC or a final TPC of a simple utterance (Figure 5). This raises the issue of insuring that the emerging auditory code is *consistent* with the motor code so that the imitative map can self-stabilize. Such consistency can be achieved through top-down motor priming which associates the compressed motor codes that represent the coordinative structures with activation patterns across the auditory feature detectors, as shown in Figure 6—an example of active internal regulation by top-down resonant feedback.

The top-down motor expectations (or priming signals) reorganize the auditory code to make it consistent with the evolving motor code. Such priming occurs during the activity of any given motor code, and hence reinforces the activity patterns across auditory feature detectors that are heard contemporaneous and consistent with such motor codes. These motorically-modified feature activity patterns are encoded in long-term memory within

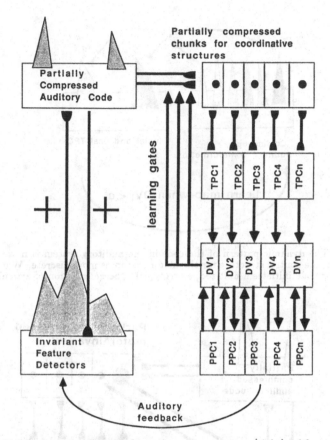

Figure 4. Circular reaction loop linking the motor system (right) with auditory system (left). Parallel motor channels for coordinative structures are shown, each with its associated learning gate, which prints (modifies the synapses for long-term memory) the imitative map between the partially-compressed auditory code and the motor code.

the auditory-to-auditory pathways to the partially compressed auditory code. Even during passive listening, these motorically-influenced auditory codes are activated. Heard speech is thereby analyzed by "how it would have been phonated." This is in agreement with the motor theory of speech perception (c.f., Studdert-Kennedy, 1984) and finds support from physiology (Ojemann, 1983). These results and the architecture of Figure 6 clarify why the concerted attempts to find purely auditory correlates of speech segments have not met with greater success (c.f., Zue, 1976; Cooper, 1980, 1983), and suggests how an artificial system capable of recognizing natural speech can incorporate motor information that human listeners employ.

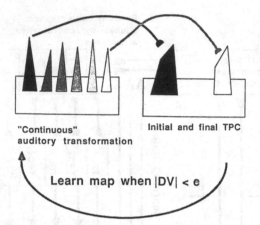

"Continuous"
auditory transformation

Initial and final TPC

Learn map when |DV| < e

Figure 5. The dynamic pattern of the code in the auditory system is more continuous, while that in the control structure for the motor system is more discrete. When activated, such a motor code initiates a unitized, stereotyped synergetic action of articulators.

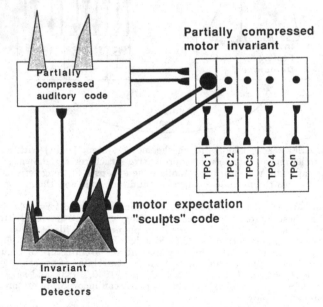

Partially compressed
motor invariant

Partially
compressed
auditory code

TPC 1 TPC 2 TPC3 TPC4 TPCn

motor expectation
"sculpts" code

Invariant
Feature
Detectors

Figure 6. Top-down priming from the motor to the auditory system reorganizes the emerging auditory code to be consistent with the motor commands. This motorically-influenced auditory code is further compressed at higher stages of the auditory system.

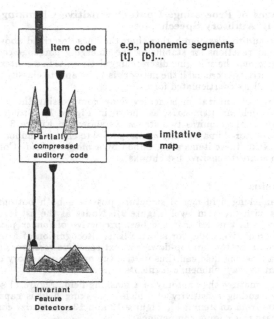

Figure 7. Unitization is achieved by compressing the partially compressed auditory code to yield an item code, which includes such units as phonemic segments.

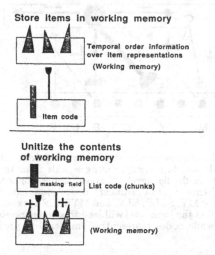

Figure 8. Context-sensitive list codes are formed via a two-level process: (top) Items are placed in *working memory*, which encodes temporal order information. Then (bottom) a masking field uses bottom-up flow and top-down priming to yield context-sensitive list codes.

5. Higher Stages of Processing: Context-Sensitive Chunking and Unitization of the Emerging Auditory Speech Code

Stages of the auditory system higher than the ones described above rely on processes other than circular reactions for stabilizing the emerging language code. Such processes *unitize*, *chunk*, or *group* the emerging discrete linguistic units in a context-sensitive manner. Such context-sensitivity is crucial if the network is to be able to classify any given phonemic segment (say) in all its coarticulated forms.

An early stage of unitization is achieved by compressing the partially compressed auditory code to yield an *item code*, as shown in Figure 7. Grouping such items into context-sensitive chunks requires two stages, as shown in Figure 8. First, sequentially occurring items are stored in a *working memory* level to encode temporal order information over the items. Next, these items are grouped by a *masking field* (Cohen and Grossberg, 1986, 1987) into context-sensitive list chunks.

6. Masking Fields

In brief, a masking field neural structure possesses both bottom-up and top-down interconnections with the item level (Figure 9). Nodes at the list level compete through mutual inhibition. List nodes that are best predictive of *longer* patterns of items will inhibit the less predictive nodes for shorter lists. Recognition of a unitized grouping of items occurs when a bottom-up top-down context-sensitive *resonance* develops. In speech networks, such a masking field can thus unitize the evolving auditory code into predictive chunks, representing, say, phonemic segments.

Figure 10 schematizes the anatomy of a masking field. Figure 11 schematizes the two primary types of coding sensitivity of which a masking field is capable in response to bottom-up inputs from an item field. Figures 12 and 13 summarize computer simulations which demonstrate this coding competence.

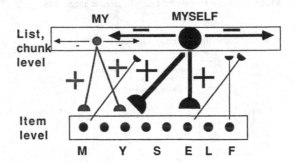

Figure 9. A masking field architecture creates context-sensitive list codes by using both bottom-up filtering signals and top-down priming signals from the list level. There is competition between units in the list level. "Larger" nodes—ones that pool information from a larger number of items—inhibit "smaller" nodes more effectively than vice versa. For instance, if list nodes for MY, SELF, ELF, and MYSELF are encoded, the presentation of the letters M-Y-S-E-L-F at the item level will lead to a resonance between the MYSELF node and the six items, while nodes representing smaller, less predictive, groupings are quickly suppressed.

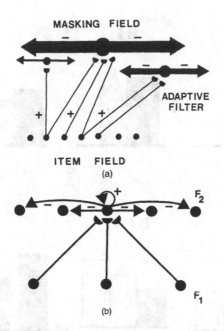

Figure 10. Masking field interactions: (a) Cells from an item field F_1 grow randomly to a masking field F_2 along positionally sensitive gradients. The nodes in the masking field grow so that larger item groupings, up to some optimal size, can activate nodes with broader and stronger inhibitory interactions. Thus the $F_1 \rightarrow F_2$ connections and the $F_2 \leftrightarrow F_2$ interactions exhibit properties of self-similarity. (b) The interactions within a masking field F_2 include positive feedback from a node to itself and negative feedback from a node to its neighbors. Long term memory (LTM) traces at the ends of $F_1 \rightarrow F_2$ pathways (designated by hemidisks) adaptively tune the filter defined by these pathways to amplify the F_2 reaction to item groupings which have previously succeeded in activating their target F_2 nodes.

The interactions between these levels can explain many speech properties, including properties of temporal invariance and phonemic restoration. When designed to incorporate a "long-term memory invariance principle" (Grossberg, 1986, 1987; Grossberg and Stone, 1986a, 1986b), the spatial pattern of activation across working memory defines an invariant code, and an attentional gain control signal to the working memory stage preserves this spatial code under changes in overall speaking rate.

Phonemic restoration occurs when an ambiguous or missing sound is clearly heard when presented in the proper context. The top-down priming of a masking field can complete ambiguous elements of the item code, so long as these items can be reorganized by the 2/3 Rule properties of the prime (Carpenter and Grossberg, 1987a, 1987b). The speech code results from a resonant wave which is controlled by feedback interactions between the working memory and masking field levels. Although the list chunks which reorganize the form and grouping of item codes utilize "future" information, this resonant wave can emerge from "past" to "future" because the internal masking of unpredictive list codes within the masking field occurs much faster than the time scale for unfolding the resonant

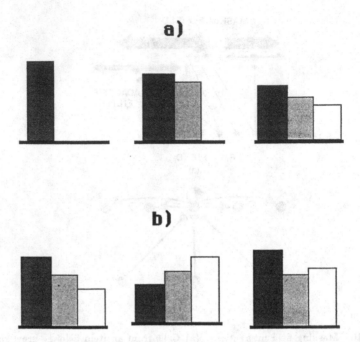

Figure 11. Two types of masking field sensitivity: (a) A masking field F_2 can automatically rescale its sensitivity to differentially react as the F_1 activity pattern expands through time to activate more F_1 cells. It hereby acts like a "multiple spatial frequency filter." (b) A masking field can differentially react to different F_1 activity patterns which activate the same set of F_1 cells. By (a) and (b), F_2 acts like a spatial pattern discriminator which can compensate for changes in overall spatial scale without losing its sensitivity to pattern changes at the finest spatial scale.

wave (Figure 14.)

The overall neural architecture employing the elements described above is shown in Figure 15.

Additional network designs are being developed for dealing with additional problems such as factoring rhythm information from linguistic information and the coding of repetitive patterns. Even as it stands, however, the architecture and design considerations described above provide a new processing architecture for understanding such issues as analysis-by-synthesis, the motor theory of speech perception, categorical perception, invariant speech perception, and phonemic restoration.

References

Bullock, D. and Grossberg, S., Neural dynamics of planned arm movements: Emergent invariants and speed-accuracy properties during trajectory formation. *Psychological Review*, in press, 1987.

Carpenter, G.A. and Grossberg, S., A massively parallel architecture for a self-organizing

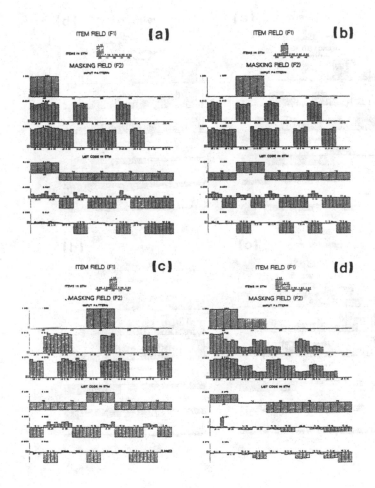

Figure 12. (a) The correct list code {0} is preferred in STM, but predictive list codes which include {0} as a part are also activated with lesser STM weights. The prediction gets less activation if {0} forms a smaller part of it. (b) The correct list code {1} is preferred in STM, but the predictive list codes which include {1} as a part are also activated with lesser STM weights. (c) The list code in response to item {2} also generates an appropriate reaction. (d) A list code of type {0, 1} is maximally activated, but part codes {0} and predictive codes which include {0, 1} as a part are also activated with lesser STM weights.

228

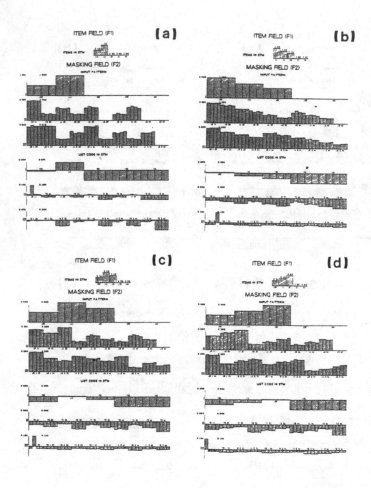

Figure 13. (a) A different list code of type $\{0,1\}$ is maximally activated, but part codes $\{1\}$ are also activated with lesser STM weight. Due to the random growth of $F_1 \rightarrow F_2$ pathways, no predictive list codes are activated (to 3 significant digits). (b)–(d) When the STM pattern across F_1 includes three items, the list code in STM strongly activates an appropriate list code. Part groupings are suppressed due to the high level of predictiveness of this list code. Comparison of Figures 12a, 12d, and 13b shows that as the item code across F_1 becomes more constraining, the list code representation becomes less distributed across F_2.

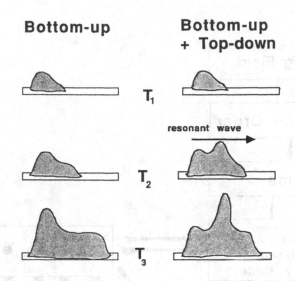

Figure 14. (Left): The activity pattern in working memory as new items enter the system, if the architecture had purely bottom-up connections. (Right): If the system has top-down priming, on the other hand, crucial features in the working memory that fit into a coherent pattern are reinforced, leading to a different distribution of neural activity. This resonant wave constitutes the speech code.

neural pattern recognition machine. *Computer Vision, Graphics, and Image Processing*, 1987, **37**, 54–115 (a).

Carpenter, G.A. and Grossberg, S., ART 2: Self-organization of stable category recognition codes for analog input patterns. *Applied Optics*, in press, 1987 (b).

Cohen, M.A. and Grossberg, S., Neural dynamics of speech and language coding: Developmental programs, perceptual grouping, and competition for short term memory. *Human Neurobiology*, 1986, **5**, 1–22.

Cohen, M.A. and Grossberg, S., Masking fields: A massively parallel neural architecture for learning, recognizing, and predicting multiple groupings of patterned data. *Applied Optics*, 1987, **26**, 1866–1891.

Cooper, F.S., Acoustics in human communication: Evolving ideas about the nature of speech. *Journal of the Acoustical Society of America*, 1980, **68**, 18–21.

Cooper, F.S., Some reflections on speech research. In P.F. MacNeilage (Ed.), **The production of speech**. New York: Springer-Verlag, 1983.

Fowler, C., Coarticulation and theories of extrinsic timing. *Journal of Phonetics*, 1980, **8**, 113–133.

Grossberg, S., **The adaptive self-organization of serial order in behavior: Speech, language, and motor control**. In E.C. Schwab and H.C. Nusbaum (Eds.), **Pattern recognition by**

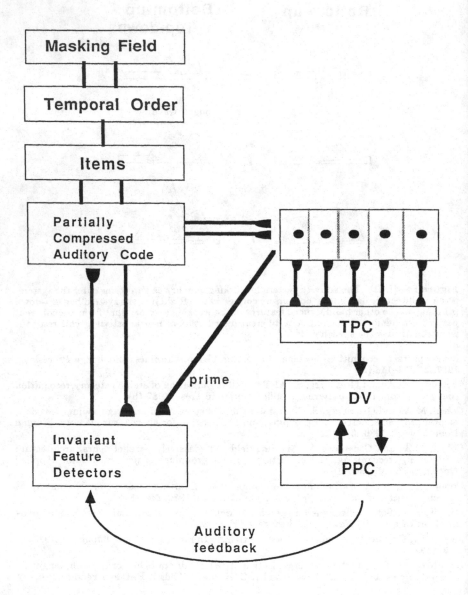

Figure 15. Global architecture for a speech recognition and synthesis system, employing the processing described above. See text for details.

humans and machines, Vol. 1: Speech perception. New York: Academic Press, 1986.

Grossberg, S. (Ed.), **The adaptive brain, II: Vision, speech, language, and motor control.** Amsterdam: Elsevier/North-Holland, 1987.

Grossberg, S. and Stone, G.O., Neural dynamics of attention switching and temporal order information in short-term memory. *Memory and Cognition*, 1986, **14**, 451–468 (a).

Grossberg, S. and Stone, G.O., Neural dynamics of word recognition and recall: Attentional priming, learning, and resonance. *Psychological Review*, 1986, **93**, 46–74 (b).

Lieberman, P., **The biology and evolution of language.** Cambridge, MA: Harvard University Press, 1984.

Ojemann, G., Brain organization for language from the perspective of electrical stimulation mapping. *Behavioral and Brain Sciences*, 1983, **2**, 189–230.

Piaget, J., **The origins of intelligence in children.** New York: Norton, 1963.

Studdert-Kennedy, M., Perceptual processing links to the motor system. In M. Studdert-Kennedy (Ed.), **Psychobiology of language.** Cambridge, MA: MIT Press, 1984, 29–39.

Zue, V.W., Acoustic characteristics of stop consonants: A controlled study. Ph.D. Dissertation, Massachusetts Institute of Technology, Electrical Engineering and Computer Science Department, 1976.

NEOCOGNITRON: A NEURAL NETWORK MODEL FOR VISUAL PATTERN RECOGNITION

Kunihiko Fukushima, Sei Miyake[*] and Takayuki Ito

NHK Science and Technical Research Laboratories

1-10-11, Kinuta, Setagaya, Tokyo 157, Japan

ABSTRACT   The "neocognitron" is a hierarchical neural network model for visual pattern recognition. It consists of many layers of neuron-like cells, and has variable connections between the cells in adjoining layers. It can acquire the ability to recognize patterns by learning, and can be trained to recognize any set of patterns. After finishing the process of learning, pattern recognition is performed on the basis of similarity in shape between patterns, and is not affected by deformation, nor by changes in size, nor by shifts in the position of the input patterns.

In the hierarchical network of the neocognitron, local features of the input pattern are extracted by the cells of a lower stage, and they are gradually integrated into more global features. The highest stage of the network is the recognition layer. Each cell of it integrates all the information of the input pattern, and responds only to one specific pattern. During this process of extracting and integrating features, errors in the relative position of local features are gradually tolerated. The operation of tolerating positional error a little at a time at each stage, rather than all in one step, plays an important role in endowing the network with an ability to recognize even distorted patterns.

We have designed a pattern-recognition system using the principle of the neocognitron, in order to demonstrate its ability. The system has been implemented on a mini-computer and has been trained to recognize handwritten numerals.

---

* On loan to ATR Auditory and Visual Perception Research Laboratories, Twin 21 Bldg. MID Tower, 2-1-61, Shiromi, Higashi-ku, Osaka 540, Japan .

1. INTRODUCTION

Modeling neural network is a promising approach not only to understanding the mechanism of the brain, but also to obtaining a design principle for new information processors. In the modeling approach, we study how to interconnect neurons in order to synthesize a brain model, which is a network with function and ability as superior as the brain. When synthesizing a model, we try to follow physiological evidence as faithfully as possible. For parts which are not yet clear, however, we construct a hypothesis and synthesize a model that follows the hypothesis. We then analyze and simulate the behavior of the model and compare it with that of the brain. If we can find any discrepancy in the behavior between the model and the brain, we change the initial hypothesis and modify the model following a new hypothesis. We then test the behavior of the model again. We repeat this procedure until the model behaves in the same way as the brain. Although we must still verify the validity of the model by physiological experiment, it is probable that the brain uses the same mechanism as the model, because both respond in the same way. Hence, modeling neural networks promises to help us to uncover the mechanism of the brain.

Once we complete a model, its simplification makes it easy to see the essential algorithm of information processing in the brain. We can use the algorithm directly as a design principle for new information processors. Thus, modeling neural networks is the most direct way for bringing the results of neurophysiological and psychological research to engineering applications.

Among many models thus obtained, a neural network model called a "neocognitron"[1)-3)] is discussed in this article. It is a hierarchical neural network model for visual pattern recognition.

2. THE STRUCTURE AND BEHAVIOR OF THE NETWORK

2.1 Physiological Background

In the visual area of the cerebrum, neurons are found to respond selectively to local features of a visual pattern, such as lines and edges in particular orientations[4)]. In the area higher than the visual cortex, it has been found that cells exist which respond selectively to certain figures like circles, triangles or squares, or even to a human face[5),6)]. Accordingly, the visual system seems to have a hierarchical structure, in which simple features are first extracted from a stimulus pattern, and then integrated into more complicated ones. In this hierarchy, a cell in a higher stage generally receives signals from a wider area of the retina, and is more insensitive to the position of the stimulus. Such neural networks in the brain are not always complete at birth, but gradually develop, adapting flexibly to circumstances after birth.

This kind of physiological evidence suggested a network structure for the neocognitron.

2.2 The Structure of the Network

The neocognitron is a hierarchical multilayered network consisting of neuron-like cells. The cells are of the analog type; that is, their inputs and output take non-negative analog values, corresponding to the instantaneous firing-frequencies of biological neurons.

The hierarchical structure of the network is illustrated in Fig. 1. There are forward connections between cells in adjoining layers. The initial stage of the network is the input layer, called U_0, and consists of a two-dimensional array of receptor cells u_0. Each of the succeeding stages has a layer of "S-cells" followed by a layer of "C-cells". Thus, in the whole network, layers of S-cells and C-cells are arranged alternately. The layer of C-cells at the highest stage is the recognition layer: the response of the cells of this layer shows the final result of pattern recognition by the neocognitron. Notation U_{Sl} and U_{Cl} are used to denote the layers of S-cells and C-cells of the l-th stage, respectively. Incidentally, each U_S-layer contains subsidiary inhibitory cells, called V-cells, but they are not drawn in Fig. 1.

S-cells or C-cells in a layer are divided into subgroups according to the kinds of feature to which they respond. Since the cells in each subgroup are arranged in a two-dimensional array, we call the subgroup a "cell-plane". In Fig. 1, each quadrangle drawn with heavy lines

236

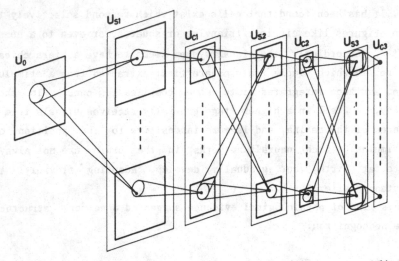

Fig. 1. Hierarchical network structure of the neocognitron[1].

represents a cell-plane, and each vertically elongated quadrangle drawn with thin lines, in which cell-planes are enclosed, represents a layer of S-cells or C-cells. Although only one cell is drawn in each cell-plane in Fig. 1, usually many cells are contained in it. Incidentally, each ellipse in the figure represents the area from which a cell receives input connections.

As schematically illustrated in Fig. 2, connections converging to cells in a cell-plane are homogeneous: All the cells in a cell-plane receive input connections of the same spatial distribution, and only

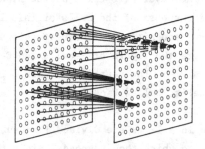

Fig. 2. Illustration showing the spatial arrangement
of the connections between cell-planes[1].

the positions of the preceding cells shift in parallel with the position of the cells in the cell-plane. This condition of homogeneity holds not only for fixed connections but also for variable connections. As discussed in Section 3.1.2, the reinforcement of the variable connections is always performed under this condition.

The density of cells in each layer is designed to decrease with the order of the stage, because each cell in a higher stage usually receives signals from larger area of the input layer and the neighboring cells come to receive similar signals. Hence, in the recognition layer at the highest stage, only one cell exists in each cell-plane.

S-cells are feature-extracting cells and somewhat resemble simple cells in the visual cortex. Figure 3 shows the input-to-output characteristics of an S-cell, in which inhibitory input reduces the effect of the excitatory inputs in a shunting manner. Connections converging to feature-extracting S-cells are variable and are reinforced during a learning (or training) process. After finishing the learning, which will be discussed later, S-cells, with the aid of the subsidiary V-cells, can extract features from the input pattern. In other words, a S-cell is activated only when a particular feature is presented at a certain position in the input layer. The features which the S-cells extract are determined during the learning process.

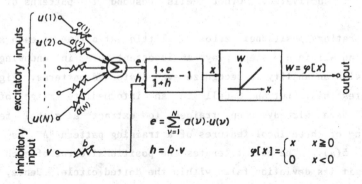

Fig. 3. Input-to-output characteristics of an S-cell: A typical example of the cells employed in the neocognitron[2].

Generally speaking, in the lower stages, local features, such as a line at a particular orientation, are extracted. In higher stages, more global features, such as a part of a training pattern, are extracted.

C-cells somewhat resemble complex cells. They are inserted in the network to allow for positional errors in the features of the stimulus. Connections from S-cells to C-cells are fixed and invariable. Each C-cell receives signals from a group of S-cells which extract the same feature, but from slightly different positions. The C-cell is activated if at least one of these S-cells is active. Even if the stimulus feature is shifted in position and another S-cell is activated instead of the first one, the same C-cell keeps responding. Hence, the C-cell's response is less sensitive to shifts in position of the input pattern.

2.3 Deformation- and Position-Invariant Recognition

In the whole network, with its alternate layers of S-cells and C-cells, the process of feature-extraction by S-cells and toleration of positional shift by C-cells are repeated as shown in Fig. 4. During this process, local features extracted in a lower stage are gradually integrated into more global features. Finally, each C-cell of the recognition layer at the highest stage integrates all the information of the input pattern, and responds only to one specific pattern. In other words, only one cell, corresponding to the category of the input pattern, is activated. Other cells respond to patterns of other category.

Tolerating positional error a little at a time at each stage, rather than all in one step, plays an important role in endowing the network with an ability to recognize even distorted patterns. Figure 5 illustrates this. Let an S-cell in an intermediate stage of the network have already been trained to extract a global feature consisting of three local features of a training pattern "A" as shown in Fig. 5(a). The cell tolerates the positional error of each local feature, if its deviation falls within the dotted circle. Hence, this S-cell responds to any of the deformed patterns shown in Fig. 5(b). That the toleration of positional errors should not be too large at this stage. If too large errors are tolerated at one step, the network may come to respond erroneously, such as recognizing a stimulus like Fig. 5(c) as an "A" pattern.

Since errors in the relative position of local features are tolerated in the process of extracting and integrating features, the same C-cell responds in the highest stage, even if the input pattern is

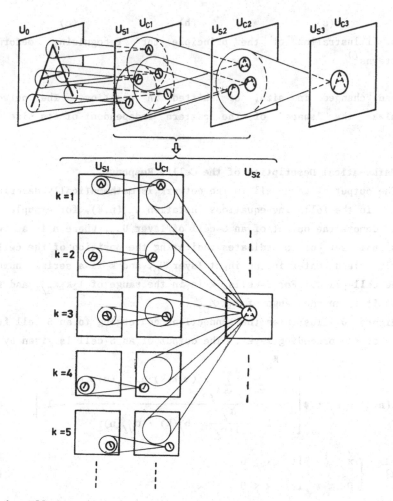

Fig. 4. Illustration of the process of pattern recognition in the neocognitron[1].

240

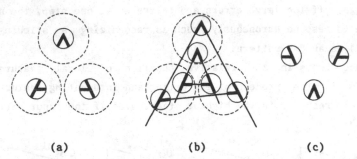

(a)	(b)	(c)

Fig. 5. Illustration of the principle for recognizing deformed
patterns[7].

deformed, changed in size, or shifted in position. The network
recognizes the "shape" of the pattern independent of its size and
position.

2.4 Mathematical Description of the Cell's Response

The output of each cell in the network is mathematically described
below. In the following equations, notation $u_{S\ell}(n,k)$, for example, is
used to denote the output of an S-cell of layer $U_{S\ell}$, where n is a two-
dimensional set of co-ordinates indicating the position of the cell's
receptive-field center in the input layer U_0, and k is a serial number
of the cell-plane. For S-cells, k is in the range of $1 \leq k \leq K_{S\ell}$, and for
C-cells it is in the range of $1 \leq k \leq K_{C\ell}$.

Figure 6 illustrates the connections converging to an S-cell from
C-cells of the preceding layer. The output of an S-cell is given by

$$u_{S\ell}(n,k) = r_\ell \cdot \phi \left[\frac{1 + \sum_{K=1}^{K_{C\ell-1}} \sum_{v \in A_\ell} a_\ell(v,\kappa,k) \cdot u_{C\ell-1}(n+v,\kappa)}{1 + \frac{r_\ell}{1+r_\ell} \cdot b_\ell(k) \cdot u_{V\ell}(n)} - 1 \right] \quad (1)$$

where

$$\phi[x] = \begin{cases} x & \text{if } x \geq 0 \\ 0 & \text{if } x < 0 . \end{cases} \quad (2)$$

In the case of $\ell=1$ in (1), $u_{C\ell-1}(n,\kappa)$ stands for $u_0(n)$ or the output of
a receptor cell of the input layer, and we have $K_{C\ell-1}=1$.

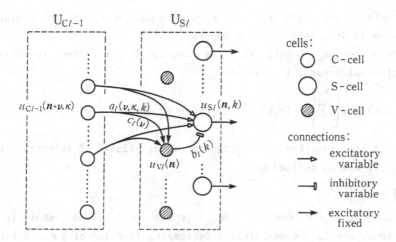

Fig. 6. Connections converging to a feature-extracting S-cell.

$a_l(\nu,\kappa,k)$ (≥ 0) is the strength of the variable excitatory connection coming from C-cell $u_{Cl-1}(n+\nu,\kappa)$ in the preceding layer. A_l denotes the summation range of ν, that is, the size of the spatial spread of the input connections to one S-cell. $b_l(k)$ (≥ 0) is the strength of the variable inhibitory connection coming from subsidiary V-cell $u_{Vl}(n)$. As discussed before in connection with Fig. 2, all the S-cells in a cell-plane have identical set of input connections. Hence, $a_l(\nu,\kappa,k)$ and $b_l(k)$ do not contain argument n representing the position of the receptive field of the cell $u_{Sl}(n,k)$.

As can be seen from (1) or from Fig. 3, the inhibitory input to this cell acts in a shunting manner. The positive constant r_l determines the efficiency of the inhibitory input to this cell.

The subsidiary V-cell which sends an inhibitory signal to this S-cell yields an output equal to the weighted root-mean-square of the signals from the preceding C-cells, that is,

$$u_{Vl}(n) = \sqrt{\sum_{\kappa=1}^{K_{Cl-1}} \sum_{\nu \in A_l} c_l(\nu) \cdot \{u_{Cl-1}(n+\nu,\kappa)\}^2} \tag{3}$$

where $c_l(\nu)$ represents the strength of the fixed excitatory connections, and is a monotonically decreasing function of $|\nu|$.

Incidentally, the role of the root-mean-square cells in feature extraction is discussed in other papers[2], [7].

The output of a C-cell, which is inserted in the network to allow for positional errors, is given by

$$u_{C\ell}(n,k) = \psi \left[\sum_{\kappa=1}^{K_{S\ell}} j_\ell(\kappa,k) \sum_{\nu \varepsilon D_\ell} d_\ell(\nu) \cdot u_{S\ell}(n+\nu,\kappa) \right] , \qquad (4)$$

where $\psi[\ ]$ is a function specifying the characteristic of saturation of the C-cell, and is defined by

$$\psi[x] = \frac{\phi[x]}{1 + \phi[x]} . \qquad (5)$$

Parameter $d_\ell(\nu)$ denotes the strength of fixed excitatory connections, and is a monotonically decreasing function of $|\nu|$. Hence, if at least one S-cell is activated in the area D_ℓ of the κ-th cell-plane, to which these connections spread, (and if $j_\ell(\kappa,k)$, which will be discussed below, is positive), this C-cell is also activated.

$j_\ell(\kappa,k)$ indicates how the output of S-cells of different cell-planes are integrated at the input of this C-cell.

In the case of learning-without-a-teacher, this process of integration is not included. Hence, $U_{S\ell}$ and $U_{C\ell}$ has the same number of cell-planes (or $K_{S\ell}=K_{C\ell}$), and

$$j_\ell(\kappa,k) = \begin{cases} 1 & \text{for } \kappa=k \\ 0 & \text{for } \kappa \neq k . \end{cases} \qquad (6)$$

In the case of learning-with-a-teacher, however, output of several number of S-cell-planes sometimes converges together to a single C-cell-plane. This condition of convergence is represented by $j_\ell(\kappa,k)$. In the case of character recognition, even characters of different styles of writing have to be correctly recognized. In other words, input characters have to be classified not only on the basis of geometrical similarity but also on the basis of customs by which some particular kinds of large deformation are admitted. Sometimes when such deformation is too large, a single S-cell-plane is not enough to extract deformed versions of a feature. In such a case, another S-cell-plane is used to extract a deformed version of the feature, and the output from these S-cell-planes are made to converge to a single C-cell-plane. It is $j_\ell(\kappa,k)$ that represents this integration process. Depending on whether or not the k-th C-cell-plane receives signals from the κ-th S-cell-plane, $j_\ell(\kappa,k)$ takes a positive value or zero, respectively. Hence, for each κ, $j_\ell(\kappa,k)$ is usually zero except for one particular value of k.

3. SELF-ORGANIZATION OF THE NETWORK

The connections converging to S-cells are variable, and are reinforced gradually in accordance with stimuli given to the network during the learning phase. Both "learning-without-a-teacher" and "learning-with-a-teacher" can be used to train the neocognitron to recognize patterns.

3.1 Learning without a Teacher

We will first discuss the case of learning-without-a-teacher (unsupervised learning)[1)-3)]. The repeated presentation of a set of training patterns is sufficient for the self-organization of the network, and it is not necessary to give any information about the categories in which these patterns should be classified. The neocognitron by itself acquires the ability to classify and recognize these patterns correctly on the basis of similarity in shape.

3.1.1 Reinforcement of maximum-output cells
Self-organization of the neocognitron is performed with two principles. The first has been introduced for the self-organization of the "cognitron"[8),9)] proposed earlier by one of the authors. According to the recent terminology, this principle can be classified under the competitive learning paradigm. Specifically, the first principle is as follows:

Among the cells situated in a certain small area, only the one which is responding the strongest has its input connections reinforced. The amount of reinforcement of each input connection to this maximum-output cell is proportional to the intensity of the response of the cell from which the relevant connection is leading.

This principle is applied to the variable input connections converging to feature-extracting S-cells. With this principle, among many S-cells in a certain small area, only the one which yields the maximum output has its input connections reinforced. Because of the "winner-takes-all" nature of this principle, the duplicated formation of cells which extract the same feature does not occur, and the formation of a redundant network can be prevented. Only the one cell giving the best response to a training stimulus is selected, and only that cell is reinforced so as to respond more appropriately to the stimulus.

Once a cell is selected and reinforced to respond to a feature, the cell usually loses its responsiveness to other features. When a different feature is presented, usually a different cell yields the maximum output and has its input connections reinforced. Thus, "division of labor" among the cells occurs automatically.

With this principle, the network also develops a self-repairing function. If a cell which has been strongly responding to a stimulus is damaged and ceases to respond, another cell, which happens to respond more strongly than others, starts to grow and substitute for the damaged cell. Until then, the larger response of the first cell had prevented the growth of a second cell.

3.1.2 Generation of a feature-extracting S-cell
Now, let's discuss how a maximum-output S-cell comes to acquire the ability to extract a feature. Figure 7 illustrates the process of reinforcement, showing only the connections converging to an S-cell. As shown in Fig. 7(a), the S-cell receives variable excitatory connections from a group of C-cells in the preceding layer. The cell also receives a variable inhibitory connection from a subsidiary inhibitory cell, called a V-cell. The V-cell receives fixed excitatory connections from the same group of C-cells as does this S-cell, and is always responds with the average intensity of the output of the C-cells. The initial strength of these variable connections is nearly zero, as discussed in Section 3.1.4.

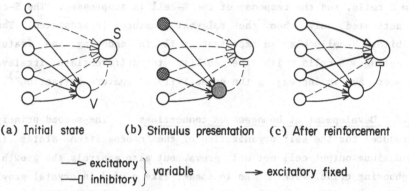

(a) Initial state   (b) Stimulus presentation   (c) After reinforcement

$\longrightarrow$ excitatory ⎫
$\longrightarrow$ inhibitory ⎬ variable $\longrightarrow$ excitatory fixed

Fig. 7. The process of reinforcement of the forward connections converging to a feature-extracting S-cell[7]. The density of the shadow in a circle represents the intensity of the response of the cell. (a) shows the initial state before training. (b) shows stimulus presentation during the training. (c) shows the connections after reinforcement.

Suppose this S-cell responds most strongly of the S-cells in its vicinity when a training stimulus is presented [see Fig. 7(b)]. According to the principle mentioned above, variable connections leading from activated C- and V-cells are reinforced as shown in Fig. 7(c). It should be noted that both excitatory and inhibitory connections are reinforced with the same principle. The variable excitatory connections to the S-cell grow into a "template" which exactly matches the spatial distribution of the response of the cells in the preceding layer. The inhibitory variable connections from the V-cell is reinforced at the same time, but not as strongly because the output of the V-cell, which is equal to the average intensity of the output of the C-cells, is not as large.

After completion of the training, the S-cell comes to acquire the ability to extract the feature of the stimulus presented during the training period. Through the excitatory connections, the S-cell receives signals indicating the existence of the relevant feature to be extracted. If an irrelevant feature is presented the inhibitory signal from the V-cell becomes stronger than the direct excitatory signals

246

from C-cells, and the response of the S-cell is suppressed. The S-cell is activated only when the relevant feature is presented. Thus, inhibitory V-cells play an important roll in endowing the feature-extracting S-cells with the ability to differentiate irrelevant features, and in increasing the selectivity of feature extraction[2].

3.1.3 Development of homogeneous connections

The second principle introduced for the self-organization of the neocognitron states that the maximum-output cell not only grows, but also controls the growth of neighboring cells, working, so to speak, like a seed in crystal growth. Neighboring cells have their input connections reinforced in the same way as the "seed cell". The process of selecting seed cells will be discussed below in more detail.

Here, we define a term "hypercolumn": a hypercolumn is a group of S-cells in a layer whose receptive fields are situated at approximately the same position. In other words, each hypercolumn contains all kinds of feature extracting cells in it, and these cells extract features from approximately the same place in the input layer. If we rearrange the cell-planes of a layer and stack them in a manner shown in Fig. 8, the cells of a hypercolumn constitutes a columnar structure. Each hypercolumn contains cells from all the cell-planes.

Now, let a training pattern be presented to the network. From each hypercolumn, the S-cell which responds the strongest is chosen as

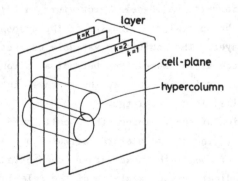

Fig. 8. Relation between cell-planes and hypercolumns within a layer[1].

a candidate for seed cells. When two candidates or more appear in one and the same cell-plane, only the one whose response is the largest is selected as the seed cell of that cell-plane. When only one candidate appears in a cell-plane, the candidate automatically becomes the seed cell of that cell-plane. If no candidate appears in a cell-plane, no seed cell is selected from that cell-plane this time.

Thus, at most one seed cell is selected from each cell-plane of S-cells at a time. Usually, a different cell becomes a seed cell when a different training pattern is given.

When a seed cell is selected from a cell-plane, all the other S-cells in the cell-plane grow so as to have input connections of the same spatial distribution as the seed cell. As the result, all the S-cells in a cell-plane grow to receive input connections of the identical spatial distribution where only the positions of the preceding C-cells are shifted in parallel from cell to cell, as illustrated in Fig. 2. In other words, connections develop homogeneously in a cell-plane. Hence, all the S-cells in the cell-plane come to respond selectively to a particular feature, and differences between these cells arise only from differences in position of the feature to be extracted.

3.1.4 Initial values of the variable connections   If the strength of all the variable connections is zero at the initial state before learning, self-organization of the network cannot start, because no cell can responds to the training pattern and maximum-output cells (or seed cells) cannot be selected. Hence, all the variable excitatory connections unconditionally get a very small value only when self-organization is going to start. In other words, each S-cell temporarily has very weak and diffused excitatory input connections only during the initial period of the self-organization. Once a reinforcement of the input connections begins, these weak and diffused initial connections are made to disappear. Incidentally, this situation coincide with the anatomical observation that, in the developing nervous system, synaptic connections between neurons are

overproduced initially and the redundant axons are eliminated gradually afterwards.

If the period of generation of these temporary weak diffused connections is delayed a little for the cells of higher stages, self-organization of the network can be performed efficiently. Specifically, it is desirable to delay it until the growth of the cells of the preceding stage has been settled.

3.1.5 Mathematical description of the reinforcement

The variable connections $a_l(\nu,\kappa,k)$ and $b_l(k)$ are reinforced depending on the intensity of the input to the seed cell. Let $u_{Sl}(\hat{n},\hat{k})$ be selected as a seed cell at a certain time. The variable connections $a_l(\nu,\kappa,\hat{k})$ and $b_l(\hat{k})$ to this seed cell, and consequently to all the S-cells in the same cell-plane as the seed cell, are reinforced by the following amount:

$$\Delta a_l(\nu,\kappa,\hat{k}) = q_l \cdot c_l(\nu) \cdot u_{Cl-1}(\hat{n}+\nu,\kappa) , \tag{7}$$

$$\Delta b_l(\hat{k}) = q_l \cdot u_{Vl}(\hat{n}) , \tag{8}$$

where q_l is a positive constant determining the speed of reinforcement. In the case of learning-with-a-teacher, which will be discussed below, a sufficiently large value is given to q_l so that the reinforcement of the input connections to each seed cell can be completed in a few steps of training-pattern presentation. In the case of learning-without-a-teacher, however, too large value of q_l is sometimes harmful.

3.2 Learning with a Teacher

As has been discussed above, in the case of learning-without-a-teacher, maximum-output cells are selected automatically as seed cells. In the case of learning-with-a-teacher (supervised learning)[3),10),11)], however, "teacher" points out which cells should be the seed cells for each training pattern. The other process of reinforcement is identical to that of the learning-without-a-teacher, and occurs automatically. It is, of course, not necessary to perform such a complicated procedure as calculating and adjusting the strength of all the connections one by one, but it is enough to point out which patterns or features should be extracted by which cells.

Efficiency of the training can be improved if training has been performed step by step from lower stages to higher stages. In other words, training of a higher stage is performed after completely finishing the training of the preceding stages.

Learning-with-a-teacher is useful when we want to train a system to recognize, for instance, hand-written characters which should be classified not only on the basis of similarity in shape but also on the basis of certain conventions. For example, the geometrical similarity between "0" and "θ" is about the same as that between "0" and "Q", but "0" and "θ" must be recognized as the same character, while "0" and "Q" must be classified into different categories. It is impossible to train the system to recognize these characters by learning-without-a-teacher only, by which characters are classified only on the basis of geometrical similarity.

4. HANDWRITTEN NUMERAL RECOGNITION

In order to demonstrate the ability of the neocognitron, we have designed a system which recognizes hand-written numerals form "0" to "9". This system[3],[12] (a modification from an old system[10]) has been implemented on a mini-computer (micro VAX-II) with an array processor (FPS-5105). The same system has also been implemented on a micro-computer (NEC PC-9801) which has a 16-bit main processor 8086 (with 384kBytes memory) and a co-processor 8087[11]. We are now implementing it in a parallel computer (NCUBE)[13].

The network has four stages of layers of S- and C-cells. Figure 9 shows how the cells of different cell-planes are spatially interconnected. This figure, in which only one cell-plane is drawn for each layer, illustrates a one-dimensional cross section of the connections between S- and C-cells.

The number of S- or C-cells in each layer is indicated at the bottom of this figure. Layer U_{C4} at the highest stage has ten cell-planes, each of which has only one C-cell. These ten C-cells corresponds to ten numeral patterns from "0" to "9".

250

$$U_0 \quad U_{S1} \quad U_{C1} \quad U_{S2} \quad U_{C2} \quad U_{S3} \quad U_{C3} \quad U_{S4} \quad U_{C4}$$

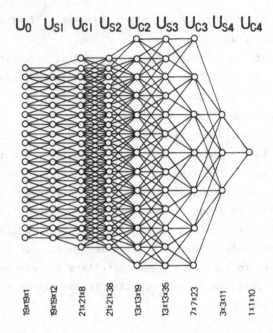

19x19x1 19x19x12 21x21x8 21x21x38 13x13x19 13x13x35 7x7x23 3x3x11 1x1x10

Fig. 9. One-dimensional view of interconnections between cells of different cell-planes[11]. Only one cell-plane is drawn in each layer. The numerals at the bottom of the figure show the total numbers of S- and C-cells in individual layers of the network.

From Fig. 9, we can read, for example, an S-cell of layer U_{S3} has 5x5 (excitatory) variable input connections from each cell-plane of layer U_{C2}. Since layer U_{C2} has 19 cell-planes, the maximum possible number of the variable input connections to each S-cell of layer U_{S3} is 5x5x19. It is important to note, however, that all of these 5x5x19 variable connections are not necessarily reinforced by learning. On the contrary, most of them usually remain at the initial state of strength of zero even after finishing learning. Since the variable connections of strength of zero need not be actually wired in the network, the effective number of connections are far less than the value directly read from this figure.

The neocognitron has been trained by learning-with-a-teacher. After finishing learning, the response of the network is tested. In this experiment, the input pattern is drawn on a magnetic tablet. Although a tablet is used to input hand-written numerals, the neocognitron does not use any temporal information about the order of the strokes of the character. The character which has already been drawn is used as the input pattern to the system. With the progress of calculation in the computer, the response of the layers of C-cells is displayed successively on a graphic terminal. Figure 10 shows an example of this display. To the input layer U_0, a numeral "2" is presented. In the recognition layer U_{C4} at the highest stage, shown at the extreme right, only cell "2" is activated. This means that the neocognitron recognizes the input pattern correctly.

Figure 11 shows some example of deformed input patterns which the neocognitron has recognized correctly. It is a matter of course that the neocognitron recognizes these patterns correctly even though they are shifted in position. When an input pattern is presented in a different position, the response of cells in intermediate layers, especially those near the input layer, varies with the shift. However, the higher the layer is, the smaller is the variation in response. The cells of the highest layer are not affected at all by a shift in position of the input pattern.

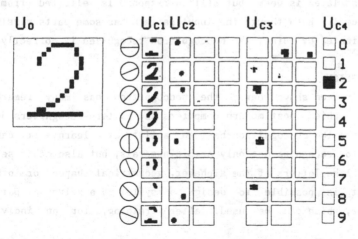

Fig. 10. An example of the response of the C-cells in the network trained to recognize handwritten numerals[12].

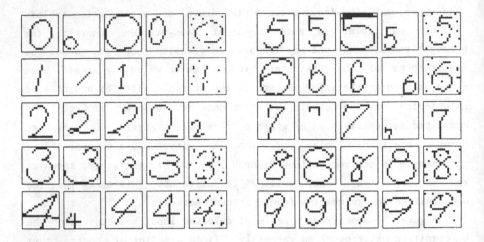

Fig. 11. Some example of deformed input patterns which the neocognitron has recognized correctly[3].

It has also been shown that even where the input pattern has been increased or diminished in size, or is skewed in shape, the response of the cells of the highest layer is not affected. Sometimes, when the input pattern has been distorted too much, the response of the cells in the highest layer is weak, but still a response is elicited from the correct cell. Even though the input pattern has some parts missing or is contaminated by noise, the neocognitron recognizes it correctly.

5. DISCUSSION

As has been shown here, the neocognitron has many remarkable properties which most modern computers and pattern-recognizers do not have. Since the neocognitron has an ability to learn, it can be trained to recognize not only Arabic numerals, but also other sets of patterns, like letters of the alphabet, geometrical shapes, or others. Hence, it is possible to design a system as a universal pattern-recognizer, which can be used, after training, for an individual purpose.

If the number of categories of the patterns to be recognized is increased, the number of cell-planes in each layer of the network also has to be increased. The number of cell-planes, however, need not be increased in proportion to the number of categories of the patterns. It is enough to increase it in less than linear proportion, because local features to be extracted at lower stages are usually contained in common in patterns of different categories.

If we want to construct a system which can recognize more complex patterns like chinese characters, it is recommended to increase the number of stages (or layers) in the network depending on the complexity of the patterns to be recognized.

The principles of the neocognitron are not restricted to the processing of visual information only, but can also be applied to other sensory information. For example, it would be possible to construct a speech-recognition system with a little modification.

Although the neocognitron has forward (i.e., afferent or bottom-up) connections only, the information-processing ability of the network can be greatly increased if backward (i.e., efferent or top-down) connections are added. The model of selective attention[7] recently proposed by one of the authors is an example of such an advanced system.

REFERENCES

1) Fukushima, K.: "Neocognitron: A Self-Organizing Neural Network Model for a Mechanism of Pattern Recognition Unaffected by Shift in Position", Biol. Cybern., **36**, 193-202 (1980).

2) Fukushima, K. and Miyake, S.: "Neocognitron: A New Algorithm for Pattern Recognition Tolerant of Deformations and Shifts in Position", Pattern Recognition, **15**, 455-469 (1982).

3) Fukushima, K.: "Neocognitron: A Hierarchical Neural Network Capable of Visual Pattern Recognition", Neural Networks, **1**, - (April 1988).

4) Hubel, D. H. and Wiesel, T. N.: "Receptive Fields, Binocular Interaction and Functional Architecture in the Cat's Visual Cortex", J. Physiol., **160**, 106-154 (1962).

5) Sato, T., Kawamura, T. and Iwai, E.: "Responsiveness of Inferotemporal Single Units to Visual Pattern Stimuli in Monkeys Performing Discrimination", Exp. Brain Res., 38, 313-319 (1980).

6) Bruce, C., Desimone, R. and Gross, C. G.: "Visual Properties of Neurons in a Polysensory Area in Superior Temporal Sulcus of the Macaque", J. Neurophysiol., 46, 369-384 (1981).

7) Fukushima, K.: "A Neural Network for Visual Pattern Recognition", Computer (IEEE Computer Society), 21[3], 65-75 (March 1988).

8) Fukushima, K.: "Cognitron: A Self-Organizing Multilayered Neural Network", Biol. Cybern., 20, 121-136 (1975).

9) K. Fukushima: "Cognitron: A Self-Organizing Multilayered Neural Network Model", NHK Technical Monograph, No. 30, Tokyo: NHK Tech. Res. Labs. (1981).

10) Fukushima, K., Miyake, S. and Ito, T.: "Neocognitron: A Neural Network Model for a Mechanism of Visual Pattern Recognition", IEEE Trans. Syst. Man Cybern., SMC-13, 826-834 (1983).

11) Fukushima, K., Miyake, S., Ito, T. and Kouno, T.: "Handwritten Numeral Recognition by the Algorithm of the Neocognitron -- An Experimental System Using a Microcomputer" (in Japanese), Trans. Information Processing Soc. Japan, 28, 627-635 (1987).

12) Fukushima, K., Miyake, S. and Ito, T.: "Neocognitron: A Biocybernetic Approach to Visual Pattern Recognition", NHK Laboratories Note, No. 336 (1986).

13) Ito, T., Fukushima, K. and Miyake, S.: "An Implementation of Neocognitron on a Parallel Computer" (in Japanese), 1988 Fall National Convention, IEICE, Japan (1988).

Part Three
APPLICATIONS

LEARNING TO PREDICT THE SECONDARY STRUCTURE
OF GLOBULAR PROTEINS

Ning Qian and Terrence J. Sejnowski

Department of Biophysics, The Johns Hopkins University
Baltimore, MD 21218, USA

Most of our knowledge of protein structure comes from the X-ray diffraction patterns of crystallized proteins. This method can be very accurate, but many steps are uncertain and the procedure is time-consuming. Recent developments in genetic engineering have vastly increased the number of known protein sequences. In addition, it is now possible to selectively alter protein sequences by site-directed mutagenesis. But to take full advantage of these techniques it would be helpful if one could predict the structure of a protein from its primary sequence of amino acids. The general problem of predicting the tertiary structure of folded proteins is unsolved.

Information about the secondary structure of a protein can be helpful in determining its structural properties. The best way to predict the structure of an new protein is to find a homologous protein whose structure has already been determined. Even if only limited regions of conserved sequences can be found, then template matching methods are applicable (Taylor, 1986). If no homologous protein with a known structure is found, existing methods for predicting secondary structures can be used but are not always reliable. Three of the most commonly used methods are those of Robson (Robson & Pain, 1971, Garnier et al., 1978), of Chou & Fasman (1978), and Lim (1974). These methods primarily exploit, in different ways, the correlations between amino acids and the local secondary structure. By local we mean an influence on the secondary structure of an amino acid by others that are no more than about 10 residues away. The average success rate of these methods is 50-53% on three types of secondary structure (α-helix, β-sheet, and coil) (Nishikawa, 1983; Kabsch & Sander, 1983a).

We present here a new method for discovering regular patterns in data that is based on neural network models. The goal of the method is to use the available information in the database of known protein structures to help predict the secondary structure of proteins for which no homologous structures are available. The average success rate of our method on a testing set of proteins non-homologous with the corresponding training set was 64.3% on three types of secondary structure (α-helix, β-sheet, and coil), with correlation coefficients (Matthews, 1975) of $C_\alpha = 0.41$, $C_\beta = 0.31$ and $C_{coil} = 0.41$ for each type respectively. These quality indices are all higher than those of previous methods. We conclude from computational experiments that no method based solely on local information in the protein sequence is likely to produce significantly better results for non-homologous proteins.

Network design. We trained a neural network model using database of known structures obtained from the Brookhaven National Laboratory. The testing set of proteins had practically no homologies with the training set. More details can be found in Qian & Sejnowski (1988). The network design used in this study is similar to the NET-talk system (Sejnowski & Rosenberg, 1987). The network maps sequences of input symbols onto sequences of output symbols. Here, the input symbols are the 20 amino acids and a special spacer symbol for regions between proteins; the output symbols correspond to the three types of secondary structures: α-helix, β-sheet, and coil.

A schematic diagram of the basic network is shown in Fig. 1. The processing units are arranged in layers, with the input units shown on the bottom and the output units shown at the top. The units on the input layer have connections to the units on the intermediate layer of "hidden" units, which in turn have connections to the units on the output layer. In networks with a single layer of modifiable weights there are no hidden units, in which case the input units are connected directly to the output layer.

The network is given a contiguous sequence of, typically, 13 amino acids. The goal of the network is to correctly predict the secondary structure for the middle amino acid. The network can be considered a "window" with 13 positions that moves through the protein, one amino acid at a time. The input layer is arranged in 13 groups. Each group has 21 units, each unit representing one of the amino acids (or spacer). For a local encoding of the input sequence, one and only one input unit in each group, corresponding to the appropriate amino acid at each position, is given a value 1, and the rest are set to 0. This is called a local coding scheme because each unit encodes a single item, in contrast with a distributed coding scheme in which each unit participates in representing several items. In some experiments we used distributed codings in which units represented biophysical properties of residues, such as their hydrophobicity. Another coding scheme that we used was the second-order conjunctive encoding, in which each unit represented a pair of residues, one residue from the middle position and a second residue at another position.

In the basic network, the output group has 3 units each representing one of the possible secondary structures for the center amino acid. For a given input and set of weights, the output of the network will be a set of numbers between 0 and 1. The secondary structure chosen was the output unit that had the highest activity level; this was equivalent to choosing the output unit that had the least mean square error with the target outputs.

We define the first-order features as the part of the mapping that can be predicted by each individual amino acid in the input window, and the second-order features as the additional part determined by all pairs of amino acids. The first order feature can be learned by a network with a single layer of modifiable weights. To learn higher-order features, a single layer network with conjunctive input codings or multi-layer network should be used.

Network training procedure. Initially, the weights in the network were assigned randomly with values uniformly distributed in the range [-0.3, 0.3]. The initial success rate was at chance level, around 33%. The performance was gradually improved by changing the weights using the back-propagation learning algorithm (Rumelhart, et al., 1986). A different random position in the concatenated training sequence of amino acids (see Section 2A above) was chosen as the center position of the input window at each training step. The surrounding amino acids were then used to clamp the input units in the window. All the amino acids in the training set were sampled once before starting over again. This random sampling procedure was adopted to prevent erratic oscillations in the performance that occurred when the amino acids were sequentially sampled. The performance of the network on the testing set was monitored frequently during training and the set of weights was kept which achieved the best average success rate on the testing set.

Testing with non-homologous proteins. We trained standard networks (13 input groups, local coding scheme, and 3 output units) with either 0 or 40 hidden units. The learning curves for the training and testing sets are shown in Fig. 2. In all cases, the percent of correctly predicted structures for both the training and testing sets rose quickly from the chance level of 33% to around 60%. Further training improved the performance of the networks with hidden units on the training set, but performance on the testing set did not improve and in fact tended to decrease. This behavior is an indication that memorization of the details of the training set is interfering with the ability of the network to generalize. The peak performance for a network with 40 hidden units was $Q_3 = 62.7\%$, with the corresponding $C_\alpha = 0.35$, $C_\beta = 0.29$ and $C_{coil} = 0.38$. The performance with no hidden units is similar, as shown in Fig. 2 and indicated below in the section on dependence on the number of hidden units.

The values of the weights for the network with no hidden units are graphically represented by a Hinton diagram in Fig. 3. The relative contribution to the secondary structure made by each amino acid at each position is apparent in this diagram. Physical properties of the amino acids can be correlated with their contributions to each form of secondary structure; in this way hypothesis can be generated concerning the physical basis for secondary structure formation (see Fig. 3).

Dependence on the number of hidden units. A surprising result is that the peak performance on the testing set was almost independent of the number of hidden units. This suggests that the common features in the training and testing proteins are all first-order features and that all of the first-order features learned from the training set that we used were common features. The higher-order features (the information due to interactions between two or more residues) learned by the network were specific to each individual protein, at least for the proteins that were used. In the next section we show that if the training set is too small then not all the first-order features learned during training are common features.

Dependence on size of the training set. A standard network with 13 input groups and no hidden units was trained on training sets with different numbers of amino acids in them. The maximum performance of the network as a function of the training set size is presented in Fig. 4. The maximum occurred after different training times in the different networks.

The maximum performance on the training set decreases with the number of amino acids in the training set because more information is being encoded in a fixed set of weights. The testing success rate, on the other hand, increases with size because the larger the training set, the better the network is able to generalize. When the training set is small, the network is able to "memorize" the details, but this strategy is not possible when the training set is large. Another conclusion from Fig. 4 is that a further increase of the data set is unlikely to improve the performance of the network on the testing set.

Cascaded networks improve performance. For a given input sequence the output of the network is a three-dimensional vector whose components have values between 0 and 1. The secondary structure for the above networks was predicted by choosing the output unit with the largest value, as mentioned in the Methods. However, information about the certainty of the prediction is not exploited by this procedure. Neither is the information available in the correlations between neighboring secondary structure assignments, since predictions are made one residue at a time. However, by designing a second network we can take advantage of this additional information.

The inputs to the second network were sequences of outputs from the first network, trained as described above. Hence, the input layer of the second network contained 13 groups with 3 units per group, each group representing the complete information about the secondary structure assignment derived from the first network. The first network was fixed while the second network was trained on the same set of training proteins as the first network. The average performance for two cascaded networks was $Q_3 = 64.3\%$, $C_\alpha = 0.41$, $C_\beta = 0.31$ and $C_{coil} = 0.41$ with 40 hidden units in both nets. This was our best result on the testing set of non-homologous proteins.

Comparison with other methods. The performance of our method for secondary structure prediction is compared with those of Robson (1971), Chou & Fasman (1978) and Lim (1974) in Table 1. The original measures of accuracy reported by these authors were based in part on the same proteins from which they derived their methods, and these proteins are equivalent to our training set. The performance of our networks with hidden units on the training set was as high as $Q_3 = 95\%$ after sufficiently long training. However, these methods should be compared on proteins with structures that were not used in or homologous with ones in the training set. The results of testing these three methods on novel proteins is reported in Table V of Nishikawa (1983) and are listed in Table 1 along with the performance of our networks on the non-homologous testing set of proteins.

Our training and testing sets of proteins were different from those used to construct and test the previous methods. To determine how much of our improvement was due to this difference, we trained a new network using 22 of the 25 proteins found in Robson & Suzuki (1976) as the training set for a network. [Three of the proteins were missing from our database]. Our testing set was a subset of those found in Table V of Nishikawa (1983). that were found in our database. The testing success rate of Robson's method on these 10 proteins was 51.2% compared with 61.9% for our method with two cascaded networks. Thus, less than 1% of the 11% improvement in Table 1 can be attributed to differences in the training sets. The relatively small effect of the larger database available to us is consistent with the asymptotic slope of the dependence on training set size shown in Fig. 4.

The new method for predicting the secondary structure of globular proteins presented here is a significant improvement over existing methods for non-homologous proteins and should have many applications. We have emphasized the distinction between training and testing sets, between homologous and non-homologous testing sets, and the balance of the relative amount of each type of secondary structure in assessing the accuracy of our method and have provided objective measures of performance that can be compared with other methods.

However, the absolute level of performance achieved by our method is still disappointingly low. Perhaps the most surprising result was the conclusion that further improvement in local methods for predicting the secondary structure of non-homologous proteins is unlikely based on known structures. The fact that networks with no hidden units performed as well as networks with hidden units on the non- homologous training set suggests that there are little or no second- or higher-order features locally available in the training set to guide the prediction of secondary structure. We also found that the second-order conjunctive input representations of the amino acids (which make second-order features available as first-order features to the output layer) does not help either. All of these experiments are consistent with the hypothesis that little or no information is available in the data beyond the first-order features that have already been extracted. Further, the dependence of the performance on the size of the training set suggests that the addition of more protein structures to the training set will not significantly improve the method for non-homologous proteins.

One limitation of the neural network approach is that the complexity of actual physical interactions in a protein molecule cannot be represented explicitly. Also, the pattern recognition method that we used is not effective when the information contained in the statistics of the training set is global. In fact, much of the local secondary structure may depend on influences outside the local neighborhood of an amino acid. Significant improvements for non-homologous proteins would require an effective way of estimating physical interactions and better methods for taking into account long-range effects.

Acknowledgements

We thank Dr. Kevin Ullmer for helping with the database and for many discussions during the course of the research. Drs. Carl Pabo and Richard Durbin suggested important improvements in the presentation. Drs. Warner Love, Richard Cone and Evangelos Moudrianakis provided helpful advice on many aspects of protein structure. We are also grateful to Paul Kienker for discussions and the use of his network simulator. TJS was supported by a Presidential Young Investigator Award (NSF BNS-83-51331).

References

Chou, P. Y. & Fasman, G. D. (1978). *Adv. Enzymol.* **47**, 45-148.

Garnier, J., Osguthorpe, D. J. & Robson, B. (1978). *J. Mol. Biol.* **120**, 97-120.

Karplus, M. (1985). *Ann. N. Y. Acad. Sci.* **439**, 107-123.

Kabsch, W. & Sander, C. (1983a). *FEBS Lett.* **155**, 179-182.

Kabsch, W. & Sander, C. (1983b). *Biopolymers* **22**, 2577-2637.

Lim, V. I. (1974). *J. Mol. Biol.* **88**, 873-894.

Mathews, B. W. (1975). *Biochim. Biophys. Acta.* **405**, 442-451.

Nishikawa, K. (1983). *Biochim. Biophys. Acta.* **748**, 285-299.

Qian & Sejnowski (1988). *J. Mol. Biol.* (in press).

Robson, B. & Pain, R. H. (1971). *J. Mol. Biol.* **58**, 237-259.

Robson, B. & Suzuki, E. (1976). *J. Mol. Biol.* **107**, 327-356.

Rumelhart, D. E., Hinton, G. E. & Williams, R. J. (1986). in *Parallel Distributed Processing, Vol.* **1**, 318-362, MIT Press.

Staden, R. (1982). *Nucleic Acids Res.* **10**, 2951-2961.

Sejnowski, T. J. & Rosenberg, R. R. (1987). *Complex Systems* **1**, 145-168.

Taylor, W. R. (1986). *J. Mol. Biol.* **188**, 233-258.

268

Widrow, R. M. & Hoff, M. E. (1960). in *Institute of Radio Engineers, Western Electronic Show and Convention, Convention Record, Part 4*, 96-104.

Figure Legends

Fig. 1: Schematic diagram of network architecture. The standard network had 13 input groups, with 21 units per group, representing a stretch of 13 contiguous amino acids (Only 7 input groups and 7 units per groups are illustrated). Information from the input layer is transformed by an intermediate layer of "hidden" units to produce a pattern of activity in 3 output units, which represent the secondary structure prediction for the central amino acid.

Fig. 2: Learning curves for real proteins with testing on non-homologous proteins. Results for two networks are shown, one with no hidden units (direct connections between input and output units) and another with 40 hidden units. The percentage of correctly predicted secondary structure is plotted as a function of the number of amino acids presented during training.

Fig. 3: Hinton diagram showing the weights from the input units to three output units for a network with one layer of weights and a local coding scheme trained on real proteins. Three gray rectangular blocks show the weights to the three output units, each representing one of the three possible secondary structures associated with the center amino acid. A weight is represented by a white square if the weight is positive (excitatory) and a black square if it is negative (inhibitory), with the area of the square proportional to the value of the weight. The 20 amino acid types are arranged vertically and the position of each of them in the 13-residue input window is represented horizontally and is numbered relative to the center position. The weight in the upper left hand corner of each large rectangle represents the bias of the output unit. The contribution to each type of secondary structure by amino acids at each position is apparent in this diagram. For example, proline is a strong α-helix breaker while alanine, leucine and methionine are strong helix formers, especially when they are on the C-terminal side. Two basic amino acids, lysine and arginine, are helix formers when they are on the C-terminal side while glutamate, an acidic amino acid, supports helical structure when it is on the N-terminal side. Isoleucine, valine tryptophan and tyrosine are strong β-sheet formers and show no preferences toward C-terminal or N-terminal side.

Fig. 4: Dependence of the prediction accuracy on the size of the training set of non-homologous proteins. Percent correct for the training and testing sets is plotted.

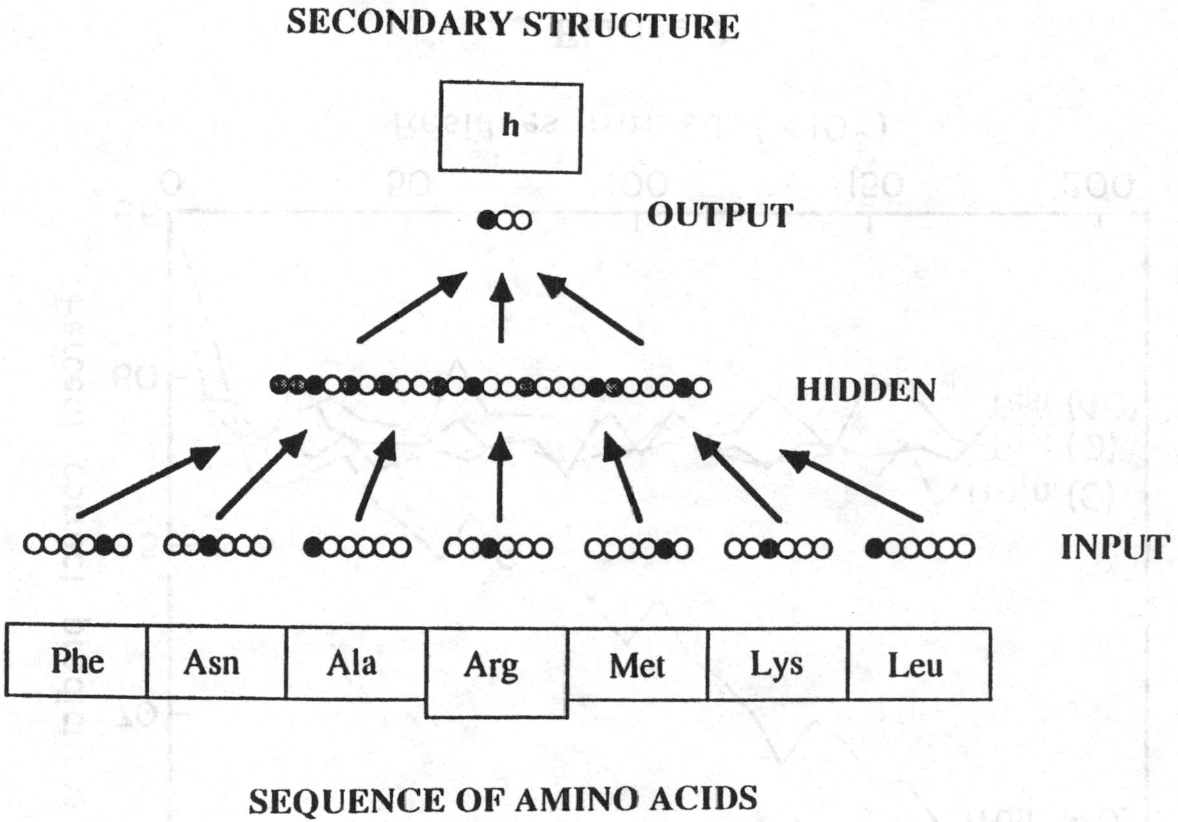

SECONDARY STRUCTURE

OUTPUT

HIDDEN

INPUT

SEQUENCE OF AMINO ACIDS

| Phe | Asn | Ala | Arg | Met | Lys | Leu |

Figure 1

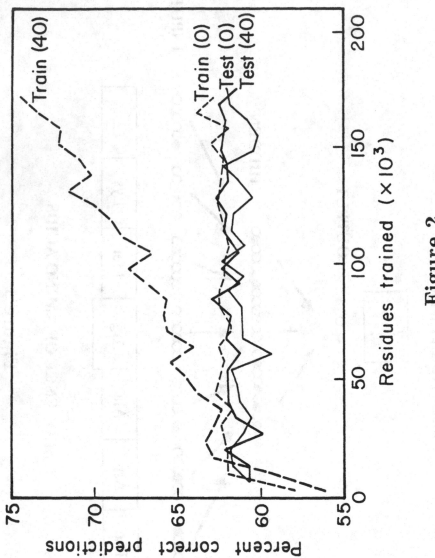

Figure 2

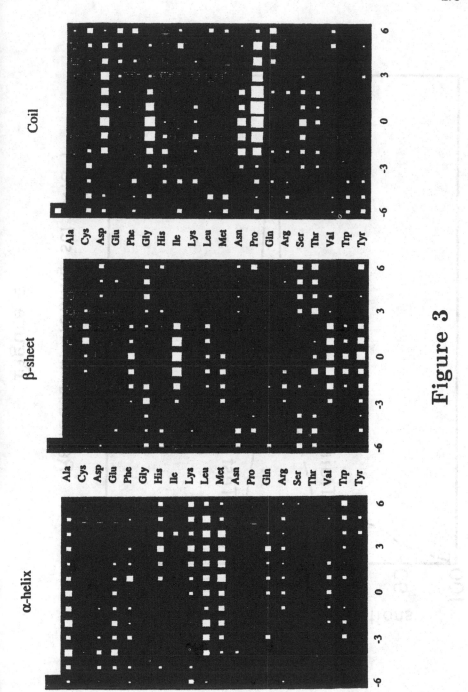

Figure 3

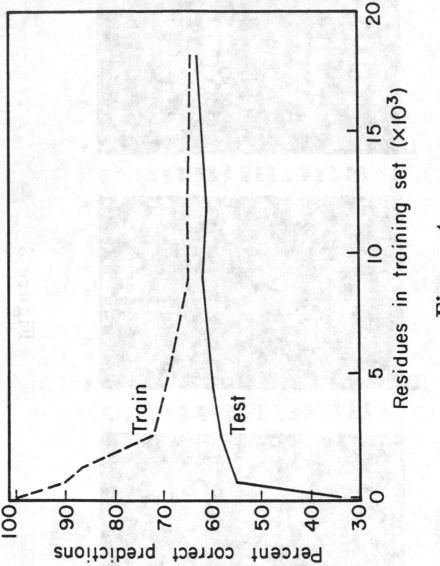

Figure 4

Table 1: Comparison of Methods

Method		$Q_3(\%)$	C_α	C_β	C_{coil}
Robson		53	0.31	0.24	0.24
Chou-Fasman		50	0.25	0.19	0.24
Lim		50	0.35	0.21	0.20
Neural	1 net	62.7	0.35	0.29	0.38
Network	2 nets	64.3	0.41	0.31	0.41

Table 1: Comparison with other methods for predicting secondary structure on a non-homologous testing set of proteins. Q_3 is the average success rate on three types of secondary structure and C_α, C_β, and C_{coil} are the corresponding correlation coefficients for the α-helix, β- sheet, and coil respectively. Results are shown for a single network (1 net) or a two network cascade (2 nets).

Exploiting Chaos to Predict the Future and Reduce Noise

J. Doyne Farmer and **John J. Sidorowich**[1]

Theoretical Division and Center for Nonlinear Studies
Los Alamos National Laboratory
Los Alamos, NM 87545.

Abstract

We discuss new approaches to forecasting, noise reduction, and the analysis of experimental data. The basic idea is to embed the data in a state space and then use straightforward numerical techniques to build a nonlinear dynamical model. We pick an *ad hoc* nonlinear representation, and fit it to the data. For higher dimensional problems we find that breaking the domain into neighborhoods using local approximation is usually better than using an arbitrary global representation. When random behavior is caused by low dimensional chaos our short term forecasts can be several orders of magnitude better than those of standard linear methods. We derive error estimates for the accuracy of approximation in terms of attractor dimension and Lyapunov exponents, the number of data points, and the extrapolation time. We demonstrate that for a given extrapolation time T iterating a short-term estimate is superior to computing an estimate for T directly.

Once we have a nonlinear dynamical model that accurately represents a data set, all the tools that were previously available only in computer experiments are extended to physical experiments. Our error estimates suggest that the use of higher order approximation techniques can give significant improvements in computing quantities such as fractal dimension or Lyapunov exponents. Furthermore, forecasting provides strong self-consistency requirements on the identification of chaotic dynamics.

We propose a nonlinear averaging scheme for separating noise from deterministic dynamics. For chaotic time series the noise reduction possible depends exponentially on the length of the time series, whereas for non-chaotic behavior it is proportional to the square root. When the equations of motion are known exactly, we can achieve noise reductions of more than ten orders of magnitude. When the equations are not known the limitation comes from prediction error, but for low dimensional systems noise reductions of several orders of magnitude are still possible.

The basic principles underlying our methods are similar to those of neural nets, but are more straightforward. For forecasting we get equivalent or better results with vastly less computer time. We suggest that these ideas can be applied to a much larger class of problems.

---

[1]Permanent address: Physics Department, UC Santa Cruz 95064

278

Contents

1 Introduction

The great promise of chaos lies in the hope that randomness might become predictable. Although chaotic dynamics puts limits on long term prediction, it implies predictability over the short term. Applications of modern nonlinear data analysis techniques indicate that chaotic dynamics is quite common, and that in many cases random behavior is due to low dimensional chaos rather than complicated dynamics involving many irreducible degrees of freedom. Until recently, however, there has been no way to exploit the presence of low dimensional chaos to actually make predictions.

In this paper we investigate straightforward but powerful approaches to this problem. We embed the data in a state space, and use simple numerical techniques to construct a nonlinear model for the dynamics. For low dimensional chaos such models can be quite accurate, allowing good forecasts and significant levels of noise reduction. They can also be used to reduce the data requirements to achieve a given level of accuracy in the computation of fractal dimension, Lyapunov exponents, or metric entropy. A good numerical model of the dynamics that can generate a data set effectively extends all the techniques that are available in a numerical experiment to physical experiments.

Most forecasting is currently done with linear methods. Linear dynamics cannot produce chaos, and linear models cannot produce good forecasts for chaotic time series. While nonlinear forecasting is an active field of investigation with a long history [33,76,64,75,63], as far as we know, the word "chaos" is not mentioned anywhere in the current forecasting literature. In this paper we re-examine forecasting problems in terms of the new paradigm that chaos offers, extending the results of a previous letter [23], where we demonstrated that local approximation can be used to make good forecasts for dynamical systems such as the Mackey-Glass differential delay equation, or chaotic convection in a He^3-He^4 mixture [43]. We have reproduced the convection results in Figure (1). For short times our forecasts are roughly 50 times as good as those of a standard linear forecasting model.

In this paper we extend our previous results to address topics such as higher order approximation, data analysis, and noise reduction, presenting derivations of some of the scaling properties we conjectured in our previous letter. A summary of some of these results will appear soon [27].

We apologize if some of our remarks are still speculative; given the current interest in these problems [35,16,18,12,49], we have decided to report work in progress along with completed work. We hope to complete a new version of this paper shortly, adding more extensive numerical results to address some of the unresolved points.

1.1 Chaos and randomness

Chaos [53,68,17,7] has caused a fundamental change in the way we think about randomness. This influence is felt strongly in physical models for random phenomena. A good example is the problem of "excess noise" in Josephson junctions. Popular models for this phenomenon [45] are now often formulated in terms of simple sys-

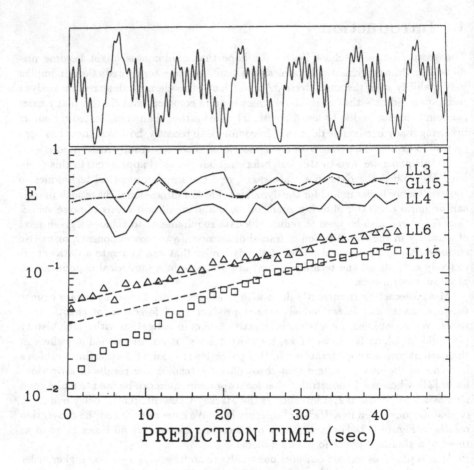

Figure 1: Top: An experimental time series obtained from Rayleigh-Benard convection in an He^3-He^4 mixture [43], with Rayleigh number $R/R_c = 12.24$ and dimension $D = 3.1$. Bottom: The normalized prediction error $\bar{E}$ (defined in Equation (15)) making forecasts with the Local Linear and Global Linear (linear autoregressive) methods. Numbers following the initial indicate the embedding dimension. The dashed lines are from Equation (34), using computed values of the metric entropy from reference [43]. Our predictions were based on $N = 30,000$ data points; with a sampling time $\Delta t = 0.07$ seconds, and a delay embedding time $\tau = 10\Delta t$. Based on the mean frequency in the power spectrum the characteristic time is roughly $t_c = 1.5$ seconds, so our database contains roughly 1400 characteristic times.

tems of deterministic differential equations, quite different from the statistical models that were the only option twenty years ago. Similarly, in this paper we show how thinking in terms of deterministic dynamical systems, and assuming that randomness arises out of chaos rather than complexity, leads to new approaches to forecasting and nonlinear modeling.

Until recently it was usually assumed that randomness was caused by extreme complication, *i.e.* the presence of many irreducible degrees of freedom. This naturally led to Kolmogorov's theory of random processes, which he defined in terms of the joint probability distribution $\mathcal{P}$ [46]. For a time series $\{x_t\}$, the dth order distribution is

$$\mathcal{P}(\xi_1,\ldots,\xi_d) = Probability\{x_1 < \xi_1,\ldots,x_d < \xi_d\}. \tag{1}$$

$\{x_t\}$ can represent events at discrete times, or samples of a continuous function. Random processes can also be discussed in terms of the probability density function P, defined as

$$\int_0^{\xi_1} \cdots \int_0^{\xi_d} P(x_1,\ldots,x_d)dx_1\ldots dx_d = \mathcal{P}(\xi_1,\ldots,\xi_d). \tag{2}$$

The process is *deterministic* if there is some value of d such that the probability density approaches a delta function in the limit of perfect measurements of $\{x_i\}$.

Many people speak of random processes as though they were a fundamental *source* of randomness. This is misleading. The theory of random processes is an empirical technique for coping with inadequate information, and makes no statements about *causes* of randomness. As far as we know, the only truly fundamental source of randomness is the uncertainty principle of quantum mechanics; everything else is deterministic, at least in principle. Nonetheless, we call many phenomena such as fluid turbulence or economics random, even though they have no obvious connection to quantum mechanics. It has traditionally been assumed that the apparent randomness of these phenomena derives solely from their complication.

We will take the practical viewpoint that randomness occurs to the extent that something cannot be predicted, which usually depends on the available information. With more data or more accurate observations, a phenomenon that previously seemed random might become more predictable, and hence less random. Randomness is in the eye of the beholder.[2] Furthermore, randomness is a matter of degree – some systems are more predictable than others.

As originally pointed out by Poincaré [60], many of the classic examples of randomness are not complicated. The dynamics of a flipping coin or a roulette ball, for example, involve only a few degrees of freedom. Their randomness comes from *sensitive dependence on initial conditions* – a small perturbation causes a much larger effect at a later time, making prediction difficult. When sensitive dependence on

---

[2]Another perhaps more fundamental notion of randomness is due to Kolmogorov and Chaitin [14]. Whether or not chaotic systems are random in this sense is controversial [78,28].

initial conditions occurs in a sustained way it is called *chaos*[3]. Since chaos is defined in the context of deterministic dynamics, in some very strict sense it might be incorrect to say that chaos is random – ultimately uncertainty originates from something external to the dynamics, such as measurement error or external "noise". But sensitive dependence exaggerates uncertainty, so that small uncertainties turn into large ones. Since chaos amplifies noise exponentially any uncertainty at all is amplified to macroscopic proportions in finite time, and short-term determinism becomes long-term randomness.

Chaos *creates* randomness by strongly amplifying what we don't know. Even with only a few degrees of freedom and a very small source of uncertainty, points in a chaotic time series that are far apart in time do not appear deterministic, unless the source of uncertainty is reduced to unreasonable proportions. Chaotic systems pass many classic "tests" of randomness; for example, some simple chaotic maps produce uncorrelated time series, with $\langle x_t x_{t+j} \rangle = 0$ unless $j = 0$. Furthermore, chaotic trajectories *look* random.

In contrast, if the dynamics are not chaotic errors grow slowly and the main requirement for determinism is that d be large enough. As long as this condition is satisfied forecasts can be made far into the future.

Dissipative dynamical systems often have the property that undisturbed trajectories approach a subset of the state space, called an *attractor*. This can cause a drastic reduction in the number of degrees of freedom. Fluid flows, for example, have an effectively infinite dimensional state space, but can have low dimensional chaotic attractors [56,9,43].

Thus, we should not distinguish chaos and randomness, but rather we should distinguish systems with low dimensional attractors from those with high dimensional attractors. If a time series is produced by motion on a very high dimensional attractor, then from a practical point of view it is impossible to gather enough information to exploit the underlying determinism. If we model the dynamics in a state space whose dimension is lower than that of the attractor, we only see a projection of the dynamics, and determinism is invisible – the dynamics look random. Even if the dimension of the model is large enough, the amount of data needed to make a good model for a high dimensional attractor may be prohibitive. This problem gets exponentially worse as the dimension increases [31,39], as is apparent in the error estimates presented in Section 3.

With many degrees of freedom, the statistical approach is probably as good as any - linear models may even be optimal. But if random behavior comes from low dimensional chaos, we can make forecasts that are much better than those of linear models. Furthermore, the resulting models can give useful diagnostic information about the nature of the underlying dynamics, aiding the search for a description in terms of first principles.

---

[3]A trajectory is *chaotic* if it has positive Lyapunov exponents, *i.e.*, if on average it is locally unstable. The motion of a roulette ball is sensitive to initial conditions, but strictly speaking not chaotic; since it comes to rest, it does not have any positive Lyapunov exponents. Chaos is a special case of sensitive dependence to initial conditions, where there is also sustained motion.

2 Model Building

2.1 State space reconstruction

Consider a time series $\{v(t_i)\}, i = 0, 1, \ldots, N$. Assume that $\{v(t_i)\}$ is stationary. This is automatic if it comes from an attractor.[4] We will assume for the moment that v is a scalar, although the extension to the case that it is a vector is straightforward. Typically $\{v(t_i)\}$ is a projection of dynamics in a higher dimensional state space, so that in order to make use of any determinism in $\{v(t_i)\}$ we must *reconstruct* a state space.

A typical example occurs in fluid flow experiments, which in principle can be modeled accurately by deterministic partial differential equations, However, in practice this may not be useful unless the data is in the correct form. For example, suppose a single probe measures a given component of the velocity at a fixed point in space. This data is simply inadequate to provide initial conditions for the Navier-Stokes equations. To build a model from the data at hand we are forced to reconstruct a state space from a single time series.

A method for doing this[5] was introduced by Packard *et al.* [59] and put on a firm mathematical foundation by Takens [71]. Suppose we create a state vector $x(t)$ by assigning coordinates

$$
\begin{aligned}
x_1(t) &= v(t), \\
x_2(t) &= v(t - \tau), \\
&\vdots \\
x_d(t) &= v(t - (d-1)\tau),
\end{aligned}
\tag{3}
$$

where τ is a delay time. If the dynamics takes place on an attractor of dimension D, then a necessary condition for determinism is $d \geq D$. If r is the dimension of a manifold containing the attractor, Takens showed that $d = 2r + 1$ is sufficient, at least in principle.

In principle, τ is arbitrary as long as it is not rationally related to $x(t)$. In practice, if τ is too small the coordinates become singular, so that $x_j \approx x_{j+1}$. If τ is too big, chaos makes x_1 and x_d causally disconnected. Taken together, these two considerations imply an effective upper bound on the embedding dimension d. In practice d is often chosen by trial and error, starting with a low value and increasing it, searching for optimal results. A more systematic procedure based on mutual information has been explored by Fraser and Swinney [30].

The use of delay coordinates to reconstruct a state space is not original to dynamical systems theory. It goes at least as far back as Yule, who in 1927 made a model for sunspot activity based on a linear combination of past values [79]. This

---

[4]We will sometimes uses subscripts to indicate coordinates, but at other times we will use them to indicate time. We hope that the context makes this clear.

[5]This was also suggested by David Ruelle

idea is also implicit in Kolmogorov's definition of a random process. The important contribution from dynamical systems theory is the demonstration that reconstruction preserves geometrical *invariants* of the dynamics, such as attractor dimension, metric entropy, and the positive Lyapunov exponents.

Delays are not the only way to embed data in a high dimensional space. Another example is derivatives, $x_1(t) = x(t), x_2(t) = x'(t), \ldots, x_d = x^{(d-1)}(t)$, which for clean data usually produce nicer embeddings than delays. But since differentiation amplifies noise, high dimensional embeddings are impractical [59,31].

Broomhead and King [11] have suggested an alternative approach. They apply the Karhunen and Loeve principal value decomposition to the delay coordinate representation, and produce embeddings that seem to have the nice properties of derivatives, but without the numerical problems. The simplest way to implement their procedure is to compute the covariance matrix $\langle x_i(t)x_j(t)\rangle_t$ and compute its eigenvectors and eigenvalues α_i. The eigenvalues α_i are the average root-mean-square projection of the d-dimensional delay coordinate time series onto the eigenvectors. Ordering them according to size, the first eigenvector has the maximum possible projection, the second has the largest possible projection for any fixed vector orthogonal to the first, and so on. The numerical calculations of Broomhead and King demonstrate that under good circumstances α_i falls off exponentially with i, until it reaches a floor determined by the noise level. The fall off steepens as the sampling time Δt decreases.

A nice feature of the Broomhead and King procedure is that a new global embedding dimension $\tilde{d}$ is computed automatically, simply by counting the number of eigenvalues above the noise floor. As long as the original embedding dimension d is sufficiently large and τ is sufficiently small, $\tilde{d}$ only depends on the *lag window* $\tau_L = \tau(d-1)$. This eliminates much of the trial and error procedure that is usually necessary to determine the dimension of the state space. If τ_L is too small, the embedding makes incomplete use of the available information, but if τ_L is too large, the determinism is overwhelmed by the amplification of noise. Our suggestion is to weight the delay coordinates before performing the principal value decomposition, according to

$$
\begin{aligned}
x_1(t) &= v(t), \\
x_2(t) &= e^{-h\tau}v(t-\tau), \\
&\vdots \\
x_d(t) &= e^{-h(d-1)\tau}v(t-(d-1)\tau),
\end{aligned}
\tag{4}
$$

where h is the metric entropy. h can be calculated by a variety of different algorithms [57]. The resulting coordinates are linearly optimal from an information theoretic point of view. This technique can also be extended for assessing the relevance of spatial samples or multivariate time series by maximizing information, as will be discussed in more detail later [25]. As pointed out by Fraser [29], while the Broomhead and King procedure has some nice features, maximizing the mutual information is a much better criterion for a good embedding than diagonalizing the covariance matrix.

Up until now we have assumed that the state $x(t)$ is constructed directly from the time series. This is is usually referred to as autoregressive (AR) modeling. For a deterministic dynamical system it is the most natural approach. An alternative is to make predictions in terms of the residuals, $\delta(t) = x(t) - \hat{x}(t)$, where $\hat{x}(t)$ is a prediction of $x(t)$. This approach, originally due to Slutsky [70], is called the moving average (MA) model. An alternative, called the ARMA model [63], extends the state space to include both. AR, MA, and ARMA models are formally equivalent, but they are not necessarily equivalent in practice. The ARMA approach may offer some advantages, even for modeling fully deterministic dynamics. We intend to investigate this further.

2.2 Learning nonlinear transformations

Once we have found a state space representation, the next task is to fit a model to the data. There are several approaches. The simplest is to make time discrete, and assume that the dynamics can be written as a map in the form

$$x(t + T) = f_T(x(t)) \tag{5}$$

where the current state is $x(t)$, and $x(t + T)$ is a future state. f and x are both d-dimensional vectors. The problem is to estimate $x(t + T)$. We will call this estimate $\hat{x}(t, T)$, and approximate the dynamics by a map $\hat{f}$ of the form

$$\hat{x}(t, T) = \hat{f}_T(x(t)). \tag{6}$$

We often iterate this equation, writing

$$\hat{x}(t, T) = \hat{f}_T(\hat{x}(t - T, T)). \tag{7}$$

An alternative approach for continuous time systems is to approximate the differential,

$$\frac{d\hat{x}}{dt} = \hat{f}(\hat{x}(t)) \tag{8}$$

and integrate. In this paper we mainly discuss the discrete-time approach, since it is faster, and we have studied it more. Differential models may offer improvements in accuracy, though, and we intend to investigate this in a future paper.

In general inaccurate measurements, external disturbances, or round off errors introduce high dimensional uncertainties that are more conveniently viewed as non-deterministic. At any given time the purely deterministic state $y(t)$ may be perturbed by $n(t)$, so that the observed state is

$$x(t) = y(t) + n(t).$$

We call $n(t)$ *noise*, and generally assume that it is small, and do our best to ignore it. For observational noise $y(t)$ is an iterate of a deterministic equation, $y(t) = f^t(y_o)$, but for external disturbances this is not possible unless the trajectory is shadowed by a purely deterministic trajectory [1,8,42]. As we demonstrate in Section 5, with a good model noise can be reduced considerably.

2.2.1 Representations

Chaotic dynamics does not occur unless f is nonlinear, so to approximate chaotic dynamics we must make nonlinear models. There are an infinite number of ways to represent nonlinear functions, and finding good nonlinear approximations is a difficult problem. As soon as we pick a representation, we make an *ad hoc* assumption, which may or may not be a good one. With prior information we may be guided to a better representation, but in the absence of any theoretical understanding we are forced to make arbitrary choices.

At this point finding a good representation is largely a matter of trial and error. We make a guess and then test to discover whether it produces a good model. Convenience plays an important role, since fitting parameters is usually time consuming unless it can be reduced to a linear problem.[6] Certain representations perform better than others across a wide class of problems, but although there are a few rigorous results [51], this is mainly a matter of numerical lore.

We now discuss a few of the representations that we are currently exploring.

Polynomials are a good representation because their parameters can be linearly fit to minimize least squares deviations, and because they arise naturally in Taylor expansions. The use of polynomials for forecasting was suggested by Wiener [76], who proposed them for moving average models, and Gabor [33,34], who proposed them for autoregressive models. Polynomials have also recently been used for fitting deterministic equations of motion by several authors [23,16,18]. The most general form for an mth degree d-dimensional polynomial is

$$A_m(x_1, \ldots, \ldots x_d) = \sum_{i_1=0, \ldots, i_d=0} a_{i_1 \ldots i_d} x_1^{i_1} \cdots x_d^{i_d}. \tag{9}$$

where $\sum_{j=1}^d i_j \leq m$. The number of parameters $a_{i_1 \ldots i_d}$ is $\frac{(d+m)!}{d!m!} \approx d^m$. Fitting this many parameters rapidly becomes impractical when m and d are large. Polynomials have the disadvantage that they do not extrapolate well beyond their domain of validity since $\|A\| \to \infty$ as $\|x\| \to \infty$.

Rational approximation by the ratio of two polynomials provides an appealing alternative [62]. Rational approximations extrapolate better than polynomials, particularly when the numerator and denominator are of the same degree, since they remain bounded as $\|x\| \to \infty$. Furthermore, like polynomials, fitting parameters by least squares is linear. The work of Bayly *et al.* [3] suggests that rational approximation does a good job of fitting some chaotic attractors, and preliminary investigations of Lee and Lee [50] suggest that they may also be a good choice for forecasting problems.

---

[6]Even when the least squares problem is linear, the solution is not well behaved if the data are singular. This can be improved through singular value decomposition [62], which is the method we use throughout this paper.

Radial basis functions [61] have been recently suggested by Casdagli [12] for nonlinear modeling. They are a global interpolation scheme with good localization properties. In their simplest form, they only depend on the distance between points.[7]

$$R(x) = \sum_{t'=1}^{N} \lambda_{t'} \phi(\|x - x_{t'}\|) \tag{10}$$

where $\| \ \|$ is the Euclidean norm, and t' is a label attached to the points in the time series. The coefficients λ_j are chosen to satisfy the interpolation conditions $x(t'+T) = R(x(t'))$. If ϕ is taken to be $\phi(r) = (r^2 + c^2)^{-\beta}$, where $\beta > -1$ and $\beta \neq 0$, (the form we use in this paper) the linear system given by Equation (10) has a unique solution as long as the $x(t')$ are distinct. A special case of radial basis functions are thin plate splines.

Neural networks provide another alternative. Although on the surface neural networks seems quite different, on closer examination it becomes clear that this difference is superficial. Neural networks can be viewed as just another class of functional representations. This was explicitly demonstrated by Lapedes and Farber [49], who recently applied a standard feed-forward neural net with two hidden layers [65] to some of the forecasting problems that we studied in reference [23]. Their neural net can be written as

$$\hat{x}(t, T) = \sum_k w_k z_k - a_o \tag{11}$$

$$z_k = \tanh\left(\sum_j w_j y_j - a_k\right), \tag{12}$$

$$y_j = \tanh\left(\sum_i w_i x_i - a_j\right), \tag{13}$$

$$x_i = x(t - i\tau), \tag{14}$$

where y_j and z_k are the values of the neurons in the two hidden layers, and x_i are the input neurons. Thus, the neurons are simply coordinates in the state space. The label "neural network" implies a particular functional representation, involving a sum over coordinates, followed by a sigmoidal function such as tanh. While this form is motivated by biological considerations, for problems in artificial intelligence or forecasting there is no reason to adhere to it unless it is better than others.

Lapedes and Farber determine the parameters a and w using back-propagation, which is essentially a nonlinear least squares algorithm. The tanh representation has several nice properties [49], but it has the disadvantage that parameters cannot be fit by solving a linear problem, and the the back propagation algorithm is time consuming. As a result, fitting parameters takes several orders of magnitude more computer

---

[7]Often a polynomial is added as well, but we do not use this.

time.

2.2.2 Local approximation

To make a good approximation for a function f, a representation must be able to conform to its variations. If f is well-behaved, any complete representation will provide a good approximation, as long as it has enough free parameters, and enough data to stably solve them. However, if f is complicated, there is no guarantee that any given representation will approximate f efficiently.

The dependence on representation can be reduced by *local approximation*. The basic idea is to break up the domain of f into local neighborhoods and fit different parameters in each neighborhood. When f is smooth the neighborhoods can be small enough so that f does not vary sharply in any of them, making the constraints of a particular representation less important.

Local approximation usually produces better fits for a given number of data points than global approximation, particularly for large data sets. It seems that most global representations reach a point of diminishing returns where adding more parameters or more data only gives a marginal improvement in accuracy. Higher order terms may oscillate, and actually cause the behavior to get worse. Past a certain point, adding more local neighborhoods is usually more efficient than adding more parameters and going to higher order. Local approximation makes it possible to use a given functional representation efficiently. The key is to choose the local neighborhood size properly, so that each neighborhood has just enough points to make the local parameter fits stable, so that adding more points would not make significant improvements.

We will define an M point *neighborhood* of x as a collection of M points $\{y_i\}$ which are in some sense close to x. Although this is perhaps a slightly unusual way to use the word "neighborhood", it is convenient in the discussion that follows. Note that although we use the language of dynamics in the following discussion, most of our remarks apply to any situation in which we wish to make a map from one set of values to another.

Although local approximation is more trouble to implement than global approximation, it is often well worth the effort, particularly for large data sets. To use local approximation in the vicinity of a point x, there are three basic steps:

1. Pick a local representation.

2. Assign neighborhoods.

3. Find a local *chart* that maps the points in each neighborhood into their future values. To make a prediction we evaluate the chart at x.

The basic idea is illustrated in Figure (2).

A simple way to assign neighborhoods is to partition the domain into disjoint sets. For example, we could use a rectangular grid. This approach is convenient, but it has

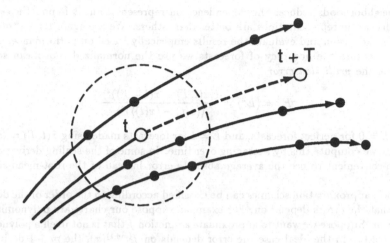

Figure 2: *Local approximation.* The current state $x(t)$ and its unknown future value $x(t + T)$ are represented by open circles. The black dots inside the dashed circle are the neighbors of $x(t)$. To make a prediction we fit a local chart with the neighbors in its domain, and the states they evolve into a time T later in its range. To make a prediction we evaluate the chart at $x(t)$.

the disadvantage that there is no overlap between the neighborhoods, and therefore no continuity between charts. A point near the boundary of its neighborhood may be poorly approximated. This is particularly true for representations such as polynomials that do a poor job of extrapolating outside their domain of validity.

One way to cope with this problem is to enforce matching conditions between adjacent neighborhoods. For many interpolation schemes, for example, this is an essential element in achieving accuracy. Unfortunately, this becomes a difficult problem for data in more than two dimensions, which is precisely the situation that we are most interested in here.

An alternative that is more accurate than disjoint partitions and more convenient than enforcing matching conditions is to overlap the neighborhoods, so that each chart is constructed from a good set of neighbors. Let $\{y_i\}$ be the n points of the neighborhood. We want to choose $\{y_i\}$ so that the predictions are as good as possible. A simple criterion is nearness: For a given metric $\| \ \|$ and a given n we will say that $\{y_i\}$ is the *nearest neighborhood* of x if it minimizes $\sum_i \|x - y_i\|$. This criterion is not optimal, as can be seen by considering linear interpolation in two dimensions; if the triangle defined by the three nearest neighbors does not enclose x, then the interpolation may be poor. In practice we find that choosing good neighborhoods makes a big difference in the quality of our predictions.

Once we have chosen neighborhoods, the next step is to fit charts to them. To do this we must pick a representation. While we anticipate that dividing the domain into neighborhoods reduces the dependence on representation, it is nonetheless the case that some representations are better than others. We are again forced to make an *ad hoc* decision and evaluate the results empirically based on performance.

To measure the accuracy of forecasts we use the normalized root-mean square error, or the *prediction error*

$$\bar{E}^2 = \langle E^2 \rangle = \frac{\langle (\hat{x}(t,T) - x(t+T))^2 \rangle}{\langle (x(t) - \langle x(t) \rangle)^2 \rangle} \tag{15}$$

Thus $\bar{E} = 0$ for perfect forecasts, and $\bar{E} = 1$ for forecasts made using $\hat{x}(t,T) = \langle x(t) \rangle$. Usually we compute this by averaging over time. In some of the scaling derivations it is also convenient to use the average absolute error instead of the root-mean-square error.

Local approximation schemes can be classified according to the order of the derivatives that the errors depend on. For example, suppose our charts are polynomials of degree m. Suppose we want to approximate a function f that is not itself a polynomial of degree m. In the ideal case the error depends on $f^{(m+1)}(x)$, the $m+1$ derivative of f. This implies that the errors are proportional to ϵ^{m+1}, where ϵ is the spacing between data points. The average spacing between N points uniformly distributed over a D dimensional space is $\epsilon \approx N^{-\frac{1}{D}}$. Calling q the *order of approximation*, we find[8]

$$\bar{E} \sim N^{-\frac{q}{D}}, \tag{16}$$

where in this case $q = m + 1$.

Achieving the ideal case where $q = m + 1$ is difficult for large q, since in general fitting a polynomial of degree m does not produce a fit that is accurate to order $m+1$. For example, suppose the number of data points is equal to the number of free parameters. The approximation may go through each point precisely, but in between it may oscillate wildly, producing an extremely inaccurate approximation. Even when we use more neighbors this may limit the accuracy.

To avoid this confusion we will use Equation (16) to *define* the order of local approximation, taking the limit as $N \to \infty$, and letting D be the information dimension of the underlying measure of the data points.

$$q = \lim_{N \to \infty} \frac{D |\log \bar{E}|}{\log N} \tag{17}$$

In general q may depend on D, f, the way in which we choose the neighborhoods, and other factors.[9]

Moving to representations of higher degree involves a tradeoff – higher degree representations potentially promise more accuracy, but also require larger neighborhoods. A larger neighborhood usually implies that the complexity of f increases.

---

[8]We will use the symbol "$\sim$" to mean "scales as", i.e. $z \sim y(x)$ implies $z = Cy(x)$, where C includes all dependencies on variables other than x.

[9]Note that the order of approximation as we have defined it here is one larger than the definition we gave in references [23,27]. We have changed our definition to correspond to common usage.

Finding the best compromise between these two effects is a central problem in local approximation.

A trivial example of local approximation is *first order*, or *nearest neighbor* approximation. This amounts to simply looking through the data set for the nearest neighbor, and predicting that the current state will do what the neighbor did a time T later. We approximate $x(t + T)$ by $\hat{x}(t, T) = x(t' + T)$, where $x(t')$ is the nearest neighbor of $x(t)$. For example, to predict tomorrow's weather we would search the historical record and find the weather pattern most similar to that of today, and predict that tomorrow's weather pattern will be the same as the neighboring pattern one day later.[10] First order approximation can sometimes be improved by finding more neighbors and averaging their predictions, for example, by weighting according to distance from the current state.

An approach that is usually superior is *local linear* or *second order* approximation. For the neighborhood $\{x(t')\}$ we simply fit a linear polynomial to the pairs $(x(t'), x(t' + T))$. When the number of nearest neighbors $M = d + 1$ and the simplex formed by the neighbors encloses $x(t)$ this is equivalent to linear interpolation. If the data is noisy, the chart may be more stable when the number of neighbors is greater than the minimum value. Again, this procedure can be improved somewhat by weighting the contributions of the neighboring points according to their distance from the current state. Linear approximation has the nice property that the number of free parameters and consequently the neighborhood size grows slowly with the embedding dimension.

Since the accuracy increases with the order of approximation, it is obviously desirable to make the order of approximation as large as possible. Any nonlinear representation is a candidate for higher order approximation. The criteria for a good local representation are somewhat different from those for a good global representation. On one hand, getting a good fit within a local neighborhood is easier, because the variations are less extreme. Wild variations or discontinuities can be accommodated by assigning neighborhoods properly. On the other hand, a local neighborhood necessarily has less data in it, so the representation must make efficient use of data. The order of approximation actually achieved may depend on other factors, such as the choice of neighborhoods, the dimension, or peculiarities of the data set. For reasonably low dimensional problems we have found that we usually achieve third order approximation, for example using quadratic polynomials. In low dimensions (two or less) it is sometimes possible to do better. [11] In higher dimensions we often find that we cannot do any better than second order approximation with our current techniques.

Another interesting alternative is to compute charts for each data point, and then view the parameters of these charts as new states. By fitting charts to them we can

---

[10]This was attempted by E.N. Lorenz, who examined roughly 4000 weather maps [52]. The results were not very successful because it was difficult to find good nearest neighbors, apparently because of the high dimensionality of weather.

[11]Casdagli has independently reached the same conclusions [12]. In two dimensions he apparently achieves *sixth* order approximation in some cases using global radial basis functions, but this does not seem to carry over in more dimensions.

make *metacharts*. For example, we could use local linear approximation both for fitting charts to past states and for interpolating between them to find a good chart for the current state. This process can obviously be continued *ad infinitum*, at least in principle. Metacharts may be smoother than charts, resulting in a more compact or more accurate description. We intend to investigate this further.[12]

Some people find local approximation objectionable because it does not result in a "closed form" model. We do not consider this a problem. The most informative diagnostic information comes from knowing statistical properties such as the Lyapunov exponents or fractal dimension. While the coefficients of a global polynomial expansion might give some extra insight, if this results in a significantly less accurate model, we may get estimates of statistical properties that are incorrect in even qualitative terms. We feel that accuracy should be the most important criterion for model selection.

2.2.3 Trajectory segmenting

For continuous time series the strategy for finding a good neighborhood is not quite the same as it is for discrete maps. In particular, the sampling time Δt becomes an important parameter, and if it is small this must be taken into account. Local neighborhoods chop a continuous trajectory into segments. We say that the sampled values of a time series lie on the same *trajectory segment* if they can be sorted so that they are contiguous in time. For example, the local neighborhood in Figure (2) contains four trajectory segments. If the sampling rate Δt is small, finding a fixed number of neighboring data points may be very different than finding a given number of trajectory segments. For example, if Δt is really small, all the neighbors of a given point will lie on the same trajectory segment. As a result they are nearly co-linear, which results in a highly singular chart. We can attempt to compensate by increasing the number of nearest neighbors, but due to nonuniformities some regions may behave differently than others. A more reliable approach is to choose neighborhoods so that they include a given number of trajectory segments, rather than a given number of points. This gives much better control over the quality of the fits.

2.2.4 Nonstationarity

Most of the results in this paper assume that we are dealing with stationary data. The assumption that the trajectory is on an attractor guarantees this, as long as the parameters are held constant. Variations of parameters, however, can result in nonstationary behavior. The most straightforward way to deal with this is to extend the state space to include time. It can be included in the metric, so that the search for nearest neighbors favors recent data. With time as a coordinate, charts can be constructed that take into account trends and other time dependent effects. Similarly,

---

[12]Y.C. Lee, who has independently suggested this approach, makes the intriguing observation that the local approximants can be used to "re-embed" the data. If f has low complexity, this might reduce the size of the data base.

a periodic function of time can be included to cope with seasonality or other periodic behavior.

2.2.5 Discontinuities

The problems caused by discontinuities can be minimized by choosing neighborhoods properly, so that their boundaries follow and do not cross discontinuities. The worst situation occurs when the points in a neighborhood are on different sides of a discontinuity – a smooth chart will inevitably produce a poor approximation.

In order to detect the presence of a discontinuity it is necessary to examine the range as well as the domain of the transformation. Two points on opposite sides of a discontinuity may be nearby in the domain. However, by extending the metric to include the range as well as the domain, singularities become evident, since points on opposite sides of a discontinuity are far apart. By definition the future value of the current state is unknown, so we cannot use it for neighborhood relations that involve both past and future values, but this is not true for the rest of the database. By extending the metric to include the range values, we can make sure that neighborhoods are chosen so that they do not cross the discontinuity. This guarantees that each chart is fit with a consistent set of points. As long as we place the current state in the correct neighborhood, on the proper side of any discontinuities, the predictions will be good. At least we win some of the time, rather than losing all the time. This procedure also makes it possible to deal with multiple valued behavior, since points on the same branch can be distinguished from those on different branches.

2.2.6 Implementing local approximation on computers

Finding neighbors in a multidimensional data set is time consuming when done by brute force. The most straightforward approach is to compute the distance to each point, which takes roughly N steps for N points. This can be reduced to roughly $\log N$ steps by organizing the data with a decision tree, such as the $k\text{-}d$ $tree$ [6,5]. The basic principle is illustrated in Figure (3).

The data set is partitioned one coordinate at a time. Any criterion can be used for the partitions; for example, for the results reported here we find the coordinate with the largest range, and partition it at its median value. The values corresponding to each partition are stored in the tree as $keys$. A given query point can be located quickly by comparing its coordinates to the keys. If we want to find the nearest neighbors this makes it possible to eliminate many points from consideration without actually computing their distance. The k-d tree has the nice property that it flexibly partitions only the parts of the space that actually contain data, adding partitions only where they are needed. Providing the k-d tree is used to find nearest neighbors, the principal speed limitation for local approximation comes from fitting the parameters of the charts.

The best approach for a given application depends on whether it is constrained by speed, computer memory, or the availability of data. At one extreme, the most

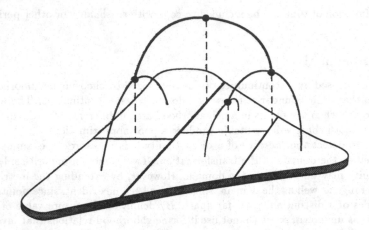

Figure 3: *The k-d tree.* The data is partitioned one coordinate at a time. Each partition is assigned a node in the tree, as indicated by the dashed lines. The partitions can be made on any coordinate, and the tree can have different depth in different regions. The result is a flexible partition of the data space, with finer partitions where there is more data. Use of the k-d tree speeds up the search for nearest neighbors, decreasing the number of steps from N to $\log N$.

accurate predictions come from finding optimal neighborhoods. This is much easier to do once we know the state, which means that we must find neighbors and fit charts as we go. At the other extreme, the fastest approach is to use fixed disjoint neighborhoods, compute charts in advance, and store them in the tree. This approach is extremely fast, requiring only the order of $\log N$ steps to locate the proper chart, plus the time needed to evaluate the chart. Even though it requires more computer memory, for real-time applications this may be the method of choice. Our numerical experiments indicate, however, that the sacrifice in accuracy can be considerable, especially in higher dimensions. Even in one dimension the difference between these two approaches for quadratic polynomial approximation can be almost an order of magnitude, as shown in Figure (4). A compromise between these two extremes is to compute charts for all the points in the data base, find the nearest neighbor of the current state, and borrow its chart to make a forecast. This procedure is fast, but consumes even more computer memory than fixed disjoint partitions.

If an application is limited by data, it may be important to update the tree after each prediction, to make use of each new data point in the next prediction. At the opposite extreme, if an unlimited supply of data is available and computer memory is at a premium, it may be desirable to find an optimal data set by selectively inserting new points where the approximation is bad and removing old points where it is already good. Regions where f changes rapidly should be covered densely, while regions where

f is relatively smooth only need to be covered sparsely. A simple criterion is to insert states that result in bad predictions, and compensate by removing states that result in good predictions. Aspects of this are discussed by Omohundro [58].

In our numerical work here, for convenience we simply use the first part of the data set to build a data base, and use the second part to make predictions. Of course, for a nonstationary process, it would be important to continually update the data base instead.

Although local approximation is certainly more trouble than global approximation, with the k-d tree it is quite fast. The k-d tree is easy to implement, as long as one avoids caveman computer languages such as FORTRAN. Local approximation does not require massive amounts of computer power. The computations for this paper were made with a SUN-3 microcomputer.

2.2.7 An historical note

At the time we wrote the paper of reference [23] we were unaware of other work using local approximation for forecasting. Now we are more informed, and a few historical comments are in order. The use of nearest neighbor (first order) approximation goes at least as far back as the work of Lorenz [52] in 1969. Linear approximation with fixed disjoint partitions was introduced by Tong and Lim [75] in 1980. Priestley gave a more general approach in the same year, and also suggested the possibility that higher order approximation might be useful. In the dynamical systems literature in 1981 local linear approximation was independently proposed as a means of computing dimension by Froehling et al. [31]. It was also proposed by Eckmann and Ruelle [21,20] and Sano and Sawada [66] to compute Lyapunov exponents, except that they omit the constant term.

More recently, several authors have independently suggested various forms of local approximation [35,18,23,12]. Linear interpolation using nearest neighbors was suggested by Peter Grassberger [35], who implemented it for the Henón map. Crutchfield and MacNamara [18] also suggested local linear approximation with disjoint neighborhoods, which they refer to as an "atlas dynamical system". In reference [23] we suggested local approximation using nearest neighbors, demonstrated its effectiveness on several experimental and numerical time series, and proposed the scaling laws that make the advantages of higher order approximation clear. Stimulated by our work, Lapedes and Farber [49] showed that neural nets could also be used for forecasting. Casdagli suggested the use of radial basis functions [12].

The use of global polynomials goes back to the work of Gabor [33] and Wiener [76]. It was also recently suggested independently by Cremers and Hübler [16], Crutchfield and McNamara [18], us [23], and Lapedes and Farber [49].

2.3 Comparison to statistically motivated methods

In this section we compare the methods discussed here, which are motivated by deterministic dynamics, to previous nonlinear forecasting methods that are based on statistical assumptions, in particular the threshold autoregressive model of Tong and Lim [75] and the local linear model of Priestley [64,41,63].

The threshold autoregressive model is formally equivalent to local linear approximation with fixed disjoint neighborhoods. A "threshold" corresponds to a partition on a given coordinate. The most important lesson learned from thinking in deterministic terms is the scale of implementation needed to get good results. For example, Tong and Lim discuss a threshold on one of the variables, splitting the state space in half, and using a different linear map for each half. While this is certainly a major improvement over a single linear map, and introduces enough nonlinearity to reproduce phenomena such as limit cycles and chaos, it is clearly inadequate to approximate a general nonlinear transformation with any accuracy. When we use fixed disjoint neighborhoods we typically partition the state space into hundreds or thousands of parts, putting thresholds on all the state space variables, in order to get good results. Also, as seen in Figure (4), using overlapping neighborhoods makes a big improvement.

Priestley's local linear model is a generalization of the threshold autoregressive model, with a different procedure for determining the charts appropriate for a given state. Instead of imposing a metric as we do, or using fixed disjoint neighborhoods as Tong and Lim do, Priestley uses an algorithm that is similar to the Kalman filter to track the free parameters in time. They vary in time as a "random walk", changing according to a linear dynamical equation. His approach has several attractive features, but the use of a linear equation for the parameters is unduly restrictive for modeling general nonlinear transformations. Also, the assumption of continuity in time means that this approach will not perform well for discrete time maps. Priestley's method avoids the arbitrariness of choosing a particular metric, but only through a loss of flexibility in other respects.

The value of assuming that the randomness of a time series is caused by chaotic dynamics rather than a more conventional random process is that it gives a new perspective. The deterministic dynamical systems approach leads naturally to innovations such as trajectory segmenting or the use of an explicit metric. This makes it natural to go to higher order approximation schemes, which can lead to big improvements in accuracy, as our numerical work clearly demonstrates. The dynamical systems viewpoint is also essential for producing the error estimates described in Section 3. Ultimately, of course, all of these methods should be judged on their performance in real-world applications.

3 Scaling of Error estimates

In this section we estimate forecasting errors. The errors depend on properties of the dynamics, such as the attractor dimension D and the Lyapunov spectrum $\{\lambda_i\}$. They also depend on properties of the data set, such as the number of data points, N, and the signal-to-noise ratio S, as well as the extrapolation time T. The resulting scaling laws provide an *a priori* means of estimating the quality of forecasts, and they also suggest improvements for computing nonlinear statistical properties such as the fractal dimension, as discussed in Section 4.

Unless otherwise stated, in this section we will assume that time is discrete, and scaled so that $t = 0, 1, 2, \ldots$. We discuss some of the problems that arise when time is continuous in Section 3.3.

3.1 Dependence on number of data points

The approximation error generally depends on the number of data points. When the accuracy improves as a power law in N, as it generally does for local approximation schemes, the dependence on N is described by the order of approximation, which we defined in Equation (16) as the scaling exponent. It is clear that for good forecasts we want to make the order of approximation as large as possible. One way to achieve this is by using local polynomials. For example, in Figure (4), we use first and second degree local polynomial charts, using data generated by the sine map,

$$x_{t+1} = sin(\pi x_t), \qquad (18)$$

where $-1 < x_t < 1$. Similarly, in Figure (5) we show the scaling behavior of the Mackey-Glass delay differential equation [55],

$$\frac{dx(t)}{dt} = -0.1x(t) + \frac{0.2x(t - \Gamma)}{1 + x(t - \Gamma)^{10}}, \qquad (19)$$

with $\Gamma = 17$. The resulting slopes are roughly $\frac{(m+1)}{D}$, where m is the degree of the chart, and D is from previous independent calculations [24].

In general it is not always possible to achieve $q = m + 1$. For example, for the delay equation we were unable to improve our results significantly using cubic charts. Improving the order of approximation is the central problem in nonlinear modeling.

3.2 Dependence on extrapolation time

Naturally for a chaotic system errors depend strongly on the time that we attempt to extrapolate into the future. The rate at which errors grow depends on the way we make predictions. There are two choices: We can make *iterative forecasts* by fitting a model for $T = 1$ and iterating to make predictions for $T = 2, 3, \ldots$. Alternatively, we can make *direct forecasts* by fitting a new model for each individual T. On the surface direct forecasting might seem more accurate, since each model is "tailored" for the time it is supposed to predict, and there is no accumulation of errors due to iteration. In fact, as we shall show here, if the model is sufficiently accurate, the opposite is true. Approximation errors for iterative forecasting grow roughly according to the largest Lyapunov exponent λ_{max}, whereas for direct forecasts the errors grow as $q\lambda_{max}$.

To show this we must first introduce the new notion of higher order Lyapunov exponents.

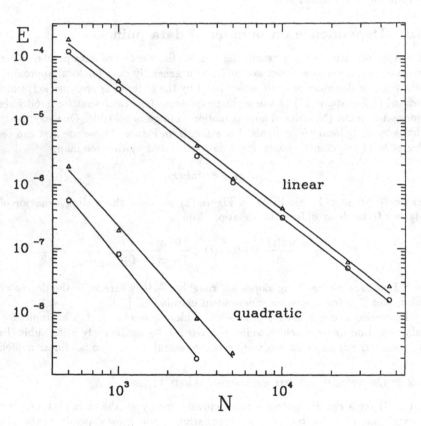

Figure 4: Local approximation by polynomials for the sine map. The slopes are roughly $-q = -(m + 1)$, where m is the degree of the polynomial, and q is the order of approximation. For the data points shown with circles we used nearest neighborhoods for the predictions, and for those indicated by triangles we used disjoint neighborhoods, constructed with the k-d tree. .

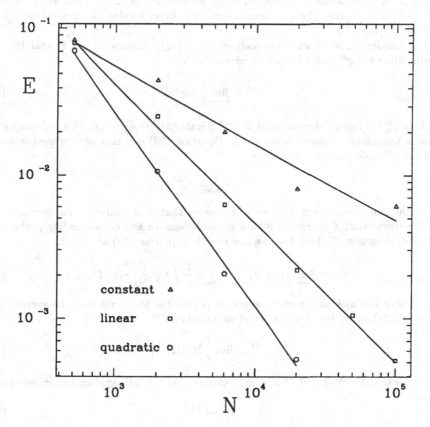

Figure 5: Local approximation by polynomials for the Mackey-Glass delay differential equation with $\Gamma = 17$, $\Delta t = 1$, $\tau = 6$, and $d = 4$ for a fixed extrapolation time $T = 85$. (The characteristic time is roughly 50.) Each value of $\bar{E}$ is based on an average of 500 forecasts. Using independently computed values for the attractor dimension D [24], we find that the measured value of $q \approx m + 1$, within the expected statistical error.

3.2.1 Higher order Lyapunov exponents

Since the accuracy of an approximation scheme depends on higher derivatives, for direct forecasting the growth rate of errors with time depends on the average growth rate of higher derivatives under iteration. This can be described in terms of a generalization of the Lyapunov exponents, which we describe in this section. For simplicity, we will only discuss the one dimensional case here, leaving the general case for a future paper [19].

Consider a one dimensional map, $x_{t+1} = f(x_t)$. Assume that f is analytic. We will define the qth order Lyapunov exponent as

$$\lambda^{(q)} = \lim_{t \to \infty} \frac{1}{t} \log |x_t^{(q)}|, \tag{20}$$

where $x_t^{(q)}$ is the qth derivative of the tth iterate, $x_t^{(q)} = \frac{d^q f^t}{dx^q}(x_0)$. This reduces to the usual Lyapunov exponent when $q = 1$. Equation (20) can also be expressed in terms of the qth order Lyapunov *number*

$$\Lambda^{(q)} = \lim_{t \to \infty} |x_t^{(q)}|^{\frac{1}{t}}.$$

For the remaining discussion we will assume that f is analytic. Furthermore, we will assume that f is ergodic with a natural measure and corresponding probability density function $P(x)$, so that we can rewrite Equation (20) as

$$\lambda^{(q)} = \lim_{t \to \infty} \frac{1}{t} \langle \log |x_t^{(q)}| \rangle = \lim_{t \to \infty} \frac{1}{t} \int \log |\frac{d^q f^t}{dx^q}(x)| P(x) dx \tag{21}$$

Using the ensemble average above, it is possible to interchange the average and the logarithm, to define a new set of exponents $l^{(q)}$[13]

$$l^{(q)} = \lim_{t \to \infty} \frac{1}{t} \log \langle |x_t^{(q)}| \rangle \tag{22}$$

It is clear that $l^{(q)} \geq \lambda^{(q)}$. For many examples, as a rough approximation we should find

$$l^{(q)} \approx \lambda^{(q)} \tag{23}$$

We will use this approximation to give us a rough idea of the relationship between first and higher order Lyapunov exponents, and also to estimate the scaling of our error estimates, acknowledging in advance that there are certainly examples for which this approximation breaks down.

When f is analytic the behavior of higher order Lyapunov exponents is at least approximately related to the first order Lyapunov exponent. To demonstrate this we will prove an equality for $l^{(2)}$ in terms of $l^{(1)}$.

---

[13]When $q = 1$ this exponent is what Fujisaka [32] calls the -1 order characteristic exponent. He generalizes the Lyapunov exponents in a different way, analogous to the definition of generalized dimensions [37].

We begin by recursively differentiating the map.

$$x_{t+1} = f(x_t) \tag{24}$$
$$x'_{t+1} = f'(x_t)x'_t \tag{25}$$
$$x''_{t+1} = f'(x_t)x''_t + f''(x_t)(x'_t)^2 \tag{26}$$

It is clear that the behavior of higher Lyapunov exponents is more complicated than that of the first order exponent. For example, $\lambda^{(1)} = \langle \log|f'(x)|\rangle_x$, but in general $\lambda^{(q)} \neq \langle \log|f^{(q)}(x)|\rangle_x$.

To get an intuitive feeling for Equation (26), suppose we neglect the first term on the right. This implies that for large t, x''_t grows roughly as the square of x'_t. However, since by definition $\langle|x'_t|\rangle$ grows as $e^{\lambda^{(1)}t}$, this suggests that the exponent for the second derivative is roughly twice that of the first.

We can prove a related inequality by making certain assumptions. Divide Equation (26) by $(x'_t f'(x_t))^2$, and rewrite it in terms of $y_t = x_t''/(x_t')^2$. After taking absolute values, averaging over x_0, and making the assumption that y_t and $f'(x_t)$ are uncorrelated, the result is

$$\langle|y_{t+1}|\rangle \leq \langle\frac{1}{|f'(x_t)|}\rangle\langle|y_t|\rangle + \langle\frac{|f''(x_t)|}{|f'(x_t)|^2}\rangle. \tag{27}$$

Let $a = \langle|f'(x_t)|^{-1}\rangle$, and $b = \langle\frac{|f''(x_t)|}{|f'(x_t)|^2}\rangle$, Equation (27) can be solved to give

$$\langle|y_t|\rangle \leq a^t\langle|y_0|\rangle + (\frac{1-a^t}{1-a})b.$$

If we also assume that $|x'_t|^{-2}$ and $|x''_t|$ are uncorrelated, then we can write $\langle|y_t|\rangle = \langle|x''_t|\rangle\langle|x'_t|^{-2}\rangle$. Dividing and taking logarithms gives

$$\log\langle|x''_t|\rangle \leq -\log\langle|x'_t|^{-2}\rangle + \log(a^t\langle|y_0|\rangle + (\frac{1-a^t}{1-a})b)$$

Since $\langle|x'_t|^{-2}\rangle > \langle|x'_t|\rangle^{-2}$ and there is a minus sign in front of the logarithm, we can make this substitution and preserve the inequality. Dividing by t gives

$$\frac{1}{t}\log\langle|x''_t|\rangle \leq \frac{2}{t}\log\langle|x'_t|\rangle + \frac{1}{t}\log(a^t\langle|y_0|\rangle + (\frac{1-a^t}{1-a})b) \tag{28}$$

By definition for a chaotic mapping $\langle\log|f'(x_t)|\rangle = \lambda^{(1)} > 0$. Since $\log\langle|f'(x_t)|\rangle \geq \langle\log|f'(x_t)|\rangle$, this also implies that $\log\langle|f'(x_t)|\rangle > 0$. If we assume that $a < 1$, when we take the limit as $t \to \infty$ the term on the right vanishes, giving

$$\lim_{t\to\infty}\frac{1}{t}\log\langle|x''_t|\rangle \leq \lim_{t\to\infty}\frac{2}{t}\log\langle|x'_t|\rangle$$

Applying the definition of $l^{(q)}$ from Equation (22) gives

$$l^{(2)} \leq 2l^{(1)} \tag{29}$$

A similar relationship holds for general q. If we assume that the approximation of Equation (23) is valid, then we have

$$\lambda^{(q)} \approx q\lambda^{(1)}. \tag{30}$$

For the numerical examples we have studied so far, this gives rough agreement. (See Figure (6) for example.)

There seems to be an analogy between higher order Lyapunov exponents and the generalized Renyi dimensions and entropies [37]. We intend to investigate this in more detail in the future [19], as well as possible connections to the higher order characteristic exponents of Fujisaka [32].

3.2.2 Direct forecasting

In this section we estimate the rate of growth of errors for direct forecasting, *i.e.*, constructing a new approximant $\hat{f}_T$ to approximate f_T at each time T. The central assumption is that the rate of growth of errors for q^{th} order approximation is dependent on the average rate of growth of the qth derivative, which can be described in terms of the qth order Lyapunov exponent.

To make this a little more concrete, it is probably worth explicitly demonstrating this for an example. Suppose that we use linear interpolation, and approximate a function F over an interval $[x_1, x_2]$ as

$$F(x) \approx \hat{F}(x) = F(x_1) + \left[\frac{F(x_2) - F(x_1)}{x_2 - x_1}\right](x - x_1). \tag{31}$$

The error that we make with this approximation is $E(x) = F(x) - \hat{F}(x)$. This can be estimated by expanding F in a Taylor's series about x_1, to get

$$E(x) = \frac{F''(x_1)}{2}(x - x_1)(x - x_2) + O(\epsilon^3),$$

where $\epsilon = x_2 - x_1$, and $O(\epsilon^n)$ indicates that the remaining terms are of order ϵ^n or smaller. The average absolute error is

$$\langle|E(x)|\rangle = \frac{1}{\epsilon}\int_{x_1}^{x_2}|E(x)|dx = \frac{1}{12}|F''(x_1)|\epsilon^2 + O(\epsilon^3) \tag{32}$$

Consider a one dimensional chaotic map, $x_{t+1} = f(x_t)$, (where the subscript now represents time). Again, assume this map is ergodic. Suppose we want to approximate the Tth iterate f^T using linear interpolation. If we use uniform knots ($\tilde{x}_i = \epsilon i$ in Equation (31)), from Equation (32) we get

$$\langle|E(x_{t+T})|\rangle = \frac{\epsilon^2}{12}\int|\frac{d^2 f^T}{dx^2}(x_t)|P(x_t)dx_t + O(\epsilon^3) = \frac{\epsilon^2}{12}\langle|x_T''|\rangle + O(\epsilon^3), \tag{33}$$

where $P(x_t)$ weights the errors on each individual forecast according to the frequency with which they occur. Taking logarithms and referring to Equations (22) and (23) shows that the mean error must scale as

$$\log \langle |E(x_{t+T})|\rangle_{x_t} \sim q l^{(1)} T \approx q \lambda^{(1)} T \qquad (34)$$

For the numerical examples that we have studied this is a fairly good approximation.

3.2.3   Iterative forecasting

In this section we derive error estimates for iterated forecasting, *i.e.*, constructing an approximant for $T = 1$ and then iterating to predict $T = 2, 3, \ldots$. Assume that at $T = 1$ we approximate f by a map g. Define the error of approximation as

$$\delta(x) = g(x) - f(x) \qquad (35)$$

where $|\delta(x)|$ is small and bounded. Similarly, define the approximation error at time T as

$$\Delta_T(x) = g^T(x) - f^T(x),$$

so by definition $\Delta_1 = \delta$. Assume that for any given T our approximation is good enough so that $\Delta_T(x)$ is small and bounded by $\Delta_{max} > |\Delta_T(x)|$ for all x. Thus, the scaling derived here will only be valid in the limit that the approximation is quite accurate.

From the previous two equations we get

$$\Delta_T(x) = f(g^{T-1}(x)) + \delta(g^{T-1}(x)) - f^T(x).$$

Expand $f(g^{T-1}(x))$ and $\delta(g^{T-1}(x))$ to first order in a Taylor series about $f^{T-1}(x)$. Assume that δ and f are both smooth, so that $f'' = O(1)$ and $\delta'' = O(1)$, where again $O(x)$ means "of the same order as x", and make use of $f' + \delta' = g'$. This implies

$$\Delta_T(x) = g'(f^{T-1}(x))\Delta_{T-1}(x) + \delta(f^{T-1}(x)) + O(\Delta_{max}^2) \qquad (36)$$

By expanding this expression for $T-2, T-3, \ldots$, it is clear that

$$\Delta_T(x) = \sum_{j=0}^{T-1} \prod_{i=j+1}^{T-1} g'(f^i(x))\delta(f^j(x)) + O(\Delta_{max}^2) \qquad (37)$$

If we also assume that $\delta' = O(\Delta_{max})$, then we can approximate g' by f', and from the chain rule we can write the product of derivatives as the derivative of the iterate, so that this becomes

$$\Delta_T(x) = \sum_{j=1}^{T} \frac{df^{T-j}}{dx}(x_j)\delta(x_{j-1}) + O(\Delta_{max}^2). \qquad (38)$$

Take absolute values and average over x, to get

$$\langle|\Delta_T(x)|\rangle = \sum_{j=1}^{T} \langle|\frac{df^{T-j}}{dx}(x_j)\delta(x_{j-1})|\rangle + O(\Delta_{max}^2). \tag{39}$$

It simplifies matters if we assume for the moment that $f(x)$ is a one dimensional map, and that $\frac{df^{T-j}}{dx}(x_j)$ and $\delta(x_{j-1})$ are uncorrelated. $\langle|\delta(x_j)|\rangle$ is independent of j. By definition (22) the average derivative $\langle|\frac{df^j}{dx}(x_0)|\rangle \approx L^j$, where $L = e^{\lambda^{(1)}}$. We can then rewrite this as

$$\langle|\Delta_T(x)|\rangle \approx \langle|\delta(x)|\rangle \sum_{i=0}^{T-1} L^i \tag{40}$$

Note that this is a natural result. It says that the cumulative error amplification from the first step is L^{T-1}, from the second step is L^{T-2}, etc. Summing the series gives

$$\langle|\Delta_T(x)|\rangle \approx \frac{L^T - 1}{L - 1}\langle|\delta(x)|\rangle. \tag{41}$$

In the limit of large T the asymptotic rate of growth is

$$\lim_{T\to\infty} \left(\frac{L^T - 1}{L - 1}\right)^{\frac{1}{T}} = L \tag{42}$$

To get an alternative view that gives a feeling for what these assumptions mean, we could return to Equation (36) and assume that

$$\delta(f^{T-1}(x)) \ll g'(f^{T-1}(x))\Delta_{T-1}(x). \tag{43}$$

Again approximating g by f and recursively substituting for Δ_{T-i}, this leads to $\Delta_T(x) = \frac{df^T}{dx}(x)\delta(x)$. Taking logarithms and absolute values, and assuming that $\delta(x)$ is uncorrelated with $\frac{df^T}{dx}(x)$, we get

$$\langle\log|\Delta_T(x)|\rangle = \langle\log|\frac{df^T}{dx}(x)|\rangle + \langle|\log\delta(x)|\rangle = T\lambda^{(1)} + \langle|\log\delta(x)|\rangle, \tag{44}$$

This agrees with previous estimates of Lorenz [54]. This also shows that the assumption of Equation (43) is roughly equivalent to the other assumptions we made leading up to Equation (42). It also suggests that the correspondence with Lyapunov exponents is more exact if we use $\langle\log E\rangle$ rather than $\log\langle E\rangle$ to evaluate the errors.

Either of these derivations depends on the assumption that $f' = O(1)$, $\delta' = O(\Delta_{max})$, and $\delta'' = O(1)$. This is not surprising: If the derivatives of the approximation are sufficiently different from those of the true dynamics, then in general the approximation will have different Lyapunov exponents. This should not be a problem for smooth approximation schemes, but this comes under question for methods such as linear interpolation, for which g is only C^1. For linear interpolation with a uniform knot spacing ϵ, it is fairly easy to show that $\delta' = O(\epsilon)$, which is larger than δ (which is $O(\epsilon^2)$), but still small. δ'' may cause problems, however, since it is a delta function. Intuitively, since we are taking an average, and the integral of the second derivative is $O(\epsilon)$, problems due to this seem unlikely, but this should be investigated in more detail. The safe case is when g and f are both smooth.

So far, we have restricted our discussion to one dimension, which is especially simple because there is only one Lyapunov exponent. In more than one dimension there can be more than one positive Lyapunov exponent, and the situation is more complicated. However, for long times, a displacement between an approximate trajectory and a true trajectory will typically line itself up along the most unstable direction, so that, as long as the largest Lyapunov exponent is sufficiently larger than the others, it will asymptotically dominate. Thus in higher dimensions we expect that the growth of errors will be dominated by the largest Lyapunov exponent. Note that this is revised from our previous paper [23]; although the metric entropy is related to the short term rate of loss of information, it does not seem to be the relevant quantity here. [14]

To conclude, for an iterative forecasting scheme satisfying the assumptions above we conjecture that the errors grow according to

$$\bar{E} \sim N^{-\frac{q}{D}} e^{\lambda_{max} T}, \tag{45}$$

where λ_{max} is the largest Lyapunov exponent. Comparing this to Equation (34), the difference is that the errors for iterative forecasts grow at an exponential rate given by λ_{max}, in contrast to $q\lambda_{max}$ for direct forecasts.

Intuitively, the superiority of iterative estimates must come from the fact that they make use of the regular structure of the higher iterates. The time series that we are trying to approximate are generated by iterating a dynamical system, and so iterative approximations are more natural. The power of the iterative procedure is reminiscent of Barnsley's methods for constructing complicated fractals from the recursive application of simple affine mappings [2]. Some preliminary results indicate the advantages of iterated forecasting in neural nets [49].

The validity of these estimates for a simple example is demonstrated in Figure (6), where we show the approximation error as a function of the extrapolation time T. As predicted, the error grows roughly according to Equation (34) for direct approximation and according to Equation (45) for iterative approximation.

For iterated forecasts we should expect that the distribution of errors will have long tails. As we have seen here, the cumulative error after many iterations is dominated by the product of the errors along the way. As long as the second moment exists, a corollary of the central limit theorem is that the probability density function of the product of many random numbers is log-normal. We expect that this will also be true for iterated forecasts, at least in the limit where we have a very accurate model and iterate many times. Log-normal distributions have long tails, corresponding to occasional very poor forecasts. Thus, if we are concerned with bounds on the worst case error rather than the mean square error, we might expect that iterated forecasts

---

[14]We would like to thank Martin Casdagli for pointing out this mistake.

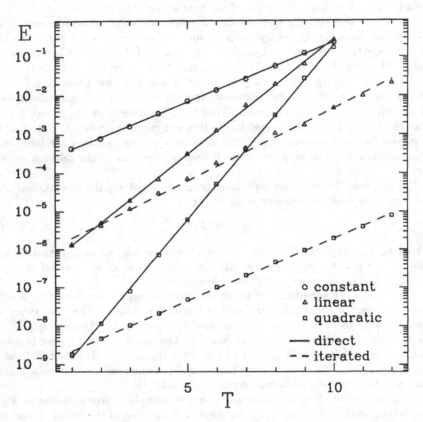

Figure 6: Forecasts made using local polynomial approximation, building the data base from a 5,000 point time series generated by Equation (18). We make 500 forecasts, and plot the average error as a function of the extrapolation time T. Letting $q = m + 1$, where m is the order of the polynomial, as expected from Equations (34) and (45), the logarithm of the error for direct forecasts grows roughly according to $q\lambda$ whereas for iterative forecasts it grows according to λ, independent of the order of approximation. λ is the Lyapunov exponent, which in this case is one bit per iteration.

will not perform as well. In fact, as discussed in Section 3.4, our numerical experiments indicate that iterated forecasts have long tails, but that direct forecasts produce tails that are even longer.

We wish to emphasize that the above results are for scaling in the limit as $\bar{E} \to 0$. In situations where the model is not good enough to achieve this limit, these scaling laws cannot be expected to hold.

3.2.4 Temporal scaling with noise

So far in this section we have assumed that the accuracy of predictions is limited by the quality of the approximation, which for a given data set is determined by the number of data points. In other cases the accuracy of prediction may be limited by noise. Even if we know the equations of motion exactly, noise limits prediction by making initial conditions uncertain, and by perturbing deterministic trajectories. The results of the previous section make it clear that the effect of noise is very much like the effect of approximation error. In fact, if we let δ in Equation (35) represent noise rather than approximation error, the results of that section go through essentially unchanged. Thus the scaling with noise is described by Equations (41) or (44), if we substitute $\langle |n(t)| \rangle$ for $\langle |\delta(x)| \rangle$.

3.3 Continuous time

So far our error estimates have been for discrete-time maps, where each iteration of the map causes a large change in the state. Two problems occur in continuous-time systems.

The first problem has to do with the meaning of N. In a continuous time system, the number of points can be increased by simply decreasing the sampling rate Δt. If Δt is already small this may not give us any new information. In this case the amount of useful data depends more on the number of characteristic time scales rather than on the number of data points. Thus, for a continuous time system the estimates should be stated in terms of the number of characteristic times, rather than the number of data points.

Unfortunately, the notion of a characteristic time scale is ambiguous. It can be estimated in many ways, for example as the correlation time, as the inverse of the average frequency of the power spectrum, or as the average time between level crossings. Yet another way to define a characteristic time comes from trajectory segmenting; the average ratio of the number of trajectory segments to the number of data points inside local neighborhoods estimates the extent to which the data is oversampled. Note that although this ambiguity about the meaning of N introduces some absolute uncertainty into the error estimates, it does not effect the scaling for any *fixed* Δt.

The second problem occurs in iterative forecasting. Although the conclusion that iterative forecasts are superior carries over to continuous time systems, it raises the question of how to pick the composition time. Clearly the effects of noise and finite precision dictate that this time should not be zero. Thus, there is some optimal time for constructing an approximation, which should minimize the errors in iterative forecasts. This is one possible source of the discrepancies observed in the following section. A similar problem occurs for approximation of differentials (Equation (8)): There are many possible ways to approximate the derivative – in the presence of noise it is important to weight contributions from different times properly. These questions will be treated in more detail in a future paper.

3.4   Numerical results

In this section we present some numerical results, investigating the scaling properties for systems that are not simple maps. So far we have mainly studied the Mackey-Glass delay differential equation (19) and the chaotic convection data of Haucke and Ecke [43].

Figure (7) shows the results of making forecasts at several different extrapolation times for the Mackey-Glass delay differential equation with $\Gamma = 17$, where the dimension of the underlying attractor is roughly $D = 2.1$ [26]. We used a a four dimensional time delay embedding, with $\tau = 6$, using only 500 data points sampled at $\Delta t = 1$. We get the best results using approximation by quadratic polynomials, and choosing twice the minimum number of nearest neighbors, which in this case is $M = 30$. As seen in the figure, iterated forecasting is superior to direct forecasting. Using least squares to fit a line to the indicated portions of the curve we get good agreement with the known value of the largest Lyapunov exponent [26]. However, we do not understand why the initial rise is larger than this, although we have observed this behavior in several other continuous data sets.

In Figure (8) we show a similar plot, except that the number of points in the database is increased to 10,000. This improves the accuracy of the forecasts substantially, roughly as expected from Equation (16). However, when we attempt to compare the slope with the previous case, we observe something puzzling: Even in the straight portions of the curve, at the right of the figure, the slope is now larger, by a factor of roughly 1.3. This may be related to the same effect mentioned above, and needs further investigation.

Our numerical experiments indicate that while our forecasts with these methods are good on average, there are large fluctuations in accuracy, with occasional very bad predictions. The distribution of errors about the mean can be quite large, as demonstrated in the error histograms of Figure (9). As we see, the error distribution has long tails. (On this scale this is only apparent for long time forecasts.) As discussed in the previous section, this is expected for iterated forecasts. However, the tails for the direct forecasts are even longer than those of the iterated forecasts. We have found that this depends on several factors, such as the proper selection of neighborhoods. If x is not enclosed by its neighborhood, for example, we get very bad predictions. The frequency of bad predictions goes up when we use fixed disjoint partitions instead of nearest neighborhoods. There may also be other factors that cause bad predictions, for example discontinuities, singular behavior, neighborhoods

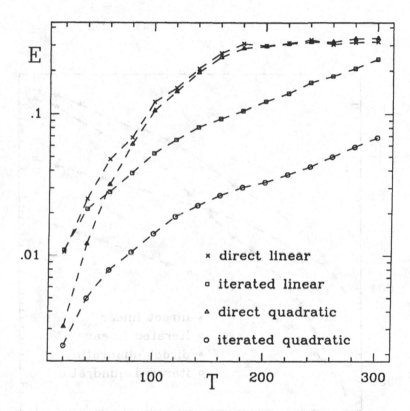

Figure 7: Prediction error for the Mackey-Glass delay differential equation as a function of extrapolation time, with $N = 500$ (roughly 10 characteristic times), $\Gamma = 17$, $\tau = 6$, $\Delta t = 1$, and $d = 4$, using nearest neighborhoods with twice the minimum number of points needed to fit the coefficients by least squares ($M = 30$ for quadratic, and $M = 10$ for linear). The iterated forecasts are made by iterating the model at the first extrapolation time shown. The fractal dimension of the underlying attractor is $D = 2.1$. To improve the statistical stability of our error estimates we have computed $\bar{E}$ by throwing out the 10% worst predictions.

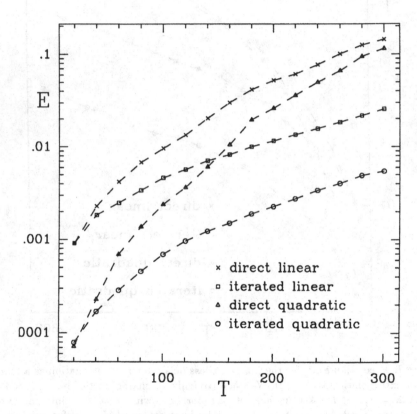

Figure 8: Same as Figure (7), except that $N = 10,000$.

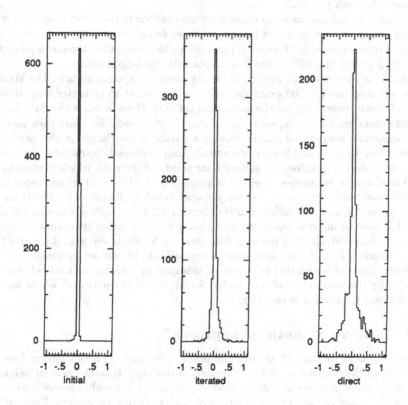

Figure 9: *Error histograms* for the data of Figure (7). The y axis shows the number of counts with a given value of the prediction error $E = x(t + T) - \hat{x}(t, T)$. The histogram on the left corresponds to short-term forecasts with $T = 20$. The middle histogram corresponds to iterated forecasts at $T = 100$, using a composition time of 20. The histogram on the right corresponds to direct forecasts, also at $T = 100$.

that for some reason or another do not make a stable fit for the parameters of their chart, or dynamical factors such as nonuniformities of the derivatives. We intend to investigate this in more detail. It should be possible to reduce bad predictions by using appropriate diagnostics and tailoring the fitting strategy for each individual neighborhood. Also, it is obviously desirable to be able to make *a priori* confidence estimates for each prediction.

Occasional bad forecasts can cause large fluctuations in the mean prediction error, so that unless we average over a large number we do not get stable results. To reduce the data requirements, for Figures (7) and (8) we have computed the mean prediction error by rejecting the 10% of the predictions with the largest errors.

Note that even when we include the worst predictions, our results for the Mackey-Glass equation using a 500 point database are equivalent to or better than those of a feed-forward neural net on the same data set [49]. If we increase the data base to 10,000 points, we see an improvement of roughly $20^{\frac{3}{2.1}} \approx 80$ for short time forecasts with quadratic polynomial charts. Using a database this large for the neural net employed by Lapedes and Farber is currently computationally intractable, even using supercomputers. In contrast, to generate one point of Figure (8), which involves fitting 1000 local charts, consumes roughly five minutes on a SUN 3/60 microcomputer.

We now return to the convection data analyzed in Figure (1). Based on our previous results, and the scaling results of Section 3.2.3, we originally assumed that we would be able to improve these results by using iterated rather than direct forecasts. We were quite surprised to discover that iterated forecasts are actually worse than direct forecasts in this case, as shown in Figure (10). We do not understand at this time why iterated forecasting is inferior in this case, or whether we could change this by altering the parameters of our model. So far, this is the only case where we have found direct forecasting is superior.

3.5   Is there an optimal approach?

These estimates make it clear that the primary limitation on short term forecasts comes from the dimension. If the dimension is too high, then the nearest neighbors of a given point may be so distant that the dynamics is poorly approximated even with a data base consisting of thousands of points. Is this an intrinsic limit, or is it possible to do better?

In fact, there is an optimal method, namely, to guess the right answer. With prior information, or a sufficiently exhaustive search, we may be able to find a model that is superior to those we would be led to by blind analysis with the techniques that we have described so far. For example, suppose a chaotic time series is produced by a differential delay equation, of the form

$$\frac{dx}{dt} = f(x(t), x(t - \Gamma)) \tag{46}$$

As demonstrated in reference [24], when Γ is large equations of this form can have attractors of very large dimension. However, suppose we know in advance that the

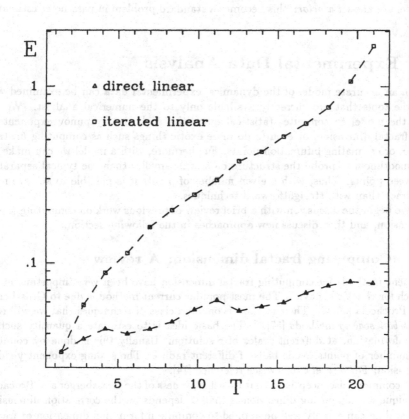

Figure 10: A comparison of iterated and direct forecasts using iterated and direct forecasting, using the same data shown in Figure (1). For reasons that we do not understand, direct forecasting is superior in this case.

dynamics is this form. If we formulate our model *as a delay equation*, the problem becomes two dimensional, irrespective of the dimension of the attractor. This obviously makes it much simpler. The error estimates we have given here apply, but D is two, irrespective of the attractor dimension. The key to this is algorithmic simplicity; if we can find a way to formulate a good model in algorithmically simple terms, even apparently complex behavior may be quite tractable. When many aspects of the model are known *a priori*, this becomes a standard problem in parameter estimation [22].

4 Experimental Data Analysis

With an accurate model of the dynamics, experimental data can be examined with all the tools that are currently available only to the numerical analyist. We can use the model to compute statistical quantities such the Lyapunov exponents or the fractal dimension, or even to do more exotic things such as computing unstable cycles or estimating bifurcation points. Furthermore, with a model we can make use of smoothness to probe the structure on a scale smaller than the typical separation between points. Thus, with a given number of points it is possible to achieve more accuracy than with straightforward techniques.

We begin the discussion with a brief review of previous work on computing fractal dimension, and then discuss new approaches in the following section.

4.1 Computing fractal dimension: A review

Efficient methods for computing fractal dimension have been very important in the search for chaotic behavior. The most popular current method is due to Grassberger and Procaccia [36,38]. Their method is one in a class of techniques that we will refer to as *ball scaling* methods [57].[15] The basic idea is to estimate a quantity such as the information, at different scales of resolution. Usually this is done by counting the number of points inside balls of different radii ϵ. The scaling exponent yields a dimension. For a recent review see reference [74].

A common misconception about the effectiveness of the Grassberger and Procaccia technique for computing dimension is that it depends on the correlation dimension. Ball scaling can equally well be applied to compute information dimension or fractal dimension [57,74]. The underlying reason for the superiority of ball scaling is the fact that it gives better estimates of probability distributions and their moments, as explained by Omohundro[16] [58].

---

[15]As far as we know the use of ball scaling to compute dimension is originally due to Pettis *et al.* (1979). Ball scaling methods independently arose from the work of several different groups, including Takens [72,73], Grassberger and Procaccia [36,38], Guckenheimer [40], and Farmer and Jen [9]. Since Grassberger and Procaccia were the first to develop and demonstrate the effectiveness of these procedures, the use of ball scaling is usually attributed to them.

[16]See page 301.

Numerical dimension computations based on ball scaling are subject to misinterpretation and must be used with care. Misapplication of the ball scaling technique has led to many false statements about the presence of chaos.

Several promising new methods to compute dimension may help solve this problem. For example, if the Broomhead and King embedding procedure is applied *locally*, to the points in a ball of radius ϵ, the embedding dimension computed from the singular values gives a good upper bound on the fractal dimension [10]. If the ball is small enough, curvature due to global structure is negligible, since inside the ball the dynamics is locally linear. The principal value decomposition automatically yields the *local* embedding dimension, which is an upper bound on the fractal dimension. Examination of the scaling with the size of balls provides a self-consistency check on the results.

Another method to compute the local embedding dimension has recently been suggested independently by Cremers and Hübler [16] and by Cenys and Pyragas [13]. This approach is similar to a method previously suggested by Packard *et al.* [59], but they examine the *scaling* of the width of conditional probability distributions with ball size and embedding dimension. If the embedding dimension is large enough the width of the distribution narrows sharply. Furthermore, if the embedding dimension is sufficiently large the width of the distribution scales linearly with ϵ, providing a self-consistency check. Yet another approach has been suggested by Bayly *et al.* [3], who propose fitting rational polynomials of different dimension, to find the dimension with minimum variance. Their method of computing dimension is analogous to that of Froehling *et al.* [31], except that they use global polynomials instead of local linear fits, which seems to produce much better results. We need further work comparing these methods to determine if they are more reliable than ball-scaling methods. Although all of these methods compute the embedding dimension rather than the fractal dimension, in practice this is often just as good.

4.2   More accurate data analysis with higher order approximation

Computations based on counting points cannot probe the dynamics to scales of resolution that are less than the typical separation distance between points. This necessarily limits the accuracy of these methods, particularly for small numbers of points in high dimensions. The accuracy of an estimate of the fractal dimension D based on N data points scales roughly as $N^{-\frac{1}{D}}$ [31].[17] Thus, in the language we have developed here, conventional ball scaling methods have the scaling properties of first order approximation.

With higher order approximation schemes it is possible to probe the dynamics to scales that are smaller than the typical separation between points. If our model is suf-

---

[17]Although this estimate was originally stated for box counting methods, it also holds for ball scaling. The difference between them comes from the constant in front, plus the difference in the demand they place on computer resources.

ficiently accurate, we can use it to obtain statistical estimates that are more accurate than those obtained directly from the data itself. There are several possibilities for doing this. For example, suppose we wish to compute the dimension of an attractor, and we are limited by the amount of available data. If we have a good model for the data, we can iterate the model to obtain new "simulated data". If the measure of the simulated data is sufficiently close to that of the true data, then we can simply apply standard algorithms to compute the dimension of the new time series. This can similarly be done to compute quantities such as the Lyapunov exponents.

A central question is how much the the data can be extended and still produce accurate estimates on a given scale. For short prediction times, the error of approximation is roughly $\bar{E} \sim N^{-\frac{1}{D}}$. If we only needed to extend the data by a few data points, we can expect these new data points to be faithful to the measure to this accuracy. As we iterate the map further, although each estimate is only for a short time, the errors in the distribution of the points may accumulate. We do not understand how to estimate this in detail, but since the evolution of the measure depends on the derivatives of the flow, we think that the accuracy will depend on the accuracy for estimating derivatives. [18] For local schemes the order of approximation for the derivative is typically one less than that of the map itself. Thus, we hope that averages computed in this way should also be accurate to this same degree of approximation. These arguments are admittedly very vague, however, and need to be made more precise, and backed up by numerical results.

Assuming that we can approximate the measure to order $q - 1$, this implies that we can generate roughly $\tilde{N} \sim N^q$ new data points and still remain faithful to the true dynamics. We can then improve our estimate of the dimension by applying the usual ball scaling algorithms to the newly generated points. Since there are now N^q of them, we can get a much better estimate. Of course, this involves more computation. Nonetheless, if the computation is limited by available data rather than computer resources, this implies that the analysis of a D dimensional attractor can be done as accurately as a $\frac{D}{q-1}$ dimensional attractor using first order methods.

Similarly, we can use this same method to improve estimates of the Lyapunov exponents. An algorithm for doing this is originally due to Wolf et al. [77], and alternatives have been suggested by Eckmann and Ruelle [21,20] and Sano and Sawada [66]. Since the Wolf algorithm involves manipulation of existing trajectories, while the Eckmann/Ruelle/Sano/Sawada algorithm employs local linear maps, on the surface it might seem to involve a higher order of approximation than the Wolf algorithm. However, since the estimate of derivatives through a linear map is only first order, in fact the order of approximation is the same.

If a higher order approximation can be found, the accuracy of these estimates can be improved in two ways. First, the estimates of local derivatives become more accurate. Second, by generating more data we can expect to get better statistics. Assuming the local derivatives are independent, the error due to statistical fluctuations goes as $\sigma \tilde{N}^{-1/2}$, where σ is the standard deviation of the local contributions,

---

[18]We thank Martin Casdagli for helping to clarify this point.

and $\tilde{N}$ is the number of simulated data points. The accuracy of approximation for the individual derivatives scales as $E \sim N^{-\frac{p}{q-1}}$, where N is the number of true data points. The dominant effect depends on the order of approximation as well as the dimension of the attractor, and whether the estimates for the derivatives are biased.

Alternatively, as we showed in reference [23], scaling properties of error estimates can be used to estimate the dimension. If the order of approximation q is known, then Equation (17) can be turned around to give a relationship for the dimension.

$$D = \lim_{N \to \infty} q \frac{\log N}{|\log \bar{E}|} \qquad (47)$$

There are several problems with this method. First, the biggest potential improvement comes about when q is large. For this method to give a reliable estimate of the dimension we must know the order of approximation *a priori*. Our numerical work so far indicates that it is difficult to achieve orders of approximation larger than two reliably when D is large. Unless we can be certain in advance of achieving a given order of approximation, we cannot estimate the dimension this way. Furthermore, the statistical stability of an estimate of $\bar{E}$ is limited by the number of points in the time series, and for a small time series statistics are not good. Thus, while checking for the proper scaling is a good measure of self consistency, we expect that iterating to generate "simulated data" will give superior estimates of the dimension.

4.3 Forecasting as a measure of self-consistency

Forecasting provides a hard test for the presence of chaos, especially when combined with the tests for the scaling properties expected from the error estimates of Section 3. It is very unlikely to make forecasts with the statistical significance that we achieve here by guessing at random. Although it is possible to construct counterexamples that would appear very much like chaos by running high dimensional noise through appropriate nonlinear filters, we doubt that such contrived examples are very likely. To paraphrase Joe Ford, "If it walks like a duck, talks like a duck, quacks like a duck, and even smells like a duck, then by golly, it seems pretty reasonable to assume it really is a duck."

5 Noise Reduction

In this section we introduce a new method for reducing noise in a dynamical system. The basic idea is nonlinear smoothing; once we can make forecasts, we can transport different points to the same point in time and average them together, to reduce the effect of noise. By applying this procedure recursively the reduction can be substantial. The basic idea is schematically illustrated in Figure (11).

We will assume that the time series $\{x_t\}$ is of the form

$$x_t = y_t + n_t, \qquad (48)$$

318

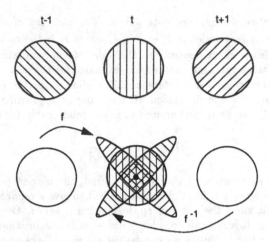

Figure 11: *Nonlinear smoothing.* The circles represent noisy measurements of a deterministic trajectory at three different times. As successive measurements are transported to the same point in time, the associated noise probability distributions distort. By weighting the values of the transported points correctly, they may be averaged together to produce a better estimate of the true value y_t. The estimates can be improved by iterating this process.

where $y_{t+1} = f(y_t)$, and the sequence n_t is uncorrelated, with $\langle n_i n_j \rangle = \sigma^2 \delta_{ij}$. We assume that n_t is unpredictable, and values at different times are independent. We will also assume that $\{n_t\}$ has a symmetric Gaussian distribution,

$$P(n_t) = \frac{1}{\sqrt{2\pi}\sigma} e^{-n_t^2/2\sigma^2}. \tag{49}$$

This assumption is convenient, but the final result does not depend on it; for example, our numerical results demonstrate that our method works quite well for noise with a bounded uniform distribution.

A standard method for reducing noise is to simply average together nearby points. For this to be effective, the average must be taken over a very short time interval, since otherwise the intrinsic dynamics of the signal dominate the noise. But if we can forecast accurately we can take the dynamics into account, and average over much longer periods of time. As we shall see, this is particularly effective in chaotic systems.

We want to transport points at different times to the same time, as illustrated in Figure (11). Conceptually, we are pulling-back and pushing-forward the probability distribution of the noise, under the action of the deterministic dynamics f. A probability distribution has an induced transformation under the mapping f that we will denote $\underline{f}$, *i.e.*,

$$P(x_{t+1} - y_{t+1}) = \underline{f}(P(x_t - y_t)).$$

Suppose we transport measurements from times $\{t - \alpha, \ldots, t + \beta\}$ to time t. Since n_t is independent, the joint probability distribution is

$$\overline{P}(f^{\alpha}(x_{t-\alpha}), \ldots, f^{-\beta}(x_{t+\beta})) = A \prod_{j=\alpha}^{j=-\beta} \underline{f}^{j}(P(x_{t-j} - y_{t-j})), \qquad (50)$$

where A is a normalization constant and $\alpha \geq 0$ and $\beta \geq 0$. If we assume that n_t is small we can linearize $\underline{f}$,

$$\underline{f}(P(x_t - y_t)) \approx A'P((f(x_t) - f(y_t))df(x_t)^{-1}) = A'P((f(x_t) - y_{t+1})df(x_t)^{-1}), \quad (51)$$

where $df(x_t)$ is the derivative of f at x_t and A' is a normalization constant. This approximation depends on the assumption that the noise is small compared to the nonlinearities of f, so that the dynamics is locally linear (so equiprobable surface remain ellipsoids under transformation, as shown in Figure (11)). Combining equations (49), (50), and (51) gives

$$\overline{P}(f^{\alpha}(x_{t-\alpha}), \ldots, f^{-\beta}(x_{t+\beta})) \approx A \prod_{j=\alpha}^{j=-\beta} e^{-\|(f^j(x_{t-j})-y_t)[df^j(x_{t-j})]^{-1}\|^2/2\sigma^2} \qquad (52)$$

We want to estimate y_t. A standard way to do this is to make a maximum likelihood estimate, which amounts to assuming that the particular sequence of fluctuations that we observe are the most likely ones, so that they lie at the maximum of the joint probability distribution. We choose our estimate Y_t to force this to be true. Since $\log \overline{P}$ has the same maximum as $\overline{P}$, we can more conveniently enforce this by setting $\frac{\partial \log \overline{P}}{\partial Y_t} = 0$. Setting $Y_t = y_t$ in Equation (52), differentiating, and solving for Y_t yields

$$Y_t = \left(\sum_{j=\alpha}^{j=-\beta} \Theta_j\right)^{-1} \sum_{j=\alpha}^{j=-\beta} \Theta_j f^j(x_{t-j}) \qquad (53)$$

where

$$\Theta_j = \left([df^j(x_{t-j})]^T df^j(x_{t-j})\right)^{-1}.$$

Θ_j is a $d \times d$ symmetric matrix that depends on x_{t-j}. It contains weighting factors that depend on local expansion and contraction rates, and take into account distortion of the noise as it is transported to different times. The directions in which the noise distribution is compressed contain more useful information, and receive higher weights.

To implement this procedure we have to estimate both $f^j(x_{t-j})$ and $df^j(x_{t-j})$. In practice, because of nonlinearities it is wise to keep the smoothing times short by keeping α and β fairly small. Further reductions are made by applying Equation (53) recursively. Since this makes it possible to keep each step short, this minimizes the effect of nonlinearities. With every pass we reduce the noise level, so that the local linear assumption of Equation (51) becomes increasingly valid. Thus once the algorithm starts to converge, further convergence is guaranteed. The recursive use of this algorithm is reminiscent of the "pull-back" algorithms for estimating Lyapunov exponents [4,69].

When we know the map exactly, with even a small amount of data and fairly large noise we can reduce the noise almost down to machine precision. To demonstrate this, we have applied this to the Henón map, as shown in Figure (12). Note that we achieve a noise reduction of roughly 10^{10}, more than 100 decibels. When the true map is not known, this is limited by forecasting accuracy. Still, even when we approximate the map we have been able to reduce the noise by a factor of 1000 or more.

As is apparent in the figure, points near either end of the time series are not smoothed nearly as accurately as those in the middle. The reason is that in a chaotic system the pulled-back values are accurate along the unstable manifold, while the push-forward values are accurate along the stable manifold. Points near the beginning of the time series have no history, and therefore noisy fluctuations along the stable manifold cannot be reduced. Similarly, points near the end of the time series have no future, and fluctuations along the unstable manifold cannot be reduced. For the Henón map, for example, as we move j steps from the beginning of the time series toward the middle, the noise is reduced along the stable manifold by a factor of roughly Λ_s^{-j}, where Λ_s is the Lyapunov number associated with the stable manifold. Similarly, as we move j steps from the end toward the center the noise is reduced by roughly Λ_u^j, where Λ_u is the Lyapunov number associated with the unstable manifold (and is greater than one).

So far we have assumed that the noise was added to an ideal trajectory, and had no effect on the dynamics. Suppose that instead the noise is coupled to the dynamics, and is included in computing the next state. In this case, there is no unique "true" trajectory. This leads to the shadowing problem: Is every noisy trajectory "shadowed" by a nearby deterministic trajectory? For hyperbolic systems, Anasov and Bowen independently proved that this true [1,8]. Although Anasov and Bowen discussed only hyperbolic systems, recently Hammel et al. [42] have applied a modified version of the Anasov-Bowen algorithm to simple systems such as the Henón map, demonstrating that they can usually find shadowing trajectories close to noisy trajectories. Although they posed the problem in a different context, their work can be viewed as noise reduction. [19] Our method is reminiscent of the Anasov-Bowen procedure, but it is applicable not only to non-hyperbolic systems, but even *non-chaotic* systems. Our method is also easy to implement in any dimension.

It is ironic that it is much easier to remove noise from a chaotic time series than from a regular time series. Without exponential expansion and contraction, according to the central limit theorem, smoothed values only become more accurate according to the square root of the number of data points, in contrast to the exponential behavior of chaotic systems. The number of points needed to achieve a given level of noise reduction for regular dynamics is therefore much larger than for chaotic dynamics.

The limit to noise reduction is usually given by the accuracy for approximating f.

---

[19] Recently this approach has been modified by E. Kostelich and J. Yorke to make it more practical (private discussion).

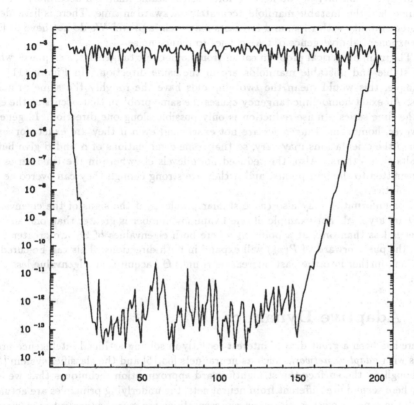

Figure 12: The noise reduction technique of Equation (53) applied to the Henón map. The map was iterated at double precision, generating a "clean" time series $\{y_t\}$, $t = 0, 1, \ldots, 200$. We then added pseudo-random numbers $\{n_t\}$ to each point, forming a noisy time series $x_t = y_t + n_t$. n_t is uniformly distributed, with a variance roughly 0.1% of that of y_t. The logarithm of $|n_t|$ is plotted at the top of the figure. Below it we plot $\log |Y_t - y_t|$, where Y_t is the smoothed value after applying Equation (53) 10,000 times to all 200 points, using $\alpha = \beta = 3$, and using the known map to compute f^j and Θ_j.

Estimates of Θ are not as important, since it is just a weighting matrix. As long as the eigendirections are roughly correct, we get a net noise reduction on every iteration. The approximation error for f is helped by the fact that the directions where we need accuracy are *contracting*; we need to forecast the stable manifold accurately forward in time, and the unstable manifold accurately backward in time. There is little decay of predictability with time, and forecasts may even improve! We do not have to fight the exponential divergence of trajectories to reduce noise.

The main numerical problem comes from homoclinic tangencies, *i.e.*, places where the stable and unstable manifolds are in the same direction. In Figure (11), for example, this would mean the two ellipsoids have the roughly the same principal axis. An exact homoclinic tangency causes the same problem that occurs at the ends of the time series – noise reduction is only possible along one direction. In general, however, homoclinic tangencies are not exact, and even if they are exact, for small α or β the orientations may vary, so that some combinations of α and β give better results than others. Also, the reduced noise levels elsewhere in the time series are transmitted to the bad points, and if they are strong enough they can overcome the instabilities.

Nonuniformities may also cause similar problems, if the signs of the eigenvalues of Θ are atypical. For example, if one Lyapunov number is greater than one and the other is less than one, at a point x_k where both eigenvalues of Θ_1 are greater than one, the push-forward of $P(x_k)$ will expand in both directions. This can be cured by reaching further into the past, increasing α until Θ_α acquires an eigenvalue less than one.

6 Adaptive Dynamics

There has been a great deal of interest recently in solving artificial intelligence problems with *adaptive networks* such as neural nets [65,15] and the classifier system [44]. Although on the surface the straightforward approximation techniques that we employ here seem quite different from neural nets, the underlying principles are actually much the same. However, since our representations are more convenient numerically, fitting parameters is hundreds of times faster.

Forecasting is an example of what is often called *learning with a teacher*. The task is to predict "outputs", based only on "inputs". For forecasting the input is the present state and the output is the future state. The record of past states provides a set of known input-output pairs which acts as a "teacher". The problem is to generalize from the teaching set and estimate unknown outputs.

We can restate the problem more formally as follows: Given an input x_i and an output y_i, we want to find maps F and G of the form

$$\hat{y}_i = F(x_i, \alpha_i) \tag{54}$$
$$\alpha_{i+1} = G(x_i, \hat{y}_i, y_i, \alpha_i) \tag{55}$$

that minimize $\|y_i - \hat{y}_i\|$, where $\hat{y}$ is an estimate of y, and the metric $\| \ \|$ provides a criterion for the estimation accuracy. α_i are parameters for F, and G is a map that changes α_i, i.e., a *learning algorithm*. x, y and i can be either continuous or discrete. For a forecasting problem, for example, i corresponds to time, x is the current state x_t, and y is a future state, $y_t = x_{t+T}$.

Neural nets correspond to a particular class of functional forms for F and G. Although this form was originally motivated by biology, there is no reason to be constrained by this in artificial intelligence problems, as reflected by many recent developments in this field. Neural nets have had success in certain problems that can be solved by learning with a teacher, for example, text to speech conversion [67] or finding gene locations on DNA sequences [48]. Lapedes and Farber have also shown that neural nets can be effective for forecasting [49].

However, as recently pointed out by Omohundro [58], alternative approaches that depart significantly from the usual form of neural nets may be computationally much more efficient. Our approach to forecasting provides a good example of this; our methods give equivalent or more accurate forecasts than the neural net of Lapedes and Farber [49], and are several orders of magnitude more efficient in terms of computer time. Furthermore, since the computations can be performed in parallel, we expect that this speed discrepancy will persist even with future parallel hardware. Omohundro has pointed out that similar methods may be employed for other problems, such as associative memory, classification, category formation, and the construction of nonlinear mappings. Many aspects of the methods that we have proposed here are applicable to this broader class of problems.

Although we have assumed that x and y are continuous, with the addition of thresholds our methods are easily converted to the discrete domain. Our work, taken together with that of Lapedes and Farber, makes it clear that the neural network solves problems by surface estimation. They show that the same is true in the discrete domain, except that answers are obtained by "rounding" the surface, truncating to a discrete value. Generalization occurs through the extrapolation of the surface to regions in which there is no data [47]. There is no *a priori* reason to constrain the functional representation to those that are currently popular in the study of neural networks.

Radial basis functions provide a particularly promising possibility. One of the key properties of the two layer tanh network is *localization*; the composition of two tanh functions forms a well-localized bump, and by adding these together it is possible to represent arbitrary functions [47]. Radial basis functions are designed to be interpolants with good localization properties, and so should be ideal replacements for the tanh. Since their parameters can be fit through linear least squares, unique solutions for radial basis functions can be found very quickly. Furthermore, as recently shown by Casdagli [12], under favorable circumstances radial basis functions can achieve orders of approximation as high as six.

Clearly these possibilities deserve more investigation.

7 Conclusions

By assuming that a random process is produced by deterministic chaos, finding a good model reduces to two parts: (1) Finding a state space embedding that maximizes determinism, and (2) fitting a nonlinear functional form to the map that sends current states to future states.

The importance of the first problem should not be underestimated. The usual time consuming procedure of searching for a good embedding by trial and error is far from optimal, both because it is time consuming and because the results are not necessarily ideal. Some improvements on this have been suggested by Fraser and Swinney [30] and by Broomhead and King [11]. We have suggested an improvement on the technique of Broomhead and King which eliminates the last free parameter. We intend to address this problem in more detail in the future.

The next problem is to approximate the dynamics from the data. The primary approach we investigate here is approximation as a discrete time map, which has the advantage of being convenient and fast. Approximation in differential terms may promise more accuracy, however, and we intend to compare these two approaches in a future paper. In either case, the problem boils down to approximating the graph of a nonlinear function. Success depends on picking a good representation. There are two basic approaches, global and local. Global approximation is convenient, but unless the representation is well matched to the map it may not produce good results, especially since many of the standard nonlinear representations undergo an explosion of parameters as the dimension increases. Local approximation has the advantage that it is less dependent on representation, and is guaranteed to get better as the number of data points increases. When used in conjunction with a data structure such as the k-d tree, it can be quite fast. The best approach depends on the details of the problem, such as the nonlinear function being approximated, the number of data points, *etc.*

An advantage of formulating the forecasting problem in the language of deterministic chaos is that it makes it possible to derive error estimates. These estimates are couched in terms of properties of the dynamics, such as the Lyapunov exponents and the dimension, the length of the data set, its signal to noise ratio, and the extrapolation time. We arrive at the conclusion that iterative forecasts are better than direct forecasts, *i.e.*, it is better to make long-term forecasts by approximating the dynamics for a short time and iterating rather than approximating directly. The iterative approach takes advantage of the recursive form of the higher iterates of dynamical systems. With the iterative approach approximation errors grow at the same rate as errors due to uncertainty in the initial state, *i.e.*, they grow exponentially according to the largest Lyapunov exponent. These results are derived in the limit as $\bar{E} \to 0$; we have observed some counterexamples. Note that in order to study the behavior of direct forecasts we had to introduce the new concept of higher order Lyapunov exponents.

Having a model of the dynamics extends all the numerical techniques that were previously available only in numerical experiments to the analysis of data in real experiments. Furthermore, when it is possible to achieve higher orders of approximation, it becomes possible to extend the available data and obtain much more accurate results than would otherwise be possible. We have given some suggestions for this, but many questions remain to be investigated.

The ability to approximate nonlinear dynamics naturally leads to a method for reducing external noise through nonlinear averaging. When the dynamics are known exactly, this technique makes noise reductions of as much as ten orders of magnitude possible. When the dynamics must be learned the limitation on this technique comes from the the accuracy of the model. However, for low dimensional systems with a modest number of data points we can produce noise reductions of several orders of magnitude. Surprisingly, noise reduction is much easier for chaotic motion than for regular motion.

All of the methods discussed above work well for low dimensional deterministic chaos. When the dimension is low they give results that are orders of magnitudes better than those of standard linear methods. However, they lose their effectiveness when the dimension is too large. The limits can be estimated through the error estimate of Equation (34). As seen from this equation, they also depend on the method used: For a given number of points higher order approximation is more accurate than low order approximation. Of course, if we have extra information, such as the functional form of the dynamics, it may be possible to overcome the constraint of large dimensions.

The methods we have described here are new and not fully explored. We anticipate that there will be considerable progress in this area in the near future.

Acknowledgements We would like to thank Martin Casdagli, Ute Dressler, Pat Hagan, Alan Lapedes, Y.C. Lee, Steve Omohundro, Norman Packard, David Rand, Rob Shaw, and Blair Swartz for valuable conversations. We are grateful for support from the Department of Energy and the Air Force Office of Scientific Research under grant AFOSR-ISSA-87-0095.

We urge the reader to use these results for peaceful purposes.

References

[1] D.V. Anasov. Geodesic flows and closed riemannian manifolds with negative curvature. *Proc. Steklov Inst. Math.*, 90, 1967.

[2] M.F. Barnsley. Making chaotic dynamical systems to order. In *Chaotic Dynamics and Fractals*, pages 53–68, Academic Press, 1986.

[3] B.J. Bayly, I. Goldhirsch, and S.A. Orszag. *Independent Degrees of Freedom of Dynamical Systems.* Technical Report, Applied and Computational Mathematics, Princeton U. preprint, 1987.

[4] G. Bennetin, L. Galgani, A. Giorgilli, and J.M. Strelcyn. *Meccanica*, 15:9, 1980.

[5] J.H. Bentley. Multidimensional binary search trees in database applications. *IEEE Transactions on Software Engineering*, SE-5(4):333–340, 1979.

[6] J.H. Bentley. Multidimensional binary search trees used for associative searching. *Communications of the ACM*, 18(9):509–517, 1975.

[7] P. Berge, Y. Pomeau, and C. Vidal. *Order in Chaos.* 1986.

[8] R. Bowen. ω-limit sets for axiom a diffeomorphisms. *Journal of Differential Equations*, 18, 1975.

[9] A. Brandstäter, J. Swift, H.L. Swinney, A. Wolf, J.D. Farmer, E. Jen, and J.P. Crutchfield. Low-dimensional chaos in a hydrodynamic system. *Physical Review Letters*, 51(16):1442, 1983.

[10] D.S. Broomhead, R. Jones, and G.P. King. Topological dimension and local coordinates from time series data. *Journal of Physics A*, 20:L563–L569, 1987.

[11] D.S. Broomhead and G.P. King. Extracting qualitative dynamics from experimental data. *Physica*, 20D:217, 1987.

[12] M. Casdagli. *Nonlinear Prediction of Chaotic Time Series.* Technical Report, Queen Mary College, London, 1988.

[13] A. Cenys and K. Pyragas. *Estimation of the Number of Degrees of Freedom from Chaotic Time Series.* Technical Report, Institute of Semiconductor Physics, Academy of Sciences of the Lithuanain SSR, Vilnius 232600, 1987.

[14] G.J. Chaitin. Randomness and mathematical proof. *Scientific American*, 232(5):47–52, 1975.

[15] J.D. Cowan and D.H. Sharp. *Neural Nets.* Technical Report LA-UR-87-4098, Los Alamos Nat. Lab., 1987.

[16] J. Cremers and A. Hübler. Construction of differential equations from experimental data. *Z. Naturforsch.*, 42a:797–802, 1987.

[17] J. P. Crutchfield, J. D. Farmer, N. H. Packard, and R. S. Shaw. Chaos. *Scientific American*, 254(12):46–57, 1986.

[18] J.P. Crutchfield and B.S. McNamara. Equations of motion from a data series. *Complex Systems*, 1:417–452, 1987.

[19] Ute Dressler, J.D. Farmer, and P. Hagen. Higher order lyapunov exponents. in preparation.

[20] J.-P. Eckmann, S. Oliffson Kamphorst, D. Ruelle, and S. Ciliberto. Lyapunov exponents from a time series. *Physical Review*, 34A(6):4971–4979, 1986.

[21] J.-P. Eckmann and D. Ruelle. Ergodic theory of chaos and strange attractors. *Reviews of Modern Physics*, 57:617, 1985.

[22] P. Eykhoff. *System Identification, Parameter and State Estimation*. Wiley, New York, 1974.

[23] J. D. Farmer and J. J. Sidorowich. Predicting chaotic time series. *Physical Review Letters*, 59(8):845–848, 1987.

[24] J.D. Farmer. Chaotic attractors of an infinite-dimensional dynamical system. *Physica*, 4D:366–393, 1982.

[25] J.D. Farmer. Discovering relevance. in preparation.

[26] J.D. Farmer. Information dimension and the probabilistic structure of chaos. *Zeitschrift Naturforschung*, 37a:1304–1325, 1982.

[27] J.D. Farmer and J.J. Sidorowich. Predicting chaotic dynamics. In J.A.S. Kelso, A.J. Mandell, and M.F. Shlesinger, editors, *Dynamic Patterns in Complex Systems*, World Scientific, Singapore, 1988.

[28] J. Ford. *Physics Today*, 36(4):40, 1983.

[29] A.M. Fraser. *Information and Entropy in Strange Attractors*. Technical Report, U. of Texas, 1987. submitted to IEEE Transactions on Information Theory.

[30] A.M. Fraser and H.L. Swinney. Independent coordinates for strange attractors from mutual information. *Physical Review*, 33A:1134–1140, 1986.

[31] H. Froehling, J.P. Crutchfield, J.D. Farmer, N.H. Packard, and R.S. Shaw. On determining the dimension of chaotic flows. *Physica*, 3D:605, 1981.

[32] H. Fujisaka. Theory of diffusion and intermittency in chaotic systems. *Progress of Theoretical Physics*, 71(3):513–523, 1984.

[33] D. Gabor. Communcation theory and cybernetics. *Transactions of the Institue of Radio Engineers*, CT-1(4):9, 1954.

[34] D. Gabor, W.P. Wilby, and R. Woodcock. A universal nonlinear filter, predictor and simulator which optimizes itself by a learning process. *Proceeding of the IEEE*, 108B:422, 1960.

[35] P. Grassberger. *Information Content and Predictability of Lumped and Distributed Dynamical Systems*. Technical Report WU-B-87-8, University of Wuppertal, 1987.

[36] P. Grassberger and I. Procaccia. *Physical Review Letters*, 50:346, 1983.

[37] P. Grassberger and I. Procaccia. Dimensions and entropies of strange attractors from a fluctuating dynamics approach. *Physica*, 13D:34–55, 1984.

[38] P. Grassberger and I. Procaccia. Measuring the strangeness of strange attractors. *Physica*, 9D:189–208, 1983.

[39] J. Guckenheimer. Noise in chaotic systems. *Nature*, 298(5872):358–361, 1982.

[40] J. Guckenheimer and G. Buzyna. *Physical Review Letters*, 51:1438, 1983.

[41] V. Haggan, S.M. Heravi, and M.B. Priestley. A study of the application of state-dependent models in non-linear time series analysis. *Journal of Time Series Analysis*, 5(2):69–102, 1984.

[42] S.M. Hammel, J.A. Yorke, and C. Grebogi. Numerical orbits of chaotic processes represent true orbits. 1987. University of Maryland preprint.

[43] H. Haucke and R. Ecke. Mode locking and chaos in rayleigh-benard convection. *Physica*, 25D:307, 1987.

[44] J. Holland. Escaping brittleness: the possibilities of general purpose machine learning algorithms applied to parallel rule-based systems. In Michalski, Carbonell, and Mitchell, editors, *Machine Learning II*, Kaufmann, 1986.

[45] B.A. Huberman, J.P. Crutchfield, and N.H. Packard. Noise phenomena in josephson junctions. *Applied Physics Letters*, 37(8):750–752, 1980.

[46] A. Kolmogorov. Grundbergriffe der wahrscheinlichkeitsrechnung. *Eng. Mat.*, 2(3), 1933.

[47] A. S. Lapedes and R. M. Farber. *How Neural Nets Work*. Technical Report, Los Alamos Nat. Lab., 1988.

[48] A. S. Lapedes and R. M. Farber. Neural net learning algorithms and genetic data analysis. in preparation.

[49] A.S. Lapedes and R. Farber. *Nonlinear Signal Processing Using Neural Networks: Prediction and System Modeling*. Technical Report LA-UR-87-, Los Alamos National Laboratory, 1987. submitted to Proc. IEEE.

[50] K. Lee and Y.C. Lee. System modeling with rational functions. in preparation.

[51] G.G. Lorentz. *Approximation of Functions*. Holt, Rinehart, and Winston, New York, 1966.

[52] E.N. Lorenz. Atmospheric predictability as revealed by naturally occurring analogues. *Journal of the Atmospheric Sciences*, 26:636–646, 1969.

[53] E.N. Lorenz. Deterministic nonperiodic flow. *Journal of Atmospheric Science*, 20:130–141, 1963.

[54] E.N. Lorenz. The growth of errors in prediction. In *Turbulence and Predictability in Geophysical Fluid Dynamics and Climate Dynamics*, pages 243–265, Soc. Italiana di Fisica, Bologna, Italy, 1985.

[55] M.C. Mackey and L. Glass. *Science*, 197:287, 1977.

[56] B. Malraison, P. Atten, P. Bergé, and M. Dubois. Dimension of strange attractors: an experimental determination for the chaotic regime of two chaotic systems. *J. Phys. Lett. (Paris)*, 44:L897, 1983.

[57] G. Mayer-Kress, editor. *Dimensions and Entropies in Chaotic Systems*, Springer-Verlag, Berlin, 1986.

[58] S. Omohundro. Efficient algorithms with neural network behavior. *Complex Systems*, 1:273–347, 1987.

[59] N.H. Packard, J.P. Crutchfield, J.D. Farmer, and R.S. Shaw. Geometry from a time series. *Physical Review Letters*, 45:712–716, 1980.

[60] H. Poincaré. *Science et Methode*. Biblioteque Scientifique, 1908. English translation by F. Maitland (Dover, 1952).

[61] M.J.D. Powell. *Radial Basis Function for Multivariable Interpolation: A Review*. Technical Report, University of Cambridge, 1985.

[62] W.H. Press, B.P. Flannery, S.A. Teukolsky, and W.T. Vettering. *Numerical Recipes*. Cambridge University Press, 1986.

[63] M.B. Priestley. New developments in time-series analysis. In M.L. Puri, J.P. Vilapiana, and W. Wertz, editors, *New Perspectives in Theoretical and Applied Statistics*, John Wiley and Sons, 1987.

[64] M.B. Priestley. State dependent models: a general approach to nonlinear time series analysis. *Journal of Time Series Analysis*, 1:47–71, 1980.

[65] D. Rummelhart and J. McClelland. *Parallel Distributed Processing*. Volume 1, MIT Press, Cambridge, MA, 1986.

[66] M. Sano and Y. Sawada. Measurement of the lyapunov spectrum from chaotic time series. *Physical Review Letters*, 55:1082, 1985.

[67] T.J. Sejnowski and C.R. Rosenberg. Nettalk. *Complex Systems*, 1, 1987.

[68] R.S. Shaw. Strange attractors, chaotic behavior, and information flow. *Z. Naturforschung*, 36a:80–112, 1981.

[69] I. Shimada and T. Nagashima. A numerical approach to ergodic problem of dissipative dynamical systems. *Progress of Theoretical Physics*, 61:1605, 1979.

[70] E. Slutsky. The summatation of random causes as the source of cyclic processes. *Problems of Economic Conditions*, 3:1, 1927. English translation in Econmetric, 5, 1937, p. 105.

[71] F. Takens. Detecting strange attractors in fluid turbulence. In D. Rand and L.-S. Young, editors, *Dynamical Systems and Turbulence*, Springer-Verlag, Berlin, 1981.

[72] F. Takens. Invariants related to dimension and entropy. In *Atas do 13°*, Colóqkio Brasiliero de Matemática, Rio de Janeiro, 1983.

[73] F. Takens. On the numerical determination of the dimension of an attractor. In *Dynamical systems and Bifurcations*, Groningen, 1984.

[74] J. Theiler. *Quantifying Chaos: Practical Estimation of the Correlation Dimension*. PhD thesis, Cal. Tech., 1987.

[75] H. Tong and K.S. Lim. Threshold autoregression, limit cycles and cyclical data. *Journal of the Royal Statistical Society B*, 42(3):245–292, 1980.

[76] N. Wiener. *Nonlinear Problems in Random Theory*. Wiley, 1958.

[77] A. Wolf, J.B. Swift, H.L. Swinney, and J.A. Vastano. Determining lyapunov exponents from a time series. *Physica*, 16D:285, 1985.

[78] S. Wolfram. Origins of randomness in physical systems. *Physical Review Letters*, 55(5), 1985.

[79] G. U. Yule. *Philos. Trans. Roy. Soc. London A*, 226:267, 1927.

How Neural Nets Work *
Alan Lapedes
Robert Farber
Theoretical Division
Los Alamos National Laboratory
Los Alamos, NM 87545
January, 1988

Abstract:
There is presently great interest in the abilities of neural networks to mimic
"qualitative reasoning" by manipulating neural incodings of symbols. Less work
has been performed on using neural networks to process floating point numbers
and it is sometimes stated that neural networks are somehow inherently inaccurate and therefore best suited for "fuzzy" qualitative reasoning. Nevertheless,
the potential speed of massively parallel operations make neural net "number
crunching" an interesting topic to explore. In this paper we discuss some of our
work in which we demonstrate that for certain applications neural networks can
achieve significantly higher numerical accuracy than more conventional techniques. In particular, prediction of future values of a chaotic time series can be
performed with exceptionally high accuracy. We analyze how a neural net is
able to predict well , and in the process show that a large class of functions from
$R^n \rightarrow R^m$ may be accurately approximated by a backpropagation neural net
with just two "hidden" layers. The network uses this functional approximation
to perform either interpolation (signal processing applications) or extrapolation
(symbol processing applications). Neural nets therefore use quite familiar methods to perform their tasks. The geometrical viewpoint advocated here seems to
be a useful approach to analyzing neural network operation and relates neural
networks to well studied topics in functional approximation.

1. Introduction

Although a great deal of interest has been displayed in neural network's
capabilities to perform a kind of qualitative reasoning, relatively little work has
been done on the ability of neural networks to process floating point numbers
in a massively parallel fashion. Clearly, this is an important ability. In this
paper we discuss some of our work in this area and show the relation between
numerical, and symbolic processing. We will concentrate on the the subject of
accurate prediction in a time series. Accurate prediction has applications in
many areas of signal processing. It is also a useful, and fascinating ability, when
dealing with natural, physical systems. Given some data from the past history
of a system, can one accurately predict what it will do in the future?

Many conventional signal processing tests, such as correlation function analysis, cannot distinguish deterministic chaotic behavior from from stochastic
noise. Particularly difficult systems to predict are those that are nonlinear and
chaotic. Chaos has a technical definition based on nonlinear, dynamical systems
theory, but intuitivly means that the system is deterministic but "random," in
a rather similar manner to deterministic, pseudo random number generators
used on conventional computers. Examples of chaotic systems in nature include
turbulence in fluids (D. Ruelle, 1971; H. Swinney, 1978), chemical reactions (K.
Tomita, 1979), lasers (H. Haken, 1975), plasma physics (D. Russel, 1980) to
name but a few. Typically, chaotic systems also display the full range of nonlinear behavior (fixed points, limit cycles etc.) when parameters are varied, and
therefore provide a good testbed in which to investigate techniques of nonlinear
signal processing. Clearly, if one can uncover the underlying, deterministic algorithm from a chaotic time series, then one may be able to predict the future

* reprinted with kind permission of the American Institute of Physics Press.

time series quite accurately.

In this paper we review and extend our work (Lapedes and Farber,1987) on predicting the behavior of a particular dynamical system, the Glass-Mackey equation. We feel that the method will be fairly general, and use the Glass-Mackey equation solely for illustrative purposes. The Glass-Mackey equation has a strange attractor with fractal dimension controlled by a constant parameter appearing in the differential equation. We present results on a neural network's ability to predict this system at two values of this parameter, one value corresponding to the onset of chaos, and the other value deeply in the chaotic regime. We also present the results of more conventional predictive methods and show that a neural net is able to achieve significantly better numerical accuracy. This particular system was chosen because of D. Farmer's and J. Sidorowich's (D. Farmer, J. Sidorowich, 1987) use of it in developing a new, non-neural net method for predicting chaos. The accuracy of this non-neural net method, and the neural net method, are roughly equivalent, with various advantages or disadvantages accruing to one method or the other depending on one's point of view. We are happy to acknowledge many valuable discussions with Farmer and Sidorowich that has led to further improvements in each method. J. Crutchfield has also developed an accurate predictive method using different methodology than that of Farmer and Sidorowich, and also different to the neural net techniques described here. See (J. Crutchfield, 1987).

We also show that a neural net never needs more than two hidden layers to solve most problems . This statement arises from a more general argument that a neural net can approximate functions from $R^n \rightarrow R^m$ with only two hidden layers, and that the accuracy of the approximation is controlled by the number of neurons in each layer. The argument assumes that the global minimum to the backpropagation minimization problem may be found, or that a local minima very close in value to the global minimum may be found. This seems to be the case in the examples we considered, and in many examples considered by other researchers, but is never guaranteed. The conclusion of an upper bound of two hidden layers is related to a similar conclusion of R. Lipman (R. Lipman, 1986) who has previously analyzed the number of hidden layers needed to form arbitrary decision regions for symbolic processing problems. Related issues are discussed by J. Denker (J. Denker et.al. 1987). It is easy to extend the argument to draw similar conclusions about an upper bound of two hidden layers for symbol processing and to place signal processing, and symbol processing in a common theoretical framework.

2. Backpropagation

Backpropagation is a learning algorithm for neural networks that seeks to find weights, T_{ij}, such that given an input pattern from a training set of pairs of Input/Output patterns, the network will produce the Output of the training set given the Input. Having learned this mapping between I and O for the training set, one then applies a new, previously unseen Input, and takes the Output as the "conclusion" drawn by the neural net based on having learned fundamental relationships between Input and Output from the training set. A popular configuration for backpropagation is a totally feedforward net (Figure 1) where Input feeds up through "hidden layers" to an Output layer.

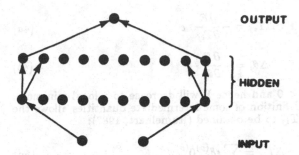

OUTPUT

HIDDEN

INPUT

Figure 1.
A feedforward neural
net. Arrows schemat-
ically indicate full
feedforward connect-
ivity

Each neuron forms a weighted sum of the inputs from previous layers to which it is connected, adds a threshold value, and produces a nonlinear function of this sum as its output value. This output value serves as input to the future layers to which the neuron is connected, and the process is repeated. Ultimately a value is produced for the outputs of the neurons in the Output layer. Thus, each neuron performs:

$$X_i^{out} = g\left(\sum_j T_{ij} X_j^{in} + \theta_i\right) \qquad (1)$$

where T_{ij} are continuous valued, positive or negative weights, θ_i is a constant, and $g(x)$ is a nonlinear function that is often chosen to be of a sigmoidal form. For example, one may choose

$$g(x) = \frac{1}{2}(1 + \tanh x) \qquad (2)$$

where tanh is the hyperbolic tangent, although the exact formula of the sigmoid is irrelevant to the results.
If $t_i^{(p)}$ are the target output values for the pth Input pattern then ones trains the network by minimizing

$$E = \sum_p \sum_i \left(t_i^{(p)} - 0_i^{(p)}\right)^2 \qquad (3)$$

where $t_i^{(p)}$ is the target output values (taken from the training set) and $0_i^{(p)}$ is the output of the network when the pth Input pattern of the training set is presented on the Input layer. i indexes the number of neurons in the Output layer.

An iterative procedure is used to minimize E. For example, the commonly used steepest descents procedure is implemented by changing T_{ij} and θ_i by ΔT_{ij} and $\Delta \theta_i$ where

$$\Delta T_{ij} = -\frac{\partial E}{\partial T_{ij}} \cdot \epsilon \tag{4a}$$

$$\Delta \theta_i = -\frac{\partial E}{\partial \theta_i} \cdot \epsilon \tag{4b}$$

This implies that $\Delta E < 0$ and hence E will decrease to a local minimum. Use of the chain rule and definition of some intermediate quantities allows the following expressions for ΔT_{ij} to be obtained (Rumelhart, 1987):

$$\Delta T_{ij} = \sum_p \epsilon \delta_i^{(p)} 0_j^{(p)} \tag{5a}$$

$$\Delta \theta_i = \epsilon \sum \delta_i^{(p)} \tag{5b}$$

where

$$\delta_i^{(p)} = \left(t_i^{(p)} - 0_i^{(p)}\right) 0_i^{(p)}(1 - 0_i^{(p)}) \tag{6}$$

if i is labeling a neuron in the Output layer; and

$$\delta_i^{(p)} = 0_i^{(p)}(1 - 0_i^{(p)}) \sum_j T_{ij}\delta_j^{(p)} \tag{7}$$

if i labels a neuron in the hidden layers. Therefore one computes $\delta_i^{(p)}$ for the Output layer first, then uses Eqn. (7) to computer $\delta_i^{(p)}$ for the hidden layers, and finally uses Eqn. (5) to make an adjustment to the weights. We remark that the steepest descents procedure in common use is extremely slow in simulation, and that a better minimization procedure, such as the classic conjugate gradient procedure (W. Press, 1986), can offer quite significant speedups. Many applications use bit representations (0,1) for symbols, and attempt to have a neural net learn fundamental relationships between the symbols. This procedure has been successfully used in converting text to speech (T. Sejnowski, 1986) and in determining whether a given fragment of DNA codes for a protein or not (A. Lapedes, R. Farber, 1987).

There is no fundamental reason, however, to use integer's as values for Input and Output. If the Inputs and Outputs are instead a collection of floating point numbers, then the network, after training, yields a specific continuous function in n variables (for n inputs) involving g(x) (i.e. hyperbolic tanh's) that provides a type of nonlinear, least mean square interpolant formula for the discrete set of data points in the training set. Use of this formula $0 = f(I_1, I_2, \ldots I_n)$ when given a new input not in the training set, is then either interpolation or extrapolation.

Since the Output values, when assumed to be floating point numbers may have a dynamic range great than $[0,1]$, one may modify the g(x) on the Output layer to be a linear function, instead of sigmoidal, so as to encompass the larger dynamic range. Dynamic range of the Input values is not so critical, however we have found that numerical problems may be avoided by scaling the Inputs (and also the Outputs) to [0,1], training the network, and then rescaling the T_{ij}, θ_i to encompass the original dynamic range. The point is that scale changes in I and O may, for feedforward networks, always be absorbed in the T_{ij}, θ_i and vice versa. We use this procedure (backpropagation, conjugate gradient, linear outputs and scaling) in the following section to predict points in a chaotic time series.

3. Prediction

Let us consider situations in Nature where a system is described by nonlinear differential equations. This is faily generic. We choose a particular nonlinear equation that has an infinite dimensional phase space, so that it is similar to other infinite dimensional systems such as partial differential equations. A differential equation with an infinite dimensional phase space (i.e. an infinite number of values are necessary to describe the initial condition) is a delay, differential equation. We choose to consider the time series generated by the Glass-Mackey equation:

$$\dot{x} = \frac{ax(t-\tau)}{1+x^{10}(t-\tau)} - bx(t) \tag{8}$$

This is a nonlinear differential, delay equation with an initial condition specified by an initial function defined over a strip of width τ (hence the infinite dimensional phase space i.e. initial functions, not initial constants are required). Choosing this function to be a constant function, and a = .2, b = .1, and $\tau = 17$ yields a time series, x(t), (obtained by integrating Eqn. (8)), that is chaotic with a fractal attractor of dimension 2.1. Increasing τ to 30 yields more complicated evolution and a fractal dimension of 3.5. The time series for 350 time steps for $\tau=30$ (time in units of τ) is plotted in Figure 2. The nonlinear evolution of the system collapses the infinite dimensional phase space down to a low (approximately 2 or 3 dimensional) fractal, attracting set. Similar chaotic systems are not uncommon in Nature. See (Farmer, 1982) for intensive analysis of the properties of this equation.

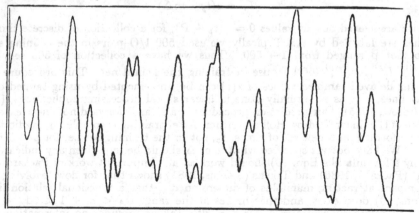

Figure 2. Example time series at tau = 30.

The goal is to take a set of values of x() at discrete times in some time window containing times less than t, and use the values to accurately predict x(t + P), where P is some prediction time step into the future. One may fix P, collect statistics on accuracy for many prediction times t (by sliding the window along the time series), and then increase P and again collect statistics on accuracy. This one may observe how an average index of accuracy changes as P is increased. In terms of Figure 2 we will select various prediction time steps, P, that correspond to attempting to predict within a "bump," to predicting a couple of "bumps" ahead. The fundamental nature of chaos dictates that prediction accuracy will decrease as P is increased. This is due to inescapable inaccuracies of finite precision in specifying the x(t) at discrete times in the past that are used for predicting the future. Thus, all predictive methods will degrade as P is increased – the question is "How rapidly does the error increase with P?" We will demonstrate that the neural net method can be orders of magnitude more accurate than conventional methods at large prediction time steps, P.

Our goal is to use backpropagation, and a neural net, to construct a function

$$0(t + P) = f(I_1(t), I_2(t - \Delta) \ldots I_m(t - m\Delta)) \qquad (9)$$

where $0(t + P)$ is the output of a single neuron in the Output layer, and $I_1 \rightarrow I_m$ are input neurons that take on values $x(t), x(t - \Delta) \ldots x(t - m\Delta)$, where Δ is a time delay. $O(t + P)$ takes on the value x(t + P). We chose the network configuation of Figure 1.

We construct a training set by selecting a set of input values:

$$I_1 = x(t_p)$$

$$I_2 = x(t_p - \Delta) \qquad (10)$$

$$I_m = x(t_p - m\Delta)$$

with associated output values $0 = x(t_p + P)$, for a collection of discrete times that are labelled by t_p. Typically we used 500 I/O pairs in the training set so that p ranged from $1 \rightarrow 500$. Thus we have a collection of 500 sets of $\{I_1^{(p)}, I_2^{(p)}, \ldots, I_m^{(p)}; 0^{(p)}\}$ to use in training the neural net. This procedure of using delayed sampled values of x(t) can be implemented by using tapped delay lines, just as is normally done in linear signal processing applications, (B. Widrow, 1985). Our prediction procedure is a straightforward nonlinear extension of the linear Widrow Hoff algorithm. After training is completed, prediction is performed on a new set of times, t_p, not in the training set i.e. for p = 500.

We have not yet specified what m or Δ should be, nor given any indication why a formula like Eqn. (9) should work at all. Important work of Packard et. al. (Packard, 1980) and Takens (Takens, 1981) shows that for flows evolving to compact attracting manifolds of dimension d_A, that a functional relation like Eqn. (9) does exist, and that m lies in the range $d_A < m + 1 < 2d_A + 1$. We therefore choose m = 4, for $\tau = 30$. Takens provides no information on Δ and we chose $\Delta = 6$ for both cases. We found that a few different choices of m and Δ can affect accuracy by a factor of 2 - a somewhat significant but not overwhelming sensitivity, in view of the fact that neural nets tend to be orders of magnitude more accurate than other methods. Takens theorem gives no information on the form of f() in Eqn. (9). It therefore is necessary to show that neural nets provide a robust approximating procedure for continuous f(), which we do in the following section. It is interesting to note that attempts to predict future values of a time series using past values of x(t) from a tapped delay line is a common procedure in signal processing, and yet there is little, if any, reference to results of nonlinear dynamical systems theory showing why any such attempt is reasonable.

After training the neural net as described above, we used it to predict 500 new values of x(t) in the future and computed the average accuracy for these points. The accuracy is defined to be the average root mean square error, divided by a constant scale factor, which we took to be the standard deviation of the data. It is necessary to remove the scale dependence of the data and dividing by the standard deviation of the data provides a scale to use. Thus the resulting "index of accuracy" is insensitive to the dynamic range of x(t).

As just described, if one wanted to use a neural net to continuously predict x(t) values at, say, 6 time steps past the last observed value (i.e. wanted to construct a net predicting x(t + 6)) then one would train one network, at P = 6, to do this. If one wanted to always predict 12 time steps past the last observed x(t) then a separate, P = 12, net would have to be trained. We, in fact, trained separate networks for P ranging between 6 and 100 in steps of 6. The index of accuracy for these networks (as obtained by computing the index of accuracy in the prediction phase) is plotted as curve D in Figure 3. There is however an alternate way to predict. If one wished to predict, say, x(t + 12) using a P = 6 net, then one can iterate the P = 6 net. That is, one uses the P = 6 net to predict the x(t +6) values, and then feeds x(t +6) back into the input line to predict x(t + 12) using the **predicted** x(t + 6) value instead of the **observed** x(t + 6) value. In fact, one can't use the observed x(t +6) value, because it hasn't been observed yet – the rule of the game is to use only data occurring at time t and before, to predict x(t +12). This procedure corresponds to iterating the map given by Eqn. (9) to perform prediction at multiples of P. Of course, the delays, Δ, must be chosen commensurate with P.

This iterative method of prediction has potential dangers. Because (in our example of iterating the P = 6 map) the predicted x(t + 6) is always made with some error, then this error is compounded in iteration, because predicted, and not observed values, are used on the input lines. However, one may predict more accurately for smaller P, so it may be the case that choosing a very accurate small P prediction, and iterating, can ultimately achieve higher accuracy at the larger P's of interest. This turns out to be true, and the iterated net method is plotted as curve E in Figure 3. It is the best procedure to use. Curves A,B,C are alternative methods (iterated polynomial, Widrow-Hoff, and non-iterated polynomial respectively. More information on these conventional methods is in (Lapedes and Farber, 1987)). Independent analysis of the iterative procedure may be found in (D. Farmer, J. Sidorowich, 1988), as well as a discussion of the effects of noise and noise reduction procedures and also a clear discussion of the Local Linear Method used by Farmer et. al.

338

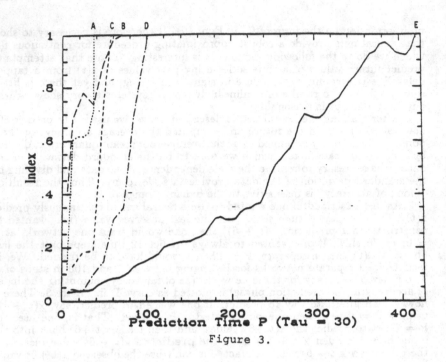

Figure 3.

4. Why It Works

Consider writing out explicitly Eqn. (9) for a two hidden layer network where the output is assumed to be a linear neuron. We consider Input connects to Hidden Layer 1, Hidden Layer 1 to Hidden Layer 2, and Hidden Layer 2 to Output. Therefore:

$$0_\ell = \sum_{k \epsilon H_2} T_{\ell k} g \left(\sum_{i \epsilon H_1} T_{ki} g \left(\sum_{j \epsilon I} T_{ij} I_j + \theta_j \right) + \theta_k \right) + \theta_\ell \qquad (11)$$

Recall that the output neurons a linear computing element so that only two g()s occur in formula (11), due to the two nonlinear hidden layers. For ease in later analysis, let us rewrite this formula as

$$0_\ell = \sum_{k \epsilon H_2} T_{\ell k} g \left(SUM_k + \theta_k \right) + \theta_\ell \qquad (12a)$$

where

$$SUM_k = \sum_{i \epsilon H_1} T_{ki} g \left(\sum_{j \epsilon I} T_{ij} I_j + \theta_j \right). \qquad (12b)$$

The T's and θ's are specific numbers specified by the training algorithm, so that after training is finished one has a relatively complicated formula (12a, 12b) that expresses the Output value as a specific, known, function of the Input values:

$$0_\ell = f(I_1, I_2, \ldots I_m).$$

A functional relation of this form, when there is only one output, may be viewed as surface in m + 1 dimensional space, in exactly the same manner one interprets the formula z = f(x,y) as a two dimensional surface in three dimensional space. The general structure of f() as determined by Eqn. (12a, 12b) is in fact quite simple. From Eqn. (12b) we see that one first forms a sum of g() functions (where g() is s sigmoidal function) and then from Eqn. (12a) one forms yet another sum involving g() functions. It may at first be thought that this special, simple form of f() restricts the type of surface that may be represented by $0_\ell = f(I_j)$. This initial thought is wrong – the special form of Eqn. (12) is actually a general representation for quite arbitrary surfaces.

To prove that Eqn. (12) is a reasonable representation for surfaces we first point out that surfaces may be approximated by adding up a series of "bumps" that are appropriately placed. An example of this occurs in familiar Fourier analysis, where wave trains of suitable frequency and amplitude are added together to approximate curves (or surfaces). Each half period of each wave of fixed wavelength is a "bump," and one adds all the bumps together to form the approximant. Let us now see how Eqn. (12) may be interpreted as adding together bumps of specified heights and positions. First consider SUM_k which is a sum of g( ) functions. In Figure (4) we plot an example of such a g() function for the case of two inputs.

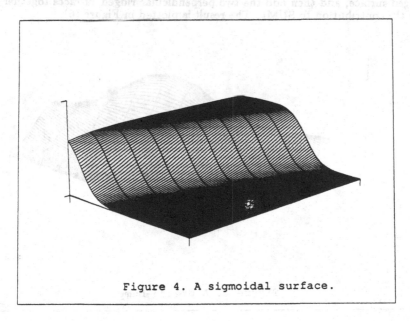

Figure 4. A sigmoidal surface.

340

The orientation of this sigmoidal surface is determined by T_{ij}, the position by θ_j, and height by T_{ki}. Now consider another g() function that occurs in SUM_k. The θ_j of the second g() function is chosen to displace it from the first, the T_{ij} is chosen so that it has the same orientation as the first, and T_{ki} is chosen to have opposite sign to the first. These two g( ) functions occur in SUM_k, and so to determine their contribution to SUM_k we sum them together and plot the result in Figure (5). The result is a ridged surface.

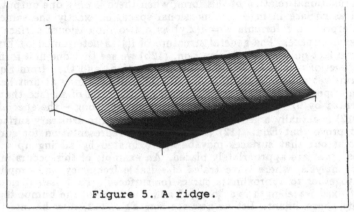

Figure 5. A ridge.

Since our goal is to obtain localized bumps we select another pair of g() functions in SUM_k, add them together to get a ridged surface perpendicular to the first ridged surface, and then add the two perpendicular ridged surfaces together to see the contribution to SUM_k. The result is plotted in Figure (6).

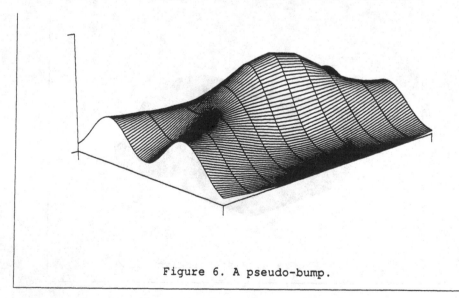

Figure 6. A pseudo-bump.

We see that this almost worked, in so much as one obtains a local maxima by this procedure. However there are also saddle-like configurations at the corners which corrupt the bump we were trying to obtain. Note that one way to fix this is to take $g(\text{SUM}_k + \theta_k)$ which will, if θ_k is chosen appropriately, depress the local minima and saddles to zero while simultaneously sending the central maximum towards 1. The result is plotted in Figure (7) and is the sought after bump.

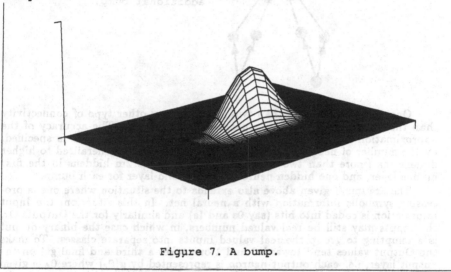

Figure 7. A bump.

Furthermore, note that the necessary $g()$ function is supplied by Eqn. (12). Therefore Eqn. (12) is a procedure to obtain localized bumps of arbitrary height and position. For two inputs, the kth bump is obtained by using four $g()$ functions from SUM_k (two $g()$ functions for each ridged surface and two ridged surfaces per bump) and then taking $g()$ of the result in Eqn. (12a). The height of the kth bump is determined by T_{θ_k} in Eqn. (12a) and the k bumps are added together by that equation as well. The general network architecture which corresponds to the above procedure of adding two $g()$ functions together to form a ridge, two perpendicular ridges together to form a pseudo-bump, and the final $g()$ to form the final bump is represented in Figure (8). To obtain any number of bumps one adds more neurons to the hidden layers by repeatedly using the connectivity of Figure (8) as a template (i.e. four neurons per bump in Hidden Layer 1, and one neuron per bump in Hidden Layer 2).

342

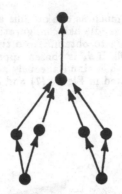

Figure 8. Connectivity needed to obtain one bump. Add four more neurons to Hidden layer 1, and one more neuron to Hidden Layer 2, for each additional bump.

One never needs more than two layers, or any other type of connectivity than that already schematically specified by Figure (8). The accuracy of the approximation depends on the number of bumps, which in turn is specified, by the number of neurons per layer. This result is easily generalized to higher dimensions (more than two Inputs) where one needs 2m hiddens in the first hidden layer, and one hidden neuron in the second layer for each bump.

The argument given above also extends to the situation where one is processing symbolic information with a neural net. In this situation, the Input information is coded into bits (say 0s and 1s) and similarly for the Output. Or, the Inputs may still be real valued numbers, in which case the binary output is attempting to group the real valued Inputs into separate classes. To make the Output values tend toward 0 and 1 one takes a third and final $g()$ on the output layer, i.e. each output neuron is represented by $g(O_\ell)$ where O_ℓ is given in Eqn. (11). Recall that up until now we have used linear neurons on the output layer. In typical backpropagation examples, one never actually achieves a hard 0 or 1 on the output layers but achieves instead some value between 0.0 and 1.0. Then typically any value over 0.5 is called 1, and values under 0.5 are called 0. This "postprocessing" step is not really outside the framework of the network formalism, because it may be performed by merely increasing the slope of the sigmoidal function on the Output layer. Therefore the only effect of the third and final $g()$ function used on the Output layer in symbolic information processing is to pass a hyperplane through the surface we have just been discussing. This plane cuts the surface, forming "decision regions," in which high values are called 1 and low values are called 0. Thus we see that the heart of the problem is to be able to form surfaces in a general manner, which is then cut by a hyperplane into general decision regions. We are therefore able to conclude that the network architecture consisting of just two hidden layers is sufficient for learning any symbol processing training set. For Boolean symbol mappings one need not use the second hidden layer to remove the saddles on the bump (c.f. Fig. 6). The saddles are lower than the central maximum so one may choose a threshold on the output layer to cut the bump at a point over the saddles to yield the correct decision region. Whether this representation is a reasonable one for subsequently achieving good prediction on a prediction set, as opposed to "memorizing" a training set, is an issue that we address below.

We also note that use of Sigma II; units (Rummelhart, 1986) or high order correlation nets (Y.-C. Lee, 1987) is an attempt to construct a surface by a general polynomial expansion, which is then cut by a hyperplane into decision regions, as in the above. Therefore the essential element of all these neural net learning algorithms are identical (i.e. surface construction), only the particular method of parameterizing the surface varies from one algorithm to another. This geometrical viewpoint, which provides a unifying framework for many neural net algorithms, may provide a useful framework in which to attempt construction of new algorithms.

Adding together bumps to approximate surfaces is a reasonable procedure to use when dealing with real valued inputs. It ties in to general approximation theory (c.f. Fourier series, or better yet, B splines), and can be quite successful as we have seen. Clearly some economy is gained by giving the neural net bumps to start with, instead of having the neural net form its own bumps from sigmoids. One way to do this would be to use multidimensional Gaussian functions with adjustable parameters.

The situation is somewhat different when processing symbolic (binary valued) data. When input symbols are encoded into N bit bit-strings then one has well defined input values in an N dimensional input space. As shown above, one can learn the training set of input patterns by appropriately forming and placing bump surfaces over this space. This is an effective method for memorizing the training set, but a very poor method for obtaining correct predictions on new input data. The point is that, in contrast to real valued inputs that come from, say, a chaotic time series, the input points in symbolic processing problems are widely separated and the bumps do not add together to form smooth surfaces. Furthermore, each input bit string is a corner of an 2^N vertex hypercube, and there is no sense in which one corner of a hypercube is surrounded by the other corners. Thus the commonly used input representation for symbolic processing problems requires that the neural net extrapolate the surface to make a new prediction for a new input pattern (i.e. new corner of the hypercube) and not interpolate, as is commonly the case for real valued inputs. Extrapolation is a farmore dangerous procedure than interpolation, and in view of the separated bumps of the training set one might expect on the basis of this argument that neural nets would fail dismally at symbol processing. This is not the case.

The solution to this apparent conundrum, of course, is that although it is sufficient for a neural net to learn a symbol processing training set by forming bumps it is not necessary for it to operate in this manner. The simplest example of this occurs in the XOR problem. One can implement the input/output mapping for this problem by duplicating the hidden layer architecture of Figure (8) appropiately for two bumps (i.e. 8 hiddens in layer 1, 2 hiddens in layer 2). As discussed above, for Boolean mappings, one can even eliminate the second hidden layer. However the architecture of Figure (9) will also suffice.

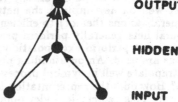

OUTPUT

Figure 9. Connectivity for XOR

HIDDEN

INPUT

344

Plotting the output of this network, Figure(9), as a function of the two inputs yields a ridge orientated to run between (0,1) and (1,0) Figure(10). Thus a neural net may learn a symbolic training set without using bumps, and a high dimensional version of this process takes place in more complex symbol processing tasks.Ridge/ravine representations of the training data are considerably more efficient than bumps (less hidden neurons and weights) and the extended nature of the surface allows reasonable predictions i.e. extrapolations.

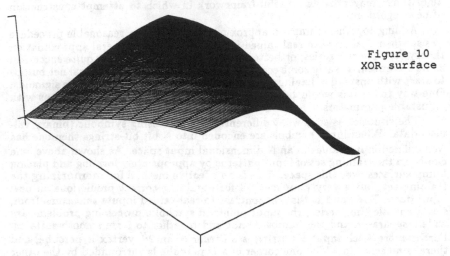

Figure 10
XOR surface

5. Conclusions

Neural nets, in contrast to popular misconception, are capable of quite accurate number crunching, with an accuracy for the prediction problem we considered that exceeds conventional methods by orders of magnitude. Neural nets work by constructing surfaces in a high dimensional space, and their operation when performing signal processing tasks on real valued inputs, is closely related to standard methods of functional approximation. One does not need more than two hidden layers for processing real valued input data, and the accuracy of the approximation is controlled by the number of neurons per layer, and not the number of layers. We emphasize that although two layers of hidden neurons are sufficient they may not be efficient. Multilayer architectures may provide very efficient networks (in the sense of number of neurons and number of weights) that can perform accurately and with minimal cost.

Effective prediction for symbolic input data is achieved by a slightly different method than that used for real value inputs. Instead of forming localized bumps (which would accurately represent the training data but would not predict well on new inputs) the network can use ridge/ravine like surfaces (and generalizations thereof) to efficiently represent the scattered input data. While neural nets generally perform prediction by interpolation for real valued data, they must perform extrapolation for symbolic data if the usual bit representations are used. An outstanding problem is why do tanh representations seem to extrapolate well in symbol processing problems? How do other functional bases do? How does the representation for symbolic inputs affect the ability to extrapolate? This geometrical viewpoint provides a unifying framework for examing many neural net algorithms, for suggesting questions about neural net operation, and for relating current neural net approaches to conventional methods.

Acknowledgments

We thank Y. C. Lee, J. D. Farmer, and J. Sidorowich for a number of valuable discussions. In particular we wish to thank D. Farmer for interesting us in the problem of prediction in chaotic time series, and for sharing numerous insights into nonlinear dynamical systems.

References

C. Barnes, C. Burks, R. Farber, A. Lapedes, K. Sirotkin, "Pattern Recognition by Neural Nets in Genetic Databases", manuscript in preparation

J. Crutchfield et. al. "Equations of Motion From A Data Series", Complex Systems 1::417-452, 1987

J. Denker et. al.,"Automatic Learning, Rule Extraction,and Generalization", ATT, Bell Laboratories preprint, 1987

D. Farmer, J.Sidorowich, Phys.Rev. Lett., 59(8), p. 845,1987

D. Farmer, Physica 4D, No.3 , p. 366, 1982

D. Farmer, J. Sidorowich, "Exploiting Chaos To Predict The Future", Los Alamos preprint, Jan. 1988

H. Haken, Phys. Lett. A53, p77 (1975)

A. Lapedes, R. Farber "Nonlinear Signal Processing Using Neural Networks: Prediction and System Modelling", LA-UR87-2662,1987

Y.C. Lee, Physica 22D,(1986)

R. Lippman, IEEE ASAP magazine,p.4, 1987

N. Packard et.al. Phys. Rev. Lett. 45, no. 9, p712 (1980)

D. Ruelle, F. Takens, Comm. Math. Phys. 20, p167 (1971)

D. Rummelhart, J. McClelland in "Parallel Distributed Processing" Vol. 1, M.I.T. Press Cambridge, MA (1986)

D. Russel et al., Phys. Rev. Lett. 45, p1175 (1980)

T. Sejnowski et al., "Net Talk: A Parallel Network that Learns to Read Aloud," Complex Systems 1, 1987

346

H. Swinney et al., Physics Today 31 (8), p41 (1978)

F. Takens, "Detecting Strange Attractor in Turbulence," Lecture Notes in Mathematics, D. Rand, L. Young (editors), Springer Berlin, p366 (1981)

K. Tomita et al., J. Stat. Phys. 21, p65 (1979)

PATTERN RECOGNITION AND SINGLE LAYER NETWORKS

by

Tom Maxwell

In this chapter we introduce the general pattern recognition problem and develop a mathematical framework for describing a class of adaptive pattern classifiers. The perceptron learning rule and the concept of order will be discussed and examples will be given. We will conclude with a demonstration of symmetry detection using high order networks.

Distinctions and Differences

At the root of all cognition is the ability to draw distinctions. Input data must be sorted into categories in order that the regularities which relate the present to past experience and models can become apparent. Since the present is never exactly the same as the past, it only by grouping similar input patterns into categories that the correspondence can be drawn between current experience and prior knowledge. Thus, as a first step in developing the tools necessary for modeling cognition, we will examine architectures for classifying input patterns.

Adaptive Pattern Classifiers

We define a classification as a mapping which takes as input a pattern vector $x = x(j)$ (bold face print will represent tensors throughout) and returns a scalar y which is typically restricted to a small range of values. We define a pattern classifier as an adaptive unit which also takes as input a pattern vector x and yields a scalar as output. If the pattern classifier yields a output which is identical to the output of the classification mapping for all inputs x of interest, then we say that the pattern classifier has implemented the classification. The components x(j) of x will be called "pixels". Usually y will be binary (+1,-1), each value representing a category into which the input vector may be placed. We define a slab of pattern classifiers as a group of classifiers which all receive the same input vector x in parallel. The output of a classifier slab will be indexed as y(i). We will be interested in adaptive pattern classifiers, whose input-output characteristics are governed by a set of parameters or "weights" W, usually represented as a set of matrices. The process of "tuning" the classifier often involves iteratively updating the weights until the desired mapping is achieved, however in some cases the weights may be calculated analytically directly from the input data. Thus our general pattern classifier can be written in the form

$$y = f(x,W) \tag{1}$$

where $y = y(i)$ is a vector of classifications, x is an input vector to be classified, f is a fixed (usually nonlinear) vector function, and W is a set of parameters which determines the structure of the categories.

Discriminant Functions

The structure of the categories generated by a pattern classifier can be described in terms of discriminant functions. Any input vector $x = x(j)$ of length Nx can be represented as a point in Nx-dimensional Euclidean space, known as pattern space. The components of the vector x(j) correspond to the rectangular coordinates of the representative point in pattern space. We shall use the notation x to denote both the input vector and the point in pattern space. Here we will consider only pattern dichotomizers ($y \in (+1,-1)$) since this is our primary interest. However, the following discussion can easily be generalized to cover the case of discrimination between many categories. A dichotomy is defined as a mapping which takes a vector x as input and yields a binary ((0,1) or (-1,1)) number as output.

The pattern dichotomizer (PD) assigns to each point in pattern space a value (+1 or -1) which indicates into which category the PD is grouping the point. Thus we can divide pattern space into two regions such that one region (denoted as R_1) maps to category 1 and the other region (denoted as R_2) maps to category 2. This division of pattern space can be represented by a function net(x,W), called a discriminant function, which has the property that, for some value of the parameter tensor W,

$$\text{net(x,W)} \begin{cases} > 0 & \text{if } x \in R1 \\ < 0 & \text{if } x \in R2. \end{cases} \qquad (2)$$

Points exactly on the boundary will be classified arbitrarily. The two regions R1 and R2 are separated from each other by a surface, defined by net(x,W) = 0, called a decision surface. The parameter tensor W controls the structure of the categorization, or equivalently, the position, shape, and orientation of the decision surface. With this definition of net(x,W) we can define a set of pattern dichotomizers by the equation

$$y(i) = \text{sgn}[\ \text{net(x,W(i))}\ ] \qquad (3)$$

where sgn[x] is the signum function defined by

$$\text{sgn}[\ x\ ] = \begin{cases} 1 & \text{if } x > 0 \\ -1 & \text{if } x \leq 0. \end{cases}$$

Within this framework, the problem of constructing a slab of pattern dichotomizers to implement a given classification has been reduced to two steps:
(1) Determine a discriminant function net(x,W) which will be suitable to the set of problems of interest.
(2) Determine an appropriate set of parameters W for the particular problem of interest.
We will now discuss these steps, beginning with the choice of a discriminant function.

Choosing A Discriminant Function

There are two factors which should be considered in the choice of a discriminant function:
   (1) The structure of the discriminant function should match the structure of the set of problems of interest. By this we mean that every classification of interest should be achievable through adjusting the parameters W.
   (2) The form of net(x,W) should be simple enough that rules can specified for the determination of the parameter values W from a given set of examples

of the desired classification.

A class of discriminate functions which has the generality to satisfy condition (1) and the simplicity to satisfy condition (2) is the Φ functions (Duda and Hart, 1973). A Φ function is a discriminate function in which the parameter values W appear only linearly. Thus an arbitrary Φ function can be represented by the equation:

$$net(x,W(i)) = \sum_j W(i,j) \, f_j(x) \qquad (4)$$

where the $f_j(x)$, which we call the "basis functions", are arbitrary fixed functions of x. The corresponding pattern dichotomizer is described by the equation

$$y(i) = sgn[\ \sum_j W(i,j) \, f_j(x)\ ]. \qquad (5)$$

The set basis functions embodies the internal representation used by the system to implement the dichotomy. If there are N parameters we say that the Φ function (or equivalently, the internal representation) has N degrees of freedom. In the computer science literature, a dichotomy is commonly called a "predicate". In the following, we will denote a predicate by the symbol r.

The constraint of linearity in the weights is required in order to define simple rules for determining the parameters W (these rules will be discussed in a later section). Due to the arbitrary nature of the $f_j(x)$ functions, Φ functions can be chosen which will perform efficiently in many different problem areas. After a short digression we will address the problem of choosing a Φ function for a particular problem area.

An important special case is the linear Φ function, described by the equation:

$$net(x,W(i)) = \sum_j W(i,j) \, x(j). \qquad (6)$$

In this case the decision surface defined by $net(x,W(i)) = 0$ is a hyperplane. Points which lie on the same side of the hyperplane will be lumped into the same category by eqn. 3. Categories which are capable of being separated by a hyperplane are said to be linearly separable.

Example: Exclusive - Or

The exclusive-or mapping, which is the simplest nonlinearly separable mapping, will be used to illustrate various concepts. This mapping can be expressed as:

Exclusive-Or:	$x(1)$	$x(2)$	τ
	1	1	1
	-1	-1	1
	-1	1	-1
	1	-1	-1

Figure 1 is a plot of the four patterns, displayed as points in pattern space. The *'s map to +1 and the o's map to -1. It is not difficult to show that there is no hyperplane (in this case, a line) which will separate the two pattern classes. Thus, this mapping is not linearly separable, and a nonlinear decision surface will be required to implement the classification. As shown in figure 2a, the categories are separated by the second order Φ function

$$net(x,W) = W(1)x(1) + W(2)x(2) + W(3)x(1)x(2)$$

with weight values $W(1) = W(2) = 0$ and $W(3) = 1$.

Another way of viewing this solution is in terms of a coordinate transformation to a higher dimensional space, which we call the "expanded space", given by:

$$x(1)' = x(1), \qquad x(2)' = x(2), \qquad x(3)' = x(1)x(2).$$

Figure 1 Exclusive-Or Mapping in Pattern Space.

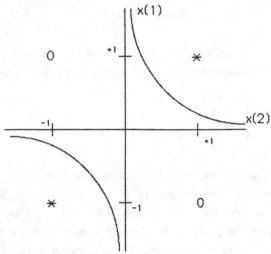

Figure 2a Nonlinear decision Surface.

The four patterns, displayed in this new space, are plotted in fig. 2b. In this space the categories are separable with a linear decision surface (eqn. 6) with weight values $W(1) = W(2) = 0$, and $W(3) = 1$ (which corresponds to the hyperplane defined by $x(3)' = 0$). Thus we see that the use of a higher order Φ function in pattern space is equivalent of the use of a linear discriminant function in an appropriately defined higher dimensional space, and that mappings which are not linearly separable in pattern space may be linearly separable when transformed to a higher dimensional expanded space. The procedure of transforming to a higher dimensional space in order to enhance linear separability is commonly called "sparse coding".

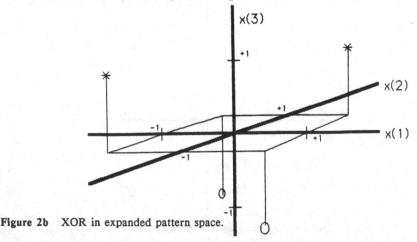

Figure 2b XOR in expanded pattern space.

The Concept of Order

The process of choosing a Φ function to implement a predicate is greatly facilitated if one can determine the order (first defined by Minsky and Papert 1969) of the desired predicate. The order of a predicate, which will be defined rigorously below, is a measure of the number of pixels that the basis functions $f_j(x)$ must depend on in order to implement the dichotomy with a system described by eqn. 5.

In order to define order rigorously, we must introduce some new notation. Let the input pattern vector be written as $x = x_R = \{ x(j) \mid j \in R \}$, where R represents the set of all points (each point is represented by a possible value of the index j) which make up the pattern vector x. A subset x_P of x is represented by $x_P = \{ x(j) \mid j \in P \}$, where P is a subset of R. We next define the support of a basis function, which intuitively represents the set of points that the function "really depends on". Define the support $S(f_j)$ of a basis function f_j as the smallest set of points S such that $f_j(x_P) = f_j(x_{S \cap P})$ for all P, where $S \cap P$ is the intersection of S and P, or the set of points contained in both S and P, and P is a subset of R. Let $|S|$ denote the number of points in S.

We can now define the order of a predicate τ as the smallest number k for which we can find a Φ function (represented by equation 4) satisfying the conditions:

(1) $| S(f_j) | \leq k$ for all j, and   (8)

(2) τ is implemented by equation (5).

The order of a predicate is a measure of its computational complexity. Our goal is to express complex predicates in terms of Φ functions whose basis functions depend on a minimal number of points. Thus if we express a predicate of order k in terms of a Φ function (eqn. 5), then we must utilize a set of basis functions which depend on at least k points. In the next section we describe the process of choosing a Φ function.

Choosing a Φ Function

For the purposes of this discussion, we will consider only Φ functions which can be expressed as polynomials. A Φ function of order k will be defined by the equation

$$net_k(x,W(i)) = W_0(i) + \sum_j W_1(i,j)x(j) + \sum_j \sum_k W_2(i,j,k)x(j)x(k) +...$$

$$(9)$$

where the high order sums are terminated after degree k in x(j). We define a kth order unit by the equation:

$$y(i) = sgn[\ net_k(x,W(i)\ ].$$

$$(10)$$

Each value of the subscript i represents a unit, the entire network of units represents a single "slab".

In some areas of application there are motivations for considering other types of Φ functions. For example, trigonometric basis functions may be most efficient in applications involving control of robot arm movements. However, we feel that with the proper coding of the inputs x(j), equation (9) should provide an efficient representation for most problems of interest.

Notice that the number of terms in a kth order Φ function is of order N^k, where N = |R| is the number of points in the input pattern x. Since N may be very large, the computation complexity of the dichotomizer will increase rapidly as the order of the Φ function increases. Thus it is advantageous to utilize a Φ function of the lowest order possible.

In order to determine the complexity of the Φ function required to implement a predicate τ, we wish to establish a correspondence between the order of the Φ function and the order of the predicate. The following theorem follows directly from a theorem proved by Minsky (1969, Theorem 1.5.3), so that interested readers are referred there for a proof. The proof depends on the assumption that the input vector x is binary (x(i) $\in$ (0,1) or (-1,1)).

Theorem 1:

A predicate τ is of order k if and only if the lowest order unit which can implement the predicate is of order k.

Thus a kth order unit will implement any kth order predicate. If we can determine the order of the predicate, then we know the order of the unit required to implement it. Minsky and Papert illustrate methods for determining the order of predicates. We will not discuss these procedures here. However, below we list some example predicates and their order. (The input vector is assumed to be binary).

Predicate	Order
And, Or	1
Exclusive-Or	2
N-Input Parity	N
Detection of Mirror Symmetries	2

In general, the order of a predicate depends on the coding of the input vector. Sparse coding can be used to reduce the order of a predicate.

Storage Capacity of a Φ Machine

Given a Φ machine with M adjustable weights and a training set of N_p patterns, we want to determine the probability $P(M,N_p)$ that a randomly selected predicate will be implementable by the unit. This result is derived by Nilsson (1965), and can be stated as (assuming M larger then about 25):

$$P(M,N_p) \sim \left\{ \begin{array}{l} 1 \text{ if } N_p < 2(M+1) \\ 0 \text{ if } N_p > 2(M+1) \end{array} \right\}. \tag{11}$$

This (approximate) step function tends to flatten out a bit as M decreases below 25. Also, since the P function never actually goes to zero, there will be a small probability of implementing a randomly selected predicate even for $N_p \gg 2(M+1)$.

We can now define the capacity C(M) of a Φ machine with M weights as the maximum size of the training set for which a randomly selected predicate can be implemented with probability near 1:

$$C(M) = 2(M+1).$$

For a polynomial unit of order r the capacity can be expressed as

$$C(r) \sim (N_x)^r$$

where N_x is the number of components (pixels) in the input vector.

Supervised Learning Problem

The supervised learning problem can be stated as follows. We are given a mapping M from a set of patterns x^s, $s \in \{1...NP\}$, to a set of classes. With each input pattern x^s is associated a class, and this class is denoted by a binary vector t^s:

$$M: \ x^s \rightarrow t^s \text{ for all } s \in \{1...NP\}. \tag{12}$$

The pair (x^s, t^s) is called an example of the map M and the full set of examples of the map is called the training set TS.

We are also given a network of adaptive pattern classifiers (which may already have been hand-crafted) designed to implement the mapping. We will denote the output of the network for input pattern x^s by $y(x^s)$.

The supervised learning problem consists of finding a set of weights which minimizes some error function E. A common example of an error function is the sum of the squares of the differences:

$$E = \sum_s (t^s - y(x^s)). \tag{13}$$

Many methods have been employed to solve this problem; several methods will be discussed below.

Optimal Associative Mappings.

Since networks of fully nonlinear units are hard to describe mathematically, let us start by examining the properties of the linear mapping network defined by:

$$y(x^s) = Wx^s, \qquad s \in \{1...NP\}, \tag{14}$$

where $x \in R^n$, $y \in R^p$, and W is a p by n matrix. Typically, we will have NP > n > p. This network can be considered to be a simple pattern dichotomizer, equivalent to equation 9 without a threshold function. Through a simple coordinate transformation to an expanded space, terms of arbitrarily high order in the pattern vector can be embedded in the input vector x^s.

With this network the supervised learning problem becomes a problem in minimizing the matrix Y - WX, where Y is the p by NP matrix whose columns are the vectors t^s and X is the n by NP matrix whose columns are the vectors x^s. As discussed extensively in (Kohonen, 1984), the best approximate solution is

$$W = YX^+ \tag{15}$$

where X^+ is the pseudo-inverse of X. The pseudo-inverse is a generalization of the concept of inverse which is applicable when the matrix to be inverted is not square. For the case in which the rows of X are linearly independent (which is to be expected in the case of NP > n) the best approximate solution can be written as

$$W = YX^T(XX^T)^{-1} = C_{yx}C_{xx}^{-1} \tag{16}$$

where X^T denotes the transpose of X and X^{-1} denotes the inverse of X. C_{yx} and C_{xx} are cross-correlation and auto-correlation matrices respectively:

$$C_{xx} = (1/m) \sum_{k=1}^{m} y^s x^{sT} \qquad C_{yx} = (1/m) \sum_{k=1}^{m} x^s x^{sT}. \tag{17}$$

In this form the solution is computationally expensive, however, since computing an inverse is expensive and there is no way to add the effect of a new example to the matrix W other then recomputing the matrix using the entire training set. However, there are iterative methods for calculating W which are of the form

$$W^k = W^{k-1} + (t^k - W^{k-1}x^k)c^{kT} \tag{18}$$

where W^k is the kth estimate of W, x^k and t^k are the kth example of the map, and c^k is a gain vector. The form of the gain vector c^k will vary depending on whether k is less then or greater then n. However, if the vectors x^s are approximately orthonormal, then in all cases the simplification occurs:

$$c^k \sim x^k \qquad \text{(if } x^k \text{ are approximately orthonormal).} \tag{19}$$

To the extent to which the vectors x^k deviate from orthonormality, the matrix

W^k will deviate from optimality. This simplification yields the training rule:

$$W^k = W^{k-1} + \propto (t^k - y^k) x^{kT}. \tag{20}$$

where $\propto$ is a learning rate parameter and $y^k = W^{k-1} x^k$.

The form of this rule is very similar to the form of many other training rules which will be discussed below (perceptron rule, back propagation, etc.). In all of these training rules, the weight matrix is updated by the product of an error vector, consisting of the difference between the desired output and the actual output, and the input vector. The differences in the various rules arise from the effects of threshold nonlinearities used in the associated networks, and the parameters used to insure convergence.

Another form of learning rule can be obtained from eqn. (16) by assuming that the x^k are approximately orthonormal. In this case we have $XX^T = C_{xx} \sim I$, the identity matrix, and eqn (16) becomes:

$$W \sim C_{yx} = YX^T = (1/m) \sum_{k=1}^{m} y^s x^{sT}. \tag{21}$$

This is the well-known outer product (or "Hebbian") learning rule which will be discussed in greater detail below.

Perceptron Learning Rule

In the previous section we discussed optimal learning rules for linear networks. In this section we discuss a generalization of one of these rules to a nonlinear pattern classifier (eqn. 5). Note that the nonlinear pattern classifier differs from the linear network only in the addition of a threshold nonlinearity. The perceptron training procedure has been discussed extensively by Rosenblatt (1962) and Minsky and Papert (1969), and is the best-known algorithm for training single layers of nonlinear pattern classifiers. It is a supervised learning procedure which consists of the following steps:

(1) Choose (sequentially) an example input x from the training set (arranged in a random periodic sequence).

(2) Generate the network output y(i) via eqn. 6.

(3) Update the weights W(i,j) via the eqn:

$$W(i,j)' = W(i,j) + \alpha(t(i)-y(i))f_j(x) \tag{22}$$

where $t(i)$ is the target, or desired output, α is a constant and the notation is slightly different from the previous section..

(4) Repeat the above procedure until the unit is able to correctly classify all patterns in the training set.

Note that the perceptron rule, eqn. (22), is identical to the iterative approximation to the optimal linear map (eqn. 20) if we (1) make a coordinate change from expanded space to pattern space $x'(j) = f_j(x)$ and (2) reinterpret y as the output of the nonlinear pattern classifier (6), instead of the linear network (14).

In order to develop an intuitive geometric interpretation of the perceptron rule, we look at a single classifier and make the coordinate change to the expanded space $x'(j) = f_j(x)$. The perceptron can then be described by the eqns. (dropping the primes on the x's):

$$y = \text{sgn}[\; \sum_j W(j)x(j)\;] = \text{sgn}[\; Wx\;] \tag{23a}$$

with the weight update rule:

$$W' = W + Ex, \quad \text{where } E = (t-y)/2 \tag{23b}$$

Here Wx denotes a vector dot product and we have taken $\alpha = 1/2$. We see that the classification of the pattern vector x is the sign of the dot product between x and the weight vector W. If x points in the same direction as W, then $y = +1$, otherwise $y = -1$. A geometric representation of these concepts is shown in fig. 3. Here we plot both x and W as vectors with tails at the origin, and construct a decision plane Γ running through the origin perpendicular to W. Then by eqn. (23), any vector x which lies on the same side of Γ as W will yield the classification $y = +1$ and any vector which lies on the opposite side will yield $y = -1$. The error term $E = (t-y)/2$ determines wether or not the classification was correct, yielding $E = 0$ if correct and $E = t$ if incorrect. Thus we see that if the classification is correct there is no

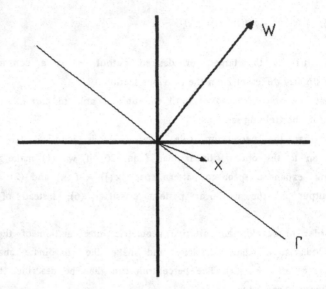

Figure 3 Decision plane in pattern space.

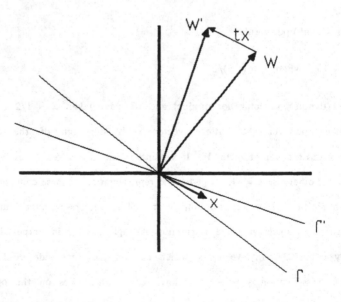

Figure 4 Weight update in pattern space.

change in the weight vector. However, if the classification is incorrect, the weight vector is updated by the amount:

$$W' = W + tx \quad \text{(if the classification of x is incorrect).} \tag{24}$$

Thus if x and W lie on the same side of the decision surface Γ and $t = -1$ (meaning that x and W should be on opposite sides of Γ for correct classification) then the weight vector W is shifted to W - x, thus swinging it away from x and making it more likely that W and x will lie on opposite sides of Γ the next time that x is encountered (see fig 4). Similarly, if x and W lie on opposite sides of Γ and $t = 1$, then W is updated to W + x, swinging it toward x and making it more likely that W and x will be on the same side of Γ the next time that x is encountered.

The procedure outlined above is repeated until a weight vector W is generated which yields $E = 0$ for all patterns in the training set. At this point we say that the learning procedure has converged to a solution of the classification problem. It can be proven that, provided the mapping is linearly separable in the expanded space, the perceptron learning procedure described above always converges in a finite number of steps. This important fact has been proven many times in many different ways; see for example Rosenblatt (1962), Nilsson (1965), or Minsky (1969).

Symmetry Detection Problem

As an illustration of the concepts previously discussed, we consider the symmetry detection problem. This problem involves the determination of which of four (or three) possible mirror symmetries exist in an N by N input pattern. The four symmetries are reflection about centered horizontal and vertical axes, and reflection about the two diagonal (corner-to-corner) axes. Every input pattern contains one of the four symmetries; the task of the network is to learn to identify which of the symmetries is present. This is a second order problem, because a comparison of two points is required in order to obtain a useful bit of information concerning the existence of a mirror symmetry. Thus the implementation of this mapping will require either a cascade of first order units or a single slab of second order units for its solution. The performance of both back propagation and simulated annealing on this problem have been investigated by other researchers, and these results are reproduced below. In this section we investigate the performance of a single slab of second order units on this problem. We will be using two supervised learning procedures: perceptron learning and outer product learning.

Simulation Description

All simulations were performed with a single slab of second order units with local interactions. These units can be described by the equation:

$$y(i) = \text{sgn}[\ net(i)\ ],$$

$$net(i) = W_0(i) + \sum_{jx}^{N} \sum_{jy}^{N} W_1(i,jx,jy)x(jx,jy) \tag{25}$$

$$+ \sum_{jx}^{N} \sum_{jy}^{N} \sum_{dx}^{2} \sum_{dy}^{2} W_2(i,jx,jy,dx,dy)x(jx,jy)x(jx+dx,jy+dy)\ ]$$

where x is the two-dimensional input vector, y is the output vector, and $W_k(..)$ are the kth order adaptive weights. The input vectors consist of a set of 10 by 10 patterns ($N = 10$) of binary ($+1,-1$) pixels. Each pattern belongs to one of the four symmetry groups listed above. The second order interactions have a range of 2 in the x and y directions, which means that any given point (pixel) will have a second order interaction with all other points within a 5x5 square centered on that point. Two output encodings were explored; a coarse coding scheme in which i varied from 1 to 2, and a sparse (unary) coding scheme, in which i ranges from 1 to 4. In the coarse coding scheme, the two output pixels were trained to encode a binary representation of the number of the symmetry group, where the order of the numbering is taken to be arbitrary. In the sparse coding scheme, each output bit is trained to be an identifier (grandmother cell) for a particular symmetry group, and the set of allowable output patterns is restricted to the set of patterns with one bit equal to one and the rest of the bits zero. Both perceptron and Hebbian learning rules were applied.

As discussed in the previous section, the perceptron learning rule is an incremental, error correcting procedure. It involves presenting training patterns sequentially, and updating the unit's weights after each presentation in an attempt to minimize the difference between the actual output of the unit on that pattern and the desired output. For this problem, the perceptron learning rule (for the second order term) can be expressed as

$$W_2(i,jx,jy,dx,dy)' = W_2(i,jx,jy,dx,dy) +$$
$$(t(i)-y(i))x(jx,jy)x(jx+dx,jy+dy) \qquad (27)$$

where $t(i)$ is the target output for the ith unit. Similar equations exist for other orders. Note that no zero or first order terms were utilized.

The Hebbian learning rule is a "one-shot" learning procedure, in which the values of the weights are generated analytically from the training set without sequential presentation of the patterns. It can be written as (for the second order terms):

$$W_2(i,jx,jy,dx,dy) = \sum_{p \in P} T(i,s(p))x^p(jx,jy)x^p(jx+dx,jy+dy) \qquad (28)$$

where the index p runs over the training set TS, and we have adopted the notation TS = { $x^p \mid p \in P$ }, so that x^p is the pth pattern in the training set. $T(i,s(p))$ is an target output matrix which depends on the output encoding scheme, and $s(p)$ is the number of the symmetry group of the pth pattern. For example, for the sparse coding scheme:

$$T(i,s(p)) = \left\{ \begin{array}{ll} 1 & \text{if } i = s(p) \\ -1/3 & \text{otherwise} \end{array} \right\} \qquad (29)$$

Similar equations exist for other orders and coding schemes. These values were chosen such that $T(i,s(p))$ satisfies the conditions:

$$\sum_i T(i,s(p)) = 0 \quad \text{and} \quad \sum_p T(i,s(p)) = 0 \qquad (30)$$

where we have assumed an equal number of training patterns in each symmetry group. These properties serve to eliminate spurious (zeroth order) correlations from eqn. 28 and play an important role in the discussion of the learning procedure below.

Simulation Results

Figures 5 through 7 summarize the simulation results for the network described above. In these figures, Nsym denotes the number of symmetry classes of the input patterns, and Noput denotes the number of output units. The weights were initialized to zero. In each iteration a randomly generated symmetric pattern was presented such that in Nsym iterations a pattern from each of the Nsym classes was presented. The weights were initialized to zero. In figures 5 thru 6 we show the learning curves for Hebbian learning with the two output encodings discussed above. In figure 7, three symmetry classes were used (all four classes were used in the other simulations). We see that the network achieves 95% accuracy on new (unseen) patterns after seeing approximately 500 training patterns. If we note that the total number of symmetric patterns is $Nsym(2^{50})$, we see that the network is generalizing very efficiently. Further simulations indicated that hebbian learning is considerable more efficient then perceptron learning on this problem.

The learning procedure can be understood as follows. Consider the ith unit, which is being trained via the Hebbian learning procedure, to detect a mirror symmetry σ_i. For simplicity of analysis, we shall examine the following form of the Hebbian learning procedure

$$W_2(i,j1,j2) \quad = \quad \sum_{p \in P} \delta(i,s(p)) \; x^p(j1) \; x^p(j2) \tag{31a}$$

where the sum runs over the training set TS, s(p) is the number of the symmetry class of the pth pattern, and

$$\delta(x,y) \quad = \quad \{ \begin{array}{ll} 1 & \text{if } x = y \\ 0 & \text{otherwise} \end{array}$$

is the delta function. Points $j1 = (jx,jy)$ and $j2 = (jx+dx,jy+dy)$ will be called σ_i-symmetric if they are mirror symmetric points under σ_i. When a second order weight for unit i represents a product between σ_i-symmetric points, then each time a pattern with that particular symmetry appears (probability 1/4), this weight is updated by +1. All other weights will be updated randomly by either +1 or -1. Thus, after the presentation of Ns patterns, The second order weights for the ith unit will be given by

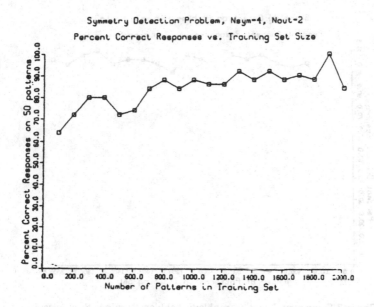

Figure 5 Symmetry detection problem, hebbian learning, coarse coding.

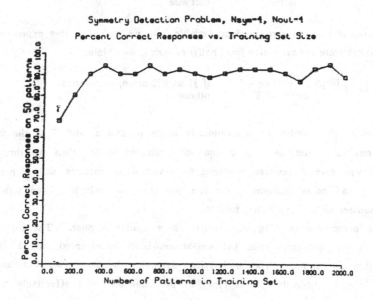

Figure 6 Symmetry detection problem, hebbian learning, sparse coding.

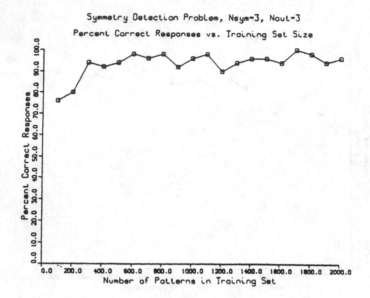

Figure 7 Symmetry detection problem, hebbian learning, three classes.

$$W_2(i,j1,j2) = \quad \{ \begin{array}{ll} Ns/4 + noise & \text{if } j1 \text{ and } j2 \text{ are } \sigma_i\text{-symmetric} \\ noise & \text{otherwise} \end{array} \qquad (32)$$

where noise is a mean-zero random variable. If we now use this expression in eqn. 26 to calculate the activation level net(i) of unit i, we obtain

$$net(i) \quad = \quad \{ \begin{array}{ll} N\sigma_i Ns/4 + noise - T & \text{if } j1 \text{ and } j2 \text{ are } \sigma_i\text{-symmetric} \\ noise - T & \text{otherwise} \end{array} \qquad (33)$$

where $N\sigma_i$ is the number of σ_i-symmetric terms in unit i, and T is the zeroth order threshold. Thus, as long as the noise remains smaller then the threshold, the unit will give a positive response to σ-symmetric patterns and a negative response to all other patterns. We can say that the unit has been sensitized to, or resonates with, σ-symmetric patterns.

The perceptron learning rules works in a similar manner. The difference is that, in the perceptron rule, the weight matrix is incremented only when the unit makes an error. The underlying dynamics is the same, but analytic analysis is more difficult. What types of problems are most effectively tackled by each of these algorithms is an open question.

Implementing Invariances

If it is known in advance that a problem possesses a certain set of invariances, it is possible to short-cut the learning process by constructing a network for solving this problem which already has these invariances imprinted. For example, one might construct a robotic brain with a priori knowledge of certain laws of physics, or a pattern recognizer with a priori knowledge of translation and rotation invariance. One method of implementing arbitrary transformation invariances in neural networks involves averaging the input of each unit over the transformation group, so that the ability to detect features which are incompatible with the imposed symmetries is washed out. All features that are equivalent under the transformation group are treated equivalently by the (invariant) units. This invariance constraint restricts the possible responses of the system to a given set of inputs, and in many cases leads to a reduction in order of the correlation matrix, thus reducing the number of high order terms required to implement a desired function.

The method of averaging over transformation groups was first presented in a neurological context by Pitts and McCulloch (1947). They discussed the hypothesis, along with copious neurological evidence possibly supporting it, that some form of this averaging process is the method by which the brain learns "universals", or invariance properties. Later Minsky and Papert (1969) used similar methods in a group-invariance theorem to demonstrate the limitations of perceptrons. Maxwell, Giles, and Lee (1986) explored the capabilities of group-invariant neural networks.

Before we introduce the general equations for implementing invariances, we will illustrate the procedure in a somewhat more intuitive fashion. Suppose we are given a second order unit described by the equation

$$y[x] = F[\ \sum_j \sum_k W_2(j,k)x(j)x(k) \ ]. \tag{34}$$

We wish to determine what constraints must be placed on the W matrix in order to ensure that the unit's output y is invariant under the translation group T. We define the translation group as the set of all translation operators: $T = \{g(m)\}$, where $g(m)x(k) = x(m+k)$. Thus the invariance constraint becomes

$$y[x] = y[g(m)x] \text{ for all } g(m) \in T. \tag{35}$$

Expanding the right hand side of eqn. 35 gives

$$y[g(m)x] = F[\sum_{j}\sum_{k} W_2(i,j,k)x(j+m)x(k+m)]$$

$$= F[\sum_{j}\sum_{k} W_2(i,j-m,k-m)x(j)x(k)] \tag{36}$$

where in the last step we have assumed either that periodic boundary conditions are appropriate or that edge effects are negligible. A comparison of eqns. 35 and 36 reveals that the constraint that must be placed on W in order to ensure translation invariance is:

$$W_2(j,k) = W_2(j-m,k-m). \text{ (Translation invariance constraint).} \tag{37}$$

In order to understand this condition, let us note that W_2 is a function of a pattern composed of two points located at j and k. The constraint of eqn. 37 stipulates that W_2 have the same value when evaluated at any pattern $(j-m,k-m)$ which is a translated version of the original pattern (j,k). In other words, in order to ensure translation invariance, W must depend only on the equivalence class of patterns under translation; the function $W(j,k)$ will not distinguish between any two patterns (j,k) and (j',k') which are translation-equivalent. This constraint can be generalized to implement invariance under an arbitrary transformation group.

It is worth noting that for the first order weight W_1, the invariance constraint is $W_1(j) = W_1(j-m)$, or $W_1(j) = $ constant. It follows the first order translation invariant term is of the form:

First order translation invariant term = $\quad W_1 \sum_{j} x(j).$

The only information contained in this term is the average pattern value, so that this term alone is of little use for pattern classification.

Another way of expressing the invariance constraint derived in the previous paragraph becomes clear if we note that the set of equivalence classes (under translation) of two point patterns is indexed by the relative coordinate $dj = j - k$. In other words, every possible value of dj corresponds to a single equivalence class and vice versa. We now note that the condition of eqn. 37 is equivalent to constraining W_2 such that it depends only on the relative coordinate dj (representing an equivalence class), and not on the absolute coordinates j and k (which would distinguish between different patterns belonging to the same equivalence class):

$$W(j,k) \rightarrow W(dj) \quad \text{(Interpreted translation invariance constraint).} \quad (38)$$

One way to get a feeling for eqn. 38 is to imagine particles located at points j and k, and envision $W(j,k)$ as representing an interparticle force or energy. If the inter-particle force depends only on the relative position dj of the particles, and not on the absolute positions j and k, then the force will be the same no matter where the particles are located in space, as long as they maintain the same relative position. This is the same as saying that the force depends only on the equivalence class (under translation) of two-particle patterns.

Implementing Invariances: General Case

We now consider the general case of implementing an arbitrary transformation invariance in a adaptive learning unit. We define an adaptive learning unit by the equation:

$$y(x) = F(net(x)) \quad (39)$$

where F is a fixed nonlinearity, and net(x) is a Φ function of x, defined by

$$net(x) = \sum_j w(j) \, f_j(x) \quad (40)$$

$f_j(x)$ is a set of fixed functions of x called the basis functions, and w(j) is a set of adaptive weights.

A unit is invariant under the action of a transformation group $G = \{g\}$ iff:

$$y(gx) = y(x) \text{ for all } g \in G. \quad (41)$$

We can now state the following theorem.

370

THEOREM: The unit defined by

$$y(x) = F(\sum_{g \in G} net(gx)) \tag{42}$$

(where the sum is over all members of the group G) is invariant under the action of the group G.

PROOF: We will demonstrate that eqn. 31 holds for units of this form.

For $h \in G$, we have:

$$y(hx) = F(\sum_{g \in G} net(ghx)) = F(\sum_{g' \in G} net(g'x)) = y(x)$$

where we have made the substitution $g' = gh$. The second step follows from the fact that a sum over $g = g'h^{-1}$ is equivalent to a sum over g', since multiplying all members of a group by a member of the same group simply results in a permutation of the members of the group. (This property of groups can be proven from the group properties listed above). QED.

The above theorem provides us with a general method for constructing group-invariant units by the method of averaging over transformation groups. If we write out the Φ function in eqn. 32 we arrive at the following form for a G-invariant unit:

$$y(x) = F(\sum_{g \in G} \sum_j f_j(gx)). \tag{33}$$

This method can be utilized to construct a wide range of invariant units.

Conclusion

The key to efficient learning in single layer networks is the choice of an appropriate Φ function. If the structure of the Φ function matches the structure of the set of problems of interest, then every classification of interest will be achievable through adjusting the parameters W, and learning will occur rapidly and efficiently. If the structure of the Φ function does not match the structure of the problem set, then there will be an important class of discriminations which will not be implementable with the network. Choosing an appropriate Φ function is not an easy task, and no general rules exist. However, within the class of polynomial networks there are some general guidelines: (1) match the order of the network to the order of the problem (if known); (2) if the problem possesses a set of invariances, implement these invariances in the network utilizing the methods described above; (3) if only local order is important, utilize a Φ function which captures only short-range correlations; etc. With methods such as these, one can adapt the structure of the network to the structure of the problem domain, resulting in efficient learning and generalization. For further discussion of these topics and other computer simulations, see Maxwell (1988).

References

Duda, RO, and Hart, PE (1973) *Pattern Classification and Scene Analysis*. New York: Wiley.

Kohonen, T (1984) *Self Organization and Associative Memory*, Springer Verlag, Berlin.

Minsky, ML and Papert, S. (1969) *Perceptrons: An Introduction to Computational Geometry*. Cambridge, MA, MIT Press.

Maxwell, T, Giles, CL, Lee, YC and Chen, HH (1986) Transformation Invariance Using High Order Correlations in Neural Network Architectures, *Proc. IEEE Intl. Conf. on Systems, Man and Cyb.*, Atlanta, Ga., p. 627.

Maxwell, T (1988) *Nonlinear Dynamics of Artificial Neural Systems*. Ph.D. dissertation, Physics Dept., Un. of Md., College Park, Md.

Nilsson, NJ (1965) *Learning Machines*. McGraw-Hill, Inc.

Rosenblatt, F (1962) *Principles of Neurodynamics*. New York: Spartan Books.

WHAT IS THE SIGNIFICANCE OF NEURAL NETWORKS FOR AI ?

by Harold H. Szu
Naval Research Laboratory, Code 5756
Washington D.C. 20375

ABSTRACT

Associative memory (AM) and attentive associative memory (AAM) have been reviewed in terms of simple neural networks (both uniform and non-uniform matched filter banks: read by inner products and write by outer products in parallel). While AM has been applied to the optical character recognition (OCR) using the set of orthogonal feature vectors deduced from image processing and computer vision, AAM can incorporate AI expert system techniques for determining the non-uniform linear combination of outer products. A rule-based system can more efficiently incorporate the frequency distribution of distorted characters according to user group profiles, say left-handed writing versus right-handed writing. Specifically in this paper, we have examined the degree of fault tolerance in AM, the ability of generalization by interpolation (auto-associative memory) and abstraction by extrapolation (hetero-associative memory). The efficiency of the closed system of rule-based knowledge representation of AI using the tuple storage has been combined with the flexibility of the non-rule based open system using the matrix knowledge representation of NI (coined for either Neural, or Network, or Natural Intelligence). Thus, the ability of generalization and abstraction becomes possible in a combined intelligent system of AI and NI.

1. INTRODUCTION

The question of the significance of neural networks for AI may be subdivided into three aspects.

(i) How can neural networks help solve AI problems ?

ANSWER: Both the well understood fault-tolerance of associative memory (AM), and the lesser understood ability of neural networks for generalization and abstraction, can be usefully incorporated into AI techniques.

(ii) How can AI help solve neural network problems ?

ANSWER: Similar to computer aided design, AI expert systems with a neural network modules can help design special purpose architectures for neural network computing.

(iii) What unsolved problems can be solved efficiently by combining AI and NI (coined for either Neural, or Network, or Natural Intelligence) techniques to utilize their respective strengths?

ANSWER: The optical character recogniton (OCR) for reading hand-written bank check and zip-codes, can be solved by combining both AI and NI techniques, as described in this paper.

Because we can only build a small neural network, we wish to endow a small set of neurons with a human-like intelligence. With present technology, whether it be electronic or optical, one cannot build a neural network of more than several hundred neurons, using existing processor elements (PE's), because of the technological difficulty associated with dense interconnectivity, about N^2 for N PE's. Thus, artificial neural networks can not yet match the size and the complexity of the human brain, that has billions of neurons and thousands of interconnects for each neuron. If we are *not*, overly ambitious in developing a **general purpose** neural computer, we can built a **special purpose** neural computer for solving special purpose problems, such as OCR.

One way to accomplish this special purpose neural computer is to combine the traditional rule-based AI wisdom with non-rule-based NI learning. This is particularly desirable in solving OCR problems because the available small neural networks can use better feature vectors obtained from other disciplines. Neural networks, built with current technology, can then provide fault tolerance for input feature vectors variations. The specific problem of hand-written character recognition, differs from the more regular, hand-printed, alphanumeric recognition problem in that it must account for such complications as connected characters and characters broken by segmentation.

Conceptually, one could solve the OCR problem using analytic, rule-based AI or neural network techniques. The OCR problem can be subdivided into character (or character string) statistics, font recognition, and character recognition; the most efficient techniques for these three subproblems are analytic (statistical), rule-based AI, and neural networks, respectively. Since the statistical techniques, applied to alphanumeric frequencies, is well known, this topic will not be discussed further. In solving the font recognition subproblem, AI rules can be set by the (statistical) frequency distribution of individual distorted characters according to user group profiles, e.g. left-handed writing versus right-handed writing. It is efficient to design AI expert system that draw upon the classical statistical pattern recognition, e.g. one stroke difference exists between "P " and "R ", or between"O " and "Q ", or in a low pass filter viewpoint only one stroke locational difference exists among four rounded letters "P " and "R ","O ", and "Q ". Furthermore, the AI rules of **pair character distortion distribution** can help solve the problem of *connected character* and *broken character after segmentation.* such as **two scripted zeros**. The **pair characer correlation matrix can be analyzed by the technique of the Karhunen-Loeve** procedure in image processing. The Karhunen-Loeve technique is compatable with AM's outer product decomposition. With the help of AI rule-based system, both the first and the second order statistics can be incorporated in the formalism of **attentive associative memory** (AAM), that processess the extra degrees of freedom in the non-uniform storage of vector outer products based on a given set of critical feature vectors.

Because the open-ended knowledge of input pattern variations may be efficiently controlled by using other disciplinary knowledge, such as AI and computer vision with a result of better combined technology, we shall review AM and AAM, and various OCR approaches by means of their specific techniques used for feature extraction and techniques used for group classification. The sooner we accept implementation limitations of the present neurocomputers, the better we can work with other disciplinary researchers. For example, we can work with researchers in AI, computer vision, image processing. Since this cross disciplinary collaboration is by nature not easy because of different trainings and languages involved, then this paper may serve a door opener for both.

Pattern recognition reseachers have been successful in machine-printed character recognition (CR) compared to optical character recognition (OCR) of hand-written bank checks or zipcodes. Difficulties of applying AI alone to an intelligent OCR may be due to the lack of non-rule-based capability of generalization and abstraction. This may be constrained by the traditional AI **one dimensional (1-D) knowledge representation,** e.g. an ordered set of tuples used in semantic networks. Similarly, difficulties of applying the neural network alone to an intelligent OCR may be in selecting critical features that is precisely one of the most challenging and unsolved problems (others are segmentations and locations). On the other hand, AI is efficient in reduce the problem to a sub-problem based on **1-D** knowledge representation of simple rules, and NI provides the fault-tolerant OCR system based on **2-D** knowlege representation. Together they give the possibility of generalization and abstraction. Thus, Szu and Tan (1988) have considered a less risky approach that consists of the traditional AI researchers who know about OCR critical features, and the neural network experts who know about AM fault tolerance. Technological developments have pointed to the readiness of such collaborations, since 2-D storage by chips or optical disks becomes cheaper than the traditional 1-D content addressable memory (CAD) processor. What's needed is a smart coprocessor such as neurocomputer. As a matter of fact, due to the 2-D nature of light, optical expert systems based on AM have been designed by Szu and Caulfield (1987) who have shown as simple replacement of 1-D tuples by 2-D matrices in a semantic network the alias problem for data fusion is solved by matrix addition and thresholding. The opto-electronical implementation of attentative associative memory model of Athale, Szu & Frielander (1986) can be expanded by means of a priori probability compiled by a pair-character correlation function of script letters. These papers may facilitate both sides the starting line of collaborations.

In this paper, we have reviewed the orthogonal subspaces of features and examined (1) the degree of fault tolerance , (2) the generalization by interpolation to other orthogonal feature vectors within the subspace, and (3) the abstraction by extrapolation to other subspaces. **AAM** may be formulated by a linear combination of outer products based on a set of orthogonal feature vectors. The combination coefficient is called the attention parameter, because it enters into the eigenvalue of **AAM** matrix that governs the recall convergence. We review briefly about the dynamics of attentive associative memory published by Szu (1988) elsewhere using arbitrary coefficients. In this paper we explicitly introduce a AI-tuple for the attention vector $a = \{a_n , n=1,...M\}$, where the inner product between the difference vector between an averaged stochastic input $|Q>$ and a fixed memory state $|m>$ is naturally used as the attention parameter defined in terms of Dirac's inner product notation: $a_m = <m|m> - <m|Q>$. Such an **AAM** matrix has non-white eigenvalue spectrum $\lambda_n \equiv a_n - (A / B)$ where the attentive memory capacity is $A \equiv \Sigma^M_{n=1} a_n$, and B is the length of the feature vectors (e.g. the number of bits). Iterative recalls are used. **Paying non-uniform attention** ($a_n \geq 1$) **increases the memory capacity** $A \geq M$ **together with a faster convergence rate proportional to the larger eigenvalue** $\lambda_m \geq \lambda$ **than a uniform attention** (i.e. $a_m = 1$). Szu's (1988) analysis has suggested that the eigenvalue spectrum and its dithering by input ensemble can play a crucial role for the convergence associated with a nonlinear dynamical system.

2. Associative Memory

Matrix associative memory works like a parallel bank of matched filters but much more efficiently in at least three counts: (1) no address coding of input and decoding for output are necessary , (2) operations are done in parallel, and (3) the connectivity matrix can be determined by itself using various adaptive (learning) algorithms.

An analytical and numerical example of AM is given as follows:

We denote M feature vectors as binary words, $U^{(m)}$, m=1,...M. Each word has B bits. The inner product of Eq(1) measures the norm, the number of bits that are one.

$$U^T \cdot U = \text{\# of one's} \qquad (1)$$

where the superscript transpose the column vector to a row vector.

The associated **bipolar words,** denoted by $V^{(m)}$, m=1, ...M, are defined as follows:

$$V = (2\,U - 1) = \text{Sgn}(\,U\,) \qquad (2)$$

where the unit vector **1** has all entries equal 1 and Sgn is the sign function that changes zero and negative quantities to -1. We prefer bipolar version to binary version because : (1) the inner product norm is always identical to the number of bits, B:

$$V^T \cdot V = B = <V|V> , \qquad (3)$$

rewritten here in terms of Dirac's bracket notation: <bra|ket> for the inner and |ket><bra| for the outer product, (2) the nature of "exclusive or" can be easily represented by bipolar multiplications:

$$+1 \times +1 = 1, -1 \times -1 = 1, +1 \times -1 = -1, -1 \times +1 = -1,$$

(3) the inner product norm is related to the Hamming distance, defined to be the number of different bits between two vectors no matter where the differences occur.

We assume an orthogonal set of feature vectors defined as follows:

$$V^{(n)T} \cdot V^{(m)} = B\,\delta_{n,m} = <n\,|\,m> \qquad (4)$$

where $\delta_{n,m}$ is the Kronecker delta. The outer product weight matrix W represents auto-associative memory:

$$[\,W\,] = \Sigma_m\,[\,V^{(m)}V^{(m)}] = \Sigma_m\,|\,m><m\,| \qquad (5)$$

Hopfield (1982, 1984) assumed the auto-associative matrix [T] to be traceless. That was used together with the symmetry property to prove convergence. Thus, the second term of Kronecker's delta matrix (1's along the main diagonal and zero elsewhere) is introduced in Eq (6) to make it traceless.

$$B[\,T\,]_{ij} = [\,W\,]_{ij} - M\,\delta_{i,j} \qquad (6)$$

B is the normalization constant, and M is the memory capacity. Using the trace operation denoted by Tr, we can easily verify Eq (6) to be traceless.

$$Tr(\,|m><m|\,) = B \qquad (7)$$

$$Tr(\,[\delta_{i,j}]\,) = B \qquad (8)$$

The tradeoff between the memory capacity and the degree of fault-tolerance has been estimated to be about 15 % of B bits [Hopfield (1982)] for pseudo-orthogonal vectors. That is,

$$M = 0.15\,B \qquad (9)$$

For orthogonal feature vectors, however, the capacity is 100 %.

$$M = B \qquad (10)$$

This fact can be demonstrated by the eigenvalue problem of the matrix which is defined to be

$$[T]\,|n> = \lambda_n\,|n> \qquad (11)$$

where the eigenvalue can be easily verified, using Eqs (4) and (6), to be *degenerate* , namely, a white spectrum for all M states,

$$\lambda_n = 1 - (M/B) \qquad (12)$$

The full capacity, M = B, corresponds to a zero eigenvalue for all B orthogonal eigenstates, one for each feature vector.

Consider a simple example where B = 4. There are 4 possible orthogonal vectors and 2^4 = 16 possible words denoted by:

0,1,2,3,4,5,6,7,8,9,10,11,12,13,14,15

We introduce orthogonal subspaces defined by the number of contiguous 1's in the binary word. The subspace consisting of words 13, 11, 7, and 14 is obviously orthogonal by shifting a "one" among 3 zeroes from the the left to the right end of the word.

Word	Binary P	Word	comple.	Binary	p	Word	Bipolar	comple.	Word	Bipolar
13	1101	2	0010	3	13	1 1 -1 1	2	-1-1 +1-1	3	
11	1 011	4	0100	3	11	1 -1 1 1	4	-1+1-1 -1	3	
7	0111	8	1000	3	7	-1 1 1 1	8	+1-1-1-1	3	
14	1110	1	0001	3	14	1 1 1 -1	1	-1-1-1+1	3	
15	1111	0	0000	4	15	1 1 1 1	0	-1-1 -1 -1	4	
6	0110	9	1001	2	6	-1 1 1-1	9	+1-1-1+1	2	
12	1100	3	0011	2	12	1 1 -1-1	3	-1-1+1+1	2	
10	1010	5	0101	1	10	1-1 1-1	5	-1+1-1+1	1	

It is readily verified that the subspace of bipolar words (13, 11, 7, 14) are mutually orthogonal to one another, as shown in Figure 1. They happen to be related to the Walsh function of periodicity p=3. The corresponding binary words have an equal angle among them [cos^{-1} (2/3)] that is not 90°. Also, the second subspace of bipolar words (15, 6, 12, 10) are also orthogonal but the two subspaces are *not* orthogonal to each other.

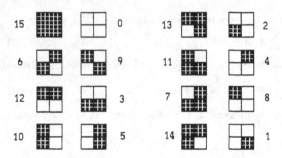

Figure 1. Two-Dimensional Representation of Walsh Base Functions Used to illustrate the fault tolerance and generalization properties of Associative Memory

We consider the storage of one word in memory.

$$4 [T_1] = [13] = | 13 > < 13 | - \delta \qquad (13)$$

If the outer product is properly normalized, it is related to the projection operator:

$$[P] = \delta - |13><13| \ (1/B) \tag{14}$$

Using Eq (4) , it can be verified that

$$[P]^2 = [P]. \tag{15}$$

We will show (1) the ability of fault tolerance, and (2) the ability for generalization.

Fault Tolerance

The following sequence of erasing (zero out) successively from the bipolar bits illustrate tolerance of missing bits.

(1) one missing bit

$$[13](0\ 1\text{-}1\ 1)^T = Sgn(3\ \text{-}2\ \text{-}2\ 3)^T = |13> \tag{16}$$

where Sgn is sign function representing the sigmoid neuron response by the point nonlinearity for extracting the algebra sign of each entries.

(2) two missing bits

$$[13](0\ 0\text{-}1\ 1)^T = Sgn(2\ 2\ \text{-}1\ 1)^T = |13> \tag{17}$$

(3) three missing bits

$$[13]\ (0\ 0\ 0\ 1)^T = Sgn(1\ 1\text{-}1\ 0)^T = |12>$$

$$[13]^2(0\ 0\ 0\ 1)^T = Sgn(1\ 1\ \text{-}1\ 3)^T = |13> \tag{18}$$

(4) four missing bits

$$[13]\ (0\ 0\ 0\ 0)^T = Sgn\ (0\ 0\ 0\ 0)^T = (\text{-}1\text{-}1\text{-}1\text{-}1)^T = |0>$$

$$[13]^2(0\ 0\ 0\ 0)^T = Sgn\ (\text{-}1\text{-}1\ 3\ \text{-}1)^T = |2>$$

$$[13]^3(0\ 0\ 0\ 0)^T = Sgn\ (\text{-}3\text{-}3+3\text{-}3)^T = |2> \tag{19}$$

which converges to a fixed point that is precisely the bipolar complement to $|13>$. In other words, the phase information is lost as an overall minus sign in the last case.

The following sequence of reversing successively from the bipolar bits illustrate tolerance of erroneous bits.

(1) one erroneous bit.

$$[13](-1 \ 1\text{-}1 \ 1)^T = \text{Sgn}(3 \ 1\text{-}1 \ 1)^T = |13> \tag{20}$$

(2) two erroneous bits.

$$[13](-1 \ \text{-}1\text{-} \ \text{-}1 \ 1)^T = \text{Sgn}(1 \ 1 \ 1 \ \text{-}1)^T = |14>$$

$$[13]^2 (\text{-}1 \ \text{-}1\text{-} \ \text{-}1 \ 1)^T = \text{Sgn} \ (\text{-}1 \ \text{-}1 \ \text{-}1 \ 1)^T = |1>$$

$$[13]^3 (\text{-}1 \ \text{-}1 \ \text{-}1 \ 1)^T = \text{Sgn} \ (1 \ 1 \ 1 \ 1)^T = |15>$$

$$[13]^4 (\text{-}1 \ \text{-}1 \ \text{-}1 \ 1)^T = \text{Sgn} \ (1 \ 1\text{-}3 \ 1)^T = |13> \tag{21}$$

(3) three erroneous bits.

$$[13](\text{-}1 \ \text{-}1\text{-} \ 1 \ 1)^T = \text{Sgn}(\text{-}1 \ \text{-}1 \ 1 \ \text{-}3)^T = | \ 2 > \tag{22}$$

which also converges to a fixed point that is also the bipolar complement of $|13>$.

Generalization within a subspace

We consider the ability to recognize a new vector that is different from the stored vector. In other words, an AM can recognize its related vectors that has not been memorized before. By recognition, we mean convergence to a different fixed point. In this sense, we say that the AM can generalize its memory to include other fixed points.

In the case of bipolar vectors, if and only if a new vector x is orthogonal to the stored vectors, associative recall "converges in a cycle of two" as defined in the following iterations:

$$\text{Sgn}([T] \ |x >) = -|x > \tag{23a}$$

$$\text{Sgn}(- [T] \ |x >) = +|x > \tag{23b}$$

This necessary and sufficient condition allows us to determine efficiently the orthogonality between a new vector and all the stored vectors.

We shall show that when a new vector $|11>$ is presented to the AM [13], due to the orthogonality between $|13>$ and $|11>$ and traceless property of [13],

$$[13] \ |11> = \text{Sgn}(- |11>) = |4>, \text{ and}$$

$$[13]^2 |11> = |11> \tag{24}$$

Once the system has acknowledged the second vector $|11>$, it is incorporated into the matrix storage.

$$4[T_2] = [13,11] = [13] + [11]$$
$$= |13><13| + |11><11| - 2\delta \tag{25}$$

If another vector, $|7>$ is presented,

$$[13,11] \; |7> = Sgn(-2 \; |7>) = |8>, \text{ and}$$

$$[13,11]^2 |7> = Sgn(4 \; |7>) = |7> \tag{26}$$

Thus, we enlarge the memory storage to have three memorized states.

$$4[T_3] = [13,11,7] = [13] + [11] + [7] =$$
$$|13><13| + |11><11| + |7><7| - 3\delta \tag{27}$$

This process is continued until the 4-bit orthogonal subspace (p=3) is filled up.

$$4[T_4] = [13] + [11] + [7] + [14] \tag{28}$$

We have demonstrated the ability to include other orthogonal vectors that have not been stored before. This example also shows the important consequence of traceless storage through its contribution to the "generalization by interpolation within the orthogonal subspace".

Given a table of orthogonal vectors, one may argue that computing inner products will also determine orthogonality. However, inner products must be done pairwise among all vectors and become inefficient as the number of vectors gets large. The above method remains efficient for all sizes.

One may furthermore argue that the difficulty is not how to construct orthogonal set, but to select critical bipolar features from gray-scale, imperfect images.

Algorithms for Construct A Critical Feature :

We shall not rely on the auto-AM to select features. One can carry out one's favorite image processing procedure to extract a set of gray-scale feature vectors, $\{|F>\}$. Bipolar feature vectors are preferred in AM because of demonstrated fault-tolerance and the special ability of traceless outer product that allow a quick convergence to a fixed point of cycle two. Given a gray-scale feature vector $|F>$, several procedures for generating a bipolar feature vector are given. The first procedure is "bipolarization", i.e. ,

$$|f> = Sgn(|F> - threshold) \tag{29}$$

The second procedure is to use the Walsh transform. We apply two-dimensional Walsh transform (as orthogonal bipolar vector space($|w_i>$)) to all gray-scale features. We select one bipolar feature vector from a specific Walsh base vector that is associated with the maximum coefficient in the Walsh transform.

$$|f> = Sgn(Max_i (\Sigma |w_i><w_i|F>) - threshold) \qquad (30)$$

where the orthonormality condition of Walsh base vectors is inserted to relate to the first method.

$$\Sigma |w_i><w_i| = [1] \qquad (31)$$

The third and the fourth procedures are to extract from the arbitrary feature vector $|G>$ the closest vector $|g>$ from either the bipolar orthogonal feature set $\{|N>\}$ or the $\{ |F>\}$ using the following traceless associative memory storage.

$$|g> = Sgn([\Sigma\Sigma|N><F|] | G> - threshold) \qquad (32)$$

$$|g> = Sgn (\Sigma c_F [|F><F|] G> - threshold) \qquad (33)$$

The linear combination coefficients $\{ c_F \}$ may be determined by the statistics of **single character distortions and variances** (similar to finding the normal modes that diagonalizes the covariance matrix and the **Karhunen-Loeve** orthogonal procedure used for outer product representation of 2-D imagery). Furthermore, the statistics of **character pair distortions**, such as **two scripted zeros**, could be used to determine the coefficients so as to resolve the problem of recognizing *connected character* and *broken character* after segmentation. We will not go into details in this approach, because of its problem-dependent nature.

The mechanism to select critical features is given as follows.

(1) Human being picks a critical feature (pictures) among the set of distorted, handwritten characters, e. g. the extra stroke among O, P, Q .

(2) Walsh transform the selected feature.

(3) Pick the Walsh function that has the largest transform value.

We choose a feature vector that is closest to the Walsh vector associated with the largest Walsh transform coefficient, and the rest follows from the procedure described in eq (24-28). We call such a set of features the critical features.

Lessons to be learned about applying associative memory to pattern recognition:

AM can only do so much. There is no way to judge the correctness of an associative recall, except by the convergence to a fixed point. One can only assign meaning to those fixed points, whether it is new or old. The proven capabilities of the AM model are (1) missing and erroneous bits recovery, and (2) the creation of new orthogonal vectors, as illustrated above. Therefore, to apply AM to pattern recognition, one must apply human interpretations to those capabilities.

Since learning is by trial and error, it is a continuous process. Suppose that a feature vector with many components representing many features (such as leg-feature and fur-feature, etc, for a tiger, coded fully as $|13>$) has been memorized by the traceless outer product. Furthermore, suppose that only certain features are known in a sequence of imperfect input vectors. (I. e., some feature values are missing. e. g. , the first in the sequence is $(0, 0, 1, 1)$). Then, the AM can fill in the missing bits. After three iterations, one finds $(-1, -1, 1, -1)^T = |2>$. One can then enlarge the traceless outer product memory to include both vectors, [13, 2]. One examines the second input vectors $(0, 0, 1, 1)$. One can verify that the enlarge memory can indeed recall the vector $|2>$, which correspond to, say, a lady, rather than a tiger. The AM "mental" capacity of recognizing other distinct objects when they show up has been demonstrated. Following this line of thought, the different subspace of different size could be assigned for different classes of objects related by a hetero-associative memory of a rectangular matrix. Such a recognition of different classes requires a complete feature set coded in the AM. It can fill all orthogonal subspaces by the "generalization procedure" illustrated in Eq(24-28).

3. ATTENTIVE ASSOCIATIVE MEMORY

Recently, Amari et al has studied the dynamics of such a system, which we will give a simple theorem. We summarize our model equations as follows:

$$< n| m > \equiv B \, \delta_{n,m} \tag{34}$$

$$[T] \, |n> = \lambda_n \, |n> \tag{35}$$

The simple model of attentive associative memory $[T]$ is a linear combination of outer products based on the set of orthogonal feature vectors, $\{ | n> , n = 1, ... M\}$, and a cue of initial state $| Q >$ that determines the set of attention parameters $\{ a_n \}$ as follows:

$$a_n = <n | n > - <n | Q > \tag{36}$$

$$B [T]_{ij} = \Sigma^M_{n=1} \, a_n \, | n_i > <n_j | - A \, [\delta_{i,j}] \tag{37}$$

that is traceless, $Tr \, \delta_{i,j} = Tr | n_i > <n_j | = B$, giving

$$A \equiv \Sigma^M_{n=1} \, a_n \tag{38}$$

and

$$\lambda_n = a_n - (A/B) \tag{39}$$

The attentive memory capacity A and eigenvalue λ_n are reduced to Hopfield's memory capacity M and a degenerate eigenvalue λ, in case of a uniform attention(i.e. $a_n = 1$),

$$\lambda \equiv 1 - r \tag{40}$$

where Amari's pattern ratio $r \equiv (M/B)$ is defined for M bipolar words (states) of B bits (neurons) each.

The dynamics is assumed to be governed by matrix-vector inner product

$$Q(t + 1) \equiv Sgn([T] Q(t)) \tag{41}$$

where a point nonlinear ity function is defined as $Sgn(x) = +1$ if $x > 0$, and -1 if $x < 0$. The succesive associative recall gives the iteration, indexed by $t = 0, 1, 2,...$, such that $Q(t) = Q$ when $t = 0$. The eigenvalue spectrum, not the distance alone, is a proper macroscopic parameter to explain the transient dynamical behaviors of the recalling process. In particular, the direction cosine

$$S_m(t)) \equiv <m| Q(t) > / <m | m > \tag{42}$$

has been derived and the logarithmic derivative is given by

$$(d/dt) \log (1 - S_m(t)) < \log (\lambda_m / 2) < 0 \tag{43}$$

Convergence to a specific m-th state is guaranteed if m-th eigenvalue (λ_m) is bounded $2 > \lambda_m > \lambda$.

Theorem 1 about the lower bound says that paying attention (i.e. non-uniform $a_n \geq 1$ always increases the memory capacity A) $\sum^M_{n=1} a_n > M$ with a faster convergence rate proportional to the eigenvalue $\lambda_m > \lambda \equiv 1 - r$

We conjecture that the statistical neurodynamics of associative memory may have similar behavior to the deterministic dynamics of attentive associative memory with a non-white eigenvalue spectrum due to random initial conditions that change with respect to the initial guess vector $|Q(t)>$, $t = 0$. The difference vector between $|Q(t)>$ from $|m>$ has an inner product norm defined as

$$2 D_m(t) \equiv <m | m > - <m|Q(t) > \tag{44}$$

If we assume that paying attention to the initial small guess error $2 D_m(0)$ amounts to choose a nonuniform and biased storage

$$a_m = 2 D_m(0) \geq 1 \tag{45}$$

and all other coefficients to be identical to 1

$$a_n = 1 \quad , n \neq m. \tag{46}$$

By definition

$$A = M + 2 D_m(0) - 1. \qquad (47)$$

Theorem 2 about the upper bound of λ_m assumes that if a small difference vector between the input $|Q>$ and the specific state $|m>$, is used as the attention parameter a_m, Eq(31a), then the critical relationship between the Amari's pattern ratio r and the initial error is analytically found for successful recalls.

$$2 D_m(0) < 2 + (M+1)/(B-1) \qquad (48)$$

The maximum permissible Hamming distance D_H, from the desired m-th state to be reached after iterative recalls, is given by the formula

$$D_H \leq (B/2) - 1 - [(M-1)/2(B+1)]((B/2) - 1 - (r/2) \qquad (49)$$

4. Conclusion

Associative memory (AM) works like a match filter , but does so efficiently. It should not be applied to image domain directly. Rather, it should be applied to feature domain so that a relatively small AM can do useful tasks at the present technology.

We shall not rely on the auto-AM to select features. Instead, features should be selected using human judgement. However, auto-AM will help us find critical features and hetero-associative memory can perform feature extraction efficiently.

There exists a large body of knowledge pertaining to features selection and extraction and pattern classification for traditional optical character recognition in the literature. This body of knowledge should be tapped and coupled with associative memory. One should not rule out the use of traditional classification techniques (such as syntactical) as extraction of high-level features which then become part of the input feature vector to an AM.

Classical pattern recognition has been demonstrated with a relatively greater success in machine-printed character recognition compared to handprinted character recognition. Difficulty may be rooted in the lack of generalization and abstraction due to machine's limited one-dimensional knowledge representation. In principle, AM should be able to complement traditional OCR with 2-D knowledge representation. Various degrees of abstraction can be achieved through a multi-layer, two-dimensional AM architecture. Note that the present technology has evolved to the point where 2-D memory (chip or optical disk) is not more expensive than 1-D memory storage with logic unit tree content addressable memory processor.

In conclusion, we can combine traditional wisdom in traditional OCR with simple AM implementable in present technology to form a human-intelligence-endowed   neural network.

Character segmentation is an important step in character recognition. Fukushima has developed neural network model (selective attention) for character segmentation in his Neocognitron [Fukushima (1987)]. The attentive associative memory model implemented opto-electronically by Athale, Szu & Friedlander (1986) can be augmented by a priori probability compiled by a character-pair correlation function of connected characters. This is an interesting area for more research.

Inputs to associative memory are linear vectors whereas inputs to OCR are rectangular arrays. Can associative memory replicate the concept of (2-D) neighborhood? The two-dimensional transform that preserves the neighborhood relationship should be used for image pre-processing before applying AM to the pattern. For example, 2-D Walsh transform can give a 1-D base Walsh vector (associated with the largest coefficient) as input feature vector to the AM.

Can AM perform syntactical parsing [Ali and Pavlidis (1977)] or rule-based structural analysis [D'Amato (1982)]? Any traditional classification technique can be used to extract high-level features for AM.

How can AM extract position and rotation invariant features? [cf. Szu (1986), Messner and Szu (1987)].

One difficulty in applying backpropagation network has been network size-scaling problem. One way to circumvent it has been to extract a small number of features as input. [cf Burr (1987), Gullichsen and Chang (1987)]. Recent advances by Ballard in 1987 permit partial connectivity between two successive layers which avoids combinatorial explosions often encountered when the input layer is directly connected to image pixels. Thus, spatial pattern relationship can be efficiently preserved in such a network while coarse-graining between successive layers can desensitize pattern variation in input images.

An AI extension of the simple AM model is attentive associative memory, (AAM), that allows us to apply AI to pay a non-uniform attention to each term of outer product storage, i.e. a linear combination of outer products in which the set of combination coefficients is determined by AI rule-based system, e.g. the frequency distribution of distorted characters according to user group profiles, e.g. left hand writing versus righthand writing. The efficiency of the closed system of rule-based knowledge representation of AI using the tuple storage is combined with the flexibility of the non-rule based open system using the matrix knowledge representation of NI (coined for either neural, or network, or natural intelligence). Thus, the ability of generalization and abstraction becomes possible for AI, and is demonstrated in a combined intelligent system of AI & NI. We can endow a simple neural network architecture based on a small set of neurons with a human-like intelligence by combining the traditional rule-based AI wisdom with non-rule-based learning. This is achievable because OCR requires better feature vectors obtained from other discipline in the sense of fault tolerance that neural networks built at the present technology can already provide with.

Appendix: Generic Definition of Neural Networks

Associative memory is a special model of neural networks. Examples of associative recalls from partial images and the success of nonlinear signal processing are recorded in the literature [cf. Kohonen (1984)]. An axiomatic definition is outlined as follows.

We shall define **three kinds of neurons**: fine-grained, medium-grained and large-grained processor elements (PEs). A fine-grained PE, represented by the lower case word *neuron* , has no internal memory analogous to neurons in the hippocampus part of the brain that is responsible for fault-tolerant associative recall. A medium-grained PE, *Neuron,* has a built-in memory analogous to Neurons in biological sensory and motor control which are responsible for reactions to approaching danger. A large-grained PE, *NEURON* , has built-in memory, control logic, and communication capabilities equivalent to a computer. NEURONs occur in nature in the form of grandmother cells or pacer/conductor cells.

These three types of neurons and their associated circuits have **four kinds of interactions:** *(1) exciting, (2) inhibiting, (3) bursting, (4) grading and delaying transmission.* In general they follow the law of the middle response or the sigmoid function (hyperbolic tangent or logistic functions) to amplify weak signals with a nonlinear quick rising function and suppress strong signals with a nonlinear tapering off saturation function. The generic definition of a Neural Network is a system which is:

1. *Non-linear* ≈ sigmoid function ≈ point non-linearity (hard limiting) shown as follows:

2. *Non-local* ≈ weighted outer product ≈ outer product (white spectrum) shown as follows:

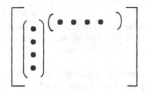

3. *Non-stationary* ≈ piecewise time stationary ≈ iterative algorithm shown as follows:

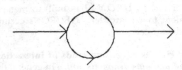

4. *Non-convex* ≈ constrained global optimization ≈ simulated annealing schematically shown as follows:

5. *Other attributes yet to be discovered* .

These successive approximations of the four *non--*principles, indicated by wiggly equality signs in (1-4), makes possible the unveiling of the complex and nonlinear neural (brain) behavior. This is possible with the use of powerful computers and more accurate models of intelligent functions. The theory is amenable to numerical simulations due to *piecewise linear, regionally local, temporarily stationary, and locally convex approximations.*

Three decades ago, Rosenblatt and co-workers built the **perceptron** solely based upon the first attribute (nonlinearity) with stochastic implementations. Thus, with hindsight, it was not surprising that Minsky and Papert could show a limited utility and propose useful alternative: **artificial intelligence (AI) rule-** based systems. *AI works in closed systems where rules govern, while neural intelligence (NI) works in open systems where rules have yet to be discovered.* Various exploitation of these efforts in neural networks are:

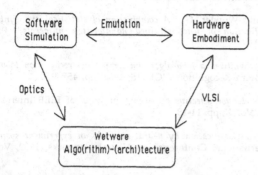

The term wet-ware, coined by Carver Mead, is neither software nor hardware, but more like a Hecht-Nielsen's net-ware based on non-programmable but trainable networks. A special version of layered neural networks has been demonstrated with the ability of phonetic interpolation in the Rumelhart, Sejnowski connectionist's networks, such as Net-Talk, Boltzmann and Cauchy Machines, and error back propagation networks.

Acknowledgement

The work has been supported by ONR under IST/SDIO program. Discussions with John Tan and Frank Polkinghorn are gratefully acknowledged.

6. REFERENCES

Ahmed, P. and Suen, C., "*Computer recognition of totally unconstrained handwritten ZIP Codes,*" International Journal of Pattern Recognition and Artificial Intelligence,Vol. 1 (1987), pp.1-15.

Ali, F. and Pavlidis, T., "*Syntactic recognition of handwritten numerals,*" IEEE Transactions on SMC, Vol. 7 (1977), pp.537-541.

Athale, R.A., Szu, H.H., & Friedlander, C.B., "*Optical implementation of associative memory with controlled nonlinearity in the correlation domain,*" Optics Letters, Vol. 11 (1986), pp. 482-484

Burr, D., "*Designing a handwriting reader,*" IEEE Trans. on PAMI, Vol.5 (1983)

Burr, D., "*Experiments with a connectionist text reader,*" Proc. of IEEE International Conference on Neural Networks, 1987, Vol. IV, pp. 717-724.

D'Amato, D., et al, "*High speed pattern recognition system for alphanumeric handprinted characters,*" Proc. of Pattern Recognition and Image Processing, 1982, pp.165-171.

390

Duda, R. and Hart, P., *Pattern Classification and Scene Analysis* Wiley-Interscience, 1973.

Duerr, B., Haettich, W., Tropf, H. and Winkler, G., *"A combination of statistical and syntactical pattern recognitionapplied to classification of unconstrained handwritten numerals,"* Pattern Recognition, Vol. 12 (1980), pp.189-199.

Fukushima, K. and Miyake, S., *"Neocognitron: a new algorithm for pattern recognition tolerant of deformations and shifts in position,"* Pattern Recognition, VOL. 15, 1982, pp. 455-469.

Fukushima, K., *"A neural network model for selective attention,"* in Proc. of IEEE International Conference on Neural Networks,1987, Vol. II, pp. 11-18.

Gullichsen, e. and Chang, E., *"Pattern classification by neural network: an experiment system for icon recognition,"* Proc. of IEEE International Conference on Neural Networks, 1987, Vol. IV, pp.725-732.

Hull, J., et al., *Optical Character Recognition Techniques in Mail Sorting: A Review of Algorithms.* Technical Report 214, State University of New York at Buffalo, Department of Computer Science, 1984.

Kohonen, T.,*"Self-Organization and Associative Memory,"*Springer-Verlag, 1984.

Stringa, L., *"LCD: a formal language for constraint)free hand-printed character recognition,"* Proc. of International Conference on Pattern Recognition, 1978, pp. 354-358.

Schurmann, J., *"Multifont word recognition system with application to postal address reading",* Proc. of Int. Conf. on Pattern Recognition, 1976, pp. 658-662.

Schurmann, J., *"Reading machines",* Proc. of Int. Conf. on Pattern Recognition, 1982, pp.1031-1044.

Shridhar, M. and Badreldin, A., *"Recognition of isolated and simply connected handwritten numerals,"* Pattern Recognition, Vol.19 (1986), pp.1-12.

Suen, C. Y., Berthod, M. and Mori, S., *"Automatic recognition of handprinted characters - the state of the art",* Proceedings of the IEEE, vol. 68 (1980), pp. 469-487.

Szu, H. H., Caulfield, H.J., *"Optical Expert Systems,"* Applied Optics, Vol. 26, pp. 1943-1947, 1987

Szu, Harold H., *"Three layers of vector outer product neuural networks for optical pattern recognition.",* In H.Szu (Ed.) *Optical and Hybrid computing* (1986) (pp. 312-330), Bellingham, WA: Society of Photo-Optical Instrumentation Engineers.

Szu, H. H. & Messner, R. A. ,*"Adaptive Invariant Novelty Filters,"* Proceedings of IEEE, Vol. 74 (1986), pp. 518-519

SELECTED BIBLIOGRAPHY ON CONNECTIONISM

Oliver G. Selfridge, Richard S. Sutton, and Charles W. Anderson*

GTE Labs, Waltham, MA.

Introduction

The topic of this annotated bibliography is connectionism, a field of computer science that has enjoyed a vast resurgence in the last ten years. Properly speaking, connectionism should be regarded as part of Artificial Intelligence, or AI, and up to some years ago it was usually so treated. The earlier entries below will make that clear. Another term for connectionism is *Neural Networks*, and it is being widely used, especially among the new efforts that are arising, including start-up companies.

Connectionism has two chief interests: one is the efficiency or novelty of certain kinds of computation; the other is models of real brains or real neural networks— the kinds made of flesh. The former is our concern here. It has been claimed that connectionism can exhibit a new kind of computing, which is "non-von-Neumann," and can thereby provide new capabilities that cannot be matched in other ways. We want to point readers towards publications that can give them background and insights about the real issues and the state of the field and its prospects. We do not cover the recent work on implementation technologies. Our audience here is anybody who wants or needs more than buzzword knowledge about the field, including researchers, students, and managers in computer science and technology.

The entries have been selected according to their relevance to learning machines that we now recognize as connectionist. Entries are ordered by date of publication.

* The authors gratefully acknowledge the helpful comments of A. Barto, M. Steenstrup, J. Franklin, and H. Klopf.

McCulloch, W.S. and Pitts, W.H., "A logical calculus of ideas immanent in nervous activity," *Bulletin of Mathematical Biophysics*, 1943, *5*, 115–133; reprinted in McCulloch's *Embodiments of Mind*, Cambridge, MA: M.I.T. Press, 1965.

> This early paper lays out the ideas behind connectionism with austere and literate precision; though in places it is not easy reading. It shows that a simple model neuron, working in discrete time and emitting a purely binary signal, can be assembled in numbers to form a Turing machine; that is, that it can compute anything that is computable at all. An awesome piece of work, considering that the junior author, who was responsible for all the mathematics and many of the ideas, was barely twenty years old.

Pitts, W.H. and McCulloch, W.S., "How we know universals: The perception of auditory and visual forms," *Bulletin of Mathematical Biophysics*, 1947, *9*, 127–147.

> A companion paper to the previous one. It shows that neural networks—that is, connectionist mechanisms—can compute features or *concept membership* in the current AI sense.

Hebb, D.O., *The Organization of Behavior: A Neurophysiological Theory*, New York: Wiley, 1949.

> This connectionist classic deals broadly with the problem of relating psychology to neurophysiology. The most lasting specific contribution has been Hebb's neurophysiological learning rule—that a synapse becomes strengthened whenever the pre- and post-synaptic neurons are simultaneously active. Hebb argued that neurons following this rule would group themselves together to form *cell assemblies*, which would then be capable of further learning and more complex behavior. One difficulty in the ideas is that the cell assemblies seemed not to behave very differently from the neurons they were assembled from. In later years, Hebb seemed to abandon his old ideas more willingly than some of his readers.

Farley, B.G. and Clark, W.A., "Simulation of self-organizing systems by digital computer," *I.R.E. Transactions on Information Theory*, 1954, vol 4, 76–84.

> The earliest publication we know of presenting simulation re-
> sults with connectionist systems. The learning problem was
> primarily one of pattern classification, but there was also discus-
> sion of what we would now recognize as reinforcement learning.
> It was presented at what is regarded as the opening guns of AI,
> the Western Joint Computer Conference session in 1954. See
> also Clark and Farley's "Generalization of pattern recognition
> in a self-organizing system" (*I.R.E. Transactions on Inf. The-
> ory*, 1955, *5*, 86–91), which includes a summary of the results
> of their earlier paper.

Rosenblatt, F., "The perceptron: A probabilistic model for information storage and organization in the brain," *Psychological Review*, 1958, *65*, 386–408. See also Rosenblatt's *Principles of Neurodynamics*, New York: Spartan, 1962.

> Rosenblatt and the perceptron are the names that today we
> most associate with the early surge and then ebbing of inter-
> est in connectionism. Probably Rosenblatt and his group at
> Cornell Aeronautical Laboratory were responsible for more hy-
> perbole per actual man-month of work than any other group
> in history—though some today may pose competition. At the
> core of the perceptron work is the *convergence theorem*, which
> states that a certain kind of perceptron will eventually learn any
> predicate it is capable of representing. As others noted later,
> the primary limitations of this result are 1) the word *eventually*,
> 2) that many predicates cannot be represented, and 3) that a
> perceptron must be explicitly told the correct behavior in order
> to learn.

Selfridge, O.G., "Pandemonium: A paradigm for learning," *The Mechanisation of Thought Processes*, London: H.M. Stationery Office, 2 vols., 1959; reprinted in *Pattern Recognition; Theory, Experiment, Computer Simulations, and Dynamic Models of Form Perception and Discovery*, Uhr, L., ed., New York: Wiley, 1966.

> A statement of the importance of features in recognition, and
> suggesting that the hierarchy of features is a dominant and nat-

ural structure; the interplay between layers has a connectionist flavor and function. This paper and Samuel's paper (below)
were the first to discuss the idea of generating new features from
combinations and mutations of old features that have already
proven useful.

Samuel, A.L., "Some studies in machine learning using the game of checkers," *IBM
Journal on Research and Development*, 1959, *3*, 210–229; reprinted in *Computers
and Thought*, Feigenbaum, E.A. and Feldman, J., eds., New York: McGraw-Hill,
1963.

Probably the most famous learning paper in AI. Although
Samuel saw his work as an alternative to the "Neural-Net Approach" that was popular at the time, it would fit well into
1980's connectionism, and is the basis for several modern learning procedures.

Widrow B. and Hoff, M.E., "Adaptive switching circuits," *1960 WESCON Convention Record Part IV*, 1960, 96–104. See also *Adaptive Signal Processing*, by
Widrow, B. & Stearns, S.D., Englewood Cliffs, NJ: Prentice-Hall, 1985.

This paper introduced the ADALINE, one of the most effective and best understood of connectionist units, and one of the
very few that have already served in useful applications. The
ADALINE continues to be widely used and to provide a theoretical base for new learning procedures (for example, *back-
propagation*). *Adaptive Signal Processing* is an excellent text-
book presentation of the ADALINE results obtained over the
years by Widrow *et al.* at M.I.T. and then at Stanford.

Minsky, M.L. and Selfridge, O.G., "Learning in random nets," *Information Theory,
Fourth London Symposium*, London: Butterworths, 1961.

This was the first real critique of Rosenblatt's perceptrons, and
pointed out that the perceptron as he had defined it, far from
being able to make general abstractions, could not even generalize towards the notion of binary parity; it also analyzed other
claims to convergence and suggested the roles that connectionist
mechanisms might play in larger systems.

Minsky, M.L., "Steps toward artificial intelligence," *Proceedings of the Institute of Radio Engineers*, 1961, *49*, 8–30; reprinted in *Computers and Thought*, Feigenbaum, E.A. & Feldman, J., eds., New York: McGraw-Hill, 1963.

> This excellent early paper in AI includes a large section on what is now termed connectionism. It should be remembered that Minsky wrote a connectionist doctorate thesis in 1954 at Princeton University ("Theory of neural-analog reinforcement systems and its application to the brain-model problem," available from University Microfilms, Ann Arbor, MI).

Nilsson, N.J., *Learning Machines*, New York: McGraw-Hill, 1965.

> This early book is a well-written exposition on linear separability in hyper-spaces; it is still one of the best teaching tools for understanding the basic processes of simple pattern recognition programs.

Minsky, M.L. and Papert, S., *Perceptrons: An Introduction to Computational Geometry*. Cambridge, MA: MIT Press, 1969.

> This properly famous book analyzed one-layer perceptrons and proved that they are inherently incapable of making some global generalizations on the basis of locally learnt examples: in particular, connectivity of a binary picture. It is a thoughtful, thorough, and well written book. However, the limitations discussed are all of perceptrons as computational mechanisms, not as learning mechanisms; that is, the limitations are on what they can compute, not on what they can learn in a practical amount of time. As Minsky and Papert note, the latter is often the more pressing concern. For example, a perceptron has no computational difficulties in learning to recognize shapes independent of their size, position, and orientation, but it can do so only after experience with each shape in all possible sizes, positions, and orientations. Although this book is often said to have killed the early perceptron work, it had already been nearly abandoned by the time the book appeared.

Mendel, J.M. and Fu, K.S., eds., *Adaptive, Learning and Pattern Recognition Systems*, New York: Academic Press, 1970.

> Though twenty years old, this thorough study of adaptive techniques in pattern recognition problems is a standard. Starting with simple linear separability, it examines gradient techniques, adaptive optimization, and reinforcement learning with good mathematical support.

Klopf, A.H., "Brain function and adaptive systems–A heterostatic theory," *Air Force Cambridge Research Laboratories Research Report*, AFCRL-72-0164, Bedford, MA., 1972. An updated version is available as Klopf's *The Hedonistic Neuron: A Theory of Memory, Learning, and Intelligence*, Washington DC: Hemisphere/Harper & Row, 1982.

> Klopf's primary contribution was to recognize that something was missing from the then-current stock of connectionist learning methods. That something was the ability to learn in environments in which you were told *how well* you were doing, but not exactly *what* you should be doing (or should have done); that is, the ability to do reinforcement learning rather than supervised or error-correction learning.

Arbib, M.A., *The Metaphorical Brain*, New York: Wiley, 1972.

> This book laid out the imperatives for modern connectionism clearly and convincingly, and helped set the stage for the current renewed interest in the area.

Sommerhoff, G., *The Logic of the Living Brain*, London: Wiley, 1974.

> A little known but excellent analysis of basic connectionist concepts and assumptions. For example, there are sections on what it means for a connectionist system to be goal-directed and to have an internal model of the world, sections on expectation, attention, and the apparent stability of the visual scene, and sections on the difference between error signals and goal signals.

Uttley, A.M., *Information Transmission in the Nervous System*, London: Academic Press, 1979.

> A compact presentation of Uttley's pioneering work in connectionism. He is best known for his early work on *conditional probability machines* (1956) and for the *informon*, a connectionist learning unit using the negative of the Hebb rule, and which he related to animal learning theories.

Fukushima, K., "Neocognitron: A self-organizing neural network model for a mechanism of pattern recognition unaffected by shift in position," *Biological Cybernetics*, 1980, *36*, 193–202.

> An often-cited problem in using connectionist networks for pattern classification is their inability to generalize to shifted or rotated versions of trained patterns. A brute-force approach to this problem is to provide sets of units that extract identical features from different parts of an input field. Fukushima's *Neocognitron* (a development of his earlier *Cognitron*) is constructed of multiple layers of such unit sets and demonstrates limited shift-invariance. The generality of this approach may be limited by the large number of units required.

Sutton, R.S. and Barto, A.G., "Toward a modern theory of adaptive networks: Expectation and prediction," *Psychological Review*, 1981, *88*, 135–170.

> A study of connectionist learning elements as models of *Pavlovian conditioning*, the simplest and best understood kind of associative learning in nature. This paper points out that while the Hebb rule is a very poor model of animal behavior, the ADALINE rule is equivalent to a popular and successful psychological theory, the *Rescorla-Wagner model*. The paper also proposes a new connectionist learning element that improves over both in some ways.

Albus, J.S., *Brains, Behavior, and Robotics*, Peterborough, NJ: McGraw-Hill/BYTE, 1981.

> This book lays out a connectionist approach to robotics, hierarchical control, and cerebellar modelling, which was pursued by the author throughout the 1970's.

Hinton, G.E. and Anderson, J.A., eds., *Parallel Models of Associative Memory*, Hillsdale, NJ: Lawrence Erlbaum, 1981.

> A collection of articles from an informal conference held at the University of California at San Diego in 1979. This conference can be said to mark the onset of renewed interest in connectionism within cognitive psychology. Included are articles by Hinton, Rumelhart, Kohonen, Anderson, Sejnowski, Feldman, Willshaw, Geman, Fahlman, and Ratcliff.

Hopfield, J.J., "Neural networks and physical systems with emergent collective computation abilities," in *Proceedings of the National Academy of Sciences, USA*, 1982, *79*, 2554–2558. See also Hopfield, J.J. and Tank, D.W., "Computing with neural circuits: A model," *Science*, 8/8/86, *233*, 625–633.

> John Hopfield has almost single-handedly created an enormous amount of activity and interest in connectionist systems among physicists and the wider lay public. In this paper, he introduced the idea of *computational energy*, a new way of understanding the computation performed by networks with feedback and effectively symmetric connections. This idea was used subsequently, for example, in the development of the Boltzmann Machine (see Ackley *et al.* below). In the Hopfield and Tank article, energy analyses are used to design networks and weight settings to solve particular problems; for instance, samples of the traveling salesman problem. This work should be taken as an excellent illustration of the energy-function design methodology, not as a demonstration of the competitiveness of such networks on combinatorial optimization problems.

Feldman, J.A. and Ballard, D.H., "Connectionist models and their properties," *Cognitive Science*, 1982, *6*, 205–254.

A scholarly argument for connectionist models as opposed to information processing models in cognitive science, and a presentation of the general "University of Rochester" connectionist model. The Rochester model emphasizes the computational advantages of massive parallelism even when *not* coupled with learning and distributed representations. Because it features a large variety of specialized types of units, the Rochester model also challenges the dogmatic assumption of many connectionists that all units should be simple and identical.

Anderson, J.A., "Cognitive and psychological computation with neural models," *IEEE Transactions on Systems, Man, and Cybernetics*, 1983, *SMC-13*, 799–814.

An excellent summary of Anderson's connectionist modelling approach to cognitive psychology, including a presentation of his "brain-state-in-a-box" model, one of the earliest (1979) and simplest extensions of associative network ideas to include feedback and nonlinearity.

Barto, A.G., Sutton R.S., and Anderson, C.W., "Neuronlike elements that can solve difficult learning control problems," *IEEE Transactions on Systems, Man, and Cybernetics*, 1983, *SMC-13*, 834–846.

This paper contains the best demonstration of the abilities of the reinforcement-learning units developed by Barto *et al.* at the University of Massachusetts. A network consisting of two units is shown to be able to solve a broomstick balancing problem that Perceptrons or ADALINEs cannot solve, and to do so much more efficiently than a non-connectionist system previously developed for this task.

Kohonen, T., *Self-Organization and Associative Memory*, Berlin: Springer-Verlag, 1984.

An excellent review of the ongoing work of this pioneer in associative memory and connectionism.

Ackley, D.H., Hinton, G.E., and Sejnowski, T.J., "A learning algorithm for Boltzmann machines," *Cognitive Science*, 1985, *9*, 147–169.

> Introduced the first effective supervised-learning algorithm applicable to networks with interior or "hidden" units.

Lee, Y.C., Doolen, G., Chen, H.H., Sun G.Z., Maxwell, T., Lee, H.Y., and Giles, C.L., "Machine learning using a higher order correlation network," in *Evolution, Games and Learning, Proceedings of the Fifth Annual International Conference of the Center for Nonlinear Studies*, 276–306, Amsterdam: North-Holland, 1985.

> Describes an approach to extending the linear computation of most connectionist units to include higher-order nonlinear terms. The extension allows the solution of nonlinear problems without resorting to multi-layer networks, and can result in spectacularly effective generalization if the selection of higher-order terms is done on a task-specific basis. However, the complexity of the individual units grows exponentially with their order and thus must be limited; multiple layers are still required to solve problems with high-order non-linearities.

Rumelhart, D.E., McClelland, J.L., and The PDP Research Group, *Parallel Distributed Processing: Explorations in the Microstructure of Cognition, Volume 1: Foundations*, Cambridge, MA: Bradford, 1986.

McClelland, J.L., Rumelhart, D.E., and The PDP Research Group, *Parallel Distributed Processing: Explorations in the Microstructure of Cognition, Volume 2: Psychological and Biological Models*, Cambridge, MA: Bradford, 1986.

> These two volumes form a splendid reference set for the field and provide a multi-disciplinary review of many of the underlying ideas. Among the many excellent articles, two must be specially mentioned: "Learning internal representations by error propagation," by Rumelhart, D.E., Hinton, G.E., & Williams, R.W., which introduces *back-propagation*, the most efficient known learning procedure for multi-layer networks; and "On learning the past tenses of English words," by Rumelhart, D.E. & McClelland, J.L., which presents a controversial connectionist model that challenges rule-oriented conceptions of language learning.

These important books share with many of the others discussed here a lack of exploration of the limits of their connectionist mechanisms. Keith Holyoak points out some of the limits of these two books in an appreciative and informative review in *Science* (5/22/87, *236*, 992–996), such as their inability to learn sequences of action in which early components receive no direct feedback.

Minsky, M.L., *Society of Mind*, New York: Simon and Schuster, 1986.

This recent popular work is at once exciting and disappointing. Minsky is articulate, witty and visionary, and he possesses a superb ability to evoke penetrating insights. Much of the discussion implicitly endorses the questions and drives that have motivated connectionists. The book as a whole seems to us to be uneven and patchy; for example, he underestimates the differences among people in how they perceive the world and behave in it. But he wisely reiterates and re-emphasizes the complexity and richness of human thought.

Sejnowski, T.E. and Rosenberg, C.R., "Parallel networks that learn to pronounce English text," *Complex Systems*, 1987, *1*, 145–168.

This paper describes *NETtalk*, a multi-layer back-propagation network that learns to convert text to its phonemic representations, using a human expert for a teacher, and with some residual error. Coupled with a commercial phoneme-to-speech box, *NETtalk*'s learning behavior makes an impressive demonstration that has captured the imagination of the public—*NETtalk* has been widely discussed in the popular press including the TODAY show. In *Proceedings of the Ninth Annual Conference of the Cognitive Science Society*, 1987, Rosenberg uses standard clustering methods to analyze the features learnt by *NETtalk*, showing one way to gain insight into the functioning of a trained network.

Carpenter, G.A. and Grossberg, S., "A massively parallel architecture for a self-organizing neural pattern recognition machine," *Computer Vision, Graphics, and Image Processing*, 1987, *37*, 54–115.

> This paper is the best presentation of Grossberg's *Adaptive Resonance Theory* (ART), which Grossberg himself introduced a decade earlier. An ART network is a mechanism for performing unsupervised clustering of input patterns. Such clustering is widely recognized as a useful technique for reducing the dimensionality of the input to a system; but additional mechanisms are required in order to relate changes in the system to its goals, a point that is not apparent from the descriptions of ART in the literature. That is, the clustering is responsive merely to the metric induced by the particular representations, and not at all to the designer's or user's purposes. Also missing in the literature are comparisons with other implementations of clustering methods. Grossberg *et al.* at Boston Unversity have analyzed this and other connectionist networks as systems of differential equations, which may facilitate their direct realization in parallel hardware.

Lippmann, R. P., "An introduction to computing with neural nets," *IEEE ASSP Magazine*, April 1987, 4–22.

> This is a gentle introduction to a few of the popular methods for using networks as pattern classifiers. They are related to standard pattern classification techniques.

Elman, J.L. and Zipser, D., "Learning the hidden structure of speech," Technical Report 8701, Insitute for Cognitive Science, University of California at San Diego, La Jolla, CA, 1987.

> This paper is a good example of the current connectionist attacks on real problems; it illustrates both what can be done and the limitations of the techniques. It is well written and clear, and it does not make claims beyond what it shows. The task is discrimination of spoken consonant/vowel pairs. The preprocessing seems to have been of over-riding importance, including sampling, A/D conversions, sophisticated normalizations, and,

in the most convincing experiments, Fourier transforms. The (few) hidden units are found to "encode the input patterns as feature types." Some feature types turn out to be easily comprehensible, but others are harder to interpret. It appears that they are, however, only simple combinations of the presence or absence of particular input values. The experimental procedure uses but a single male speaker.

Hinton, G. E., "Connectionist learning procedures," Technical Report CMU-CS-87-115, Carnegie-Mellon University, Pittsburgh, PA, 1987; to appear in *Artificial Intelligence*, 1988.

This technical report reviews learning methods for multilayer networks and some early work with single-layer networks. Research issues are briefly discussed. Examples of supervised, unsupervised, and reinforcement learning methods are described. This paper is the most concise treatment covering these three forms of learning.

Sutton, R.S., "Learning to predict by the methods of temporal differences," Technical Report TR87-509.1, GTE Laboratories, Waltham, MA, 1987; to appear in *Machine Learning*, 1988.

From the Abstract: "This article introduces and provides the first formal results in the theory of *temporal-difference methods*, a class of statistical learning procedures specialized for prediction—that is, for using past experience with an incompletely known system to predict its future behavior ... It is argued that most problems to which supervised learning is currently applied are really prediction problems of the sort to which [these] methods can be applied to advantage."

Anderson, J.A. and Rosenfeld, E., eds., *Neurocomputing: Foundations of Research*, Cambridge, MA: MIT Press, to appear in 1988.

A collection of 43 connectionist reprints, including many of those listed here, and "with a general introduction, introductions to each paper, and reference materials."

HIERTALKER: A DEFAULT HIERARCHY OF HIGH ORDER NEURAL NETWORKS THAT LEARNS TO READ ENGLISH ALOUD

Z.G. An* , Y.C. Lee , and G.D. Doolen
Center for Nonlinear Studies
Los Alamos National Laboratory

Abstract

A new learning algorithm based on a default hierarchy of high order neural networks is proposed. A simulator of such a hierarchy, HIERtalk ,is applied to the conversion of English words to phonemes. Achieved accuracy is 99% for trained words, and 96% for new words.

1. Introduction It is a problem of general interest to determine the relationships among two or more sets of discrete symbols. This problem occurs in translating languages (both computer languages such as Pascal and Lisp, and natural human languages such as English and Chinese), writing compilers for computer languages, determining the correspondence between genetic codes and their structural realizations, and in mapping ordinary spelling onto a phonetic transcription appropriate to drive a speech synthesizer.

This last problem has recently been studied by Sejnowski and Rosenberg .[1] Using a first order neural network with an iterative error correction scheme, known as back-propagation,[2] Sejnowski's NETtalk can recall the correct pronunciation of a corpus of training words and generalize to novel words. The accuracy with which the NETtalk converts English words into phonemes is typically 91-93% for the words in the training set, and 89-91% for new words. NETtalk is based on an automated learning procedure, in contrast to traditional AI approaches (for example DECtalk, a commercial product that converts text to speech), which rely on highly labor intensive entries of phonological rules. Additionally, Stanfil and Waltz also tackled this problem of text to phoneme translation using a memory-based reasoning approach (MBRtalk) which works directly from a data base or training set[3]. The accuracy that the MBRtalk achieved is 86% for new words.

In this paper we present a new learning procedure, based on a default hierarchy of high order neural networks, which exhibited an enhanced capability of generalization and a good efficiency. This new architecture is suitable for learning regularities or "building blocks" embedded in a stream of information with inherent long range correlations. Moreover, it is not plagued by a combinatoric explosion of rules in learning. When applied to the conversion of English words to phonemes, a simulator of such a hierarchy, HIERtalker, achieved an accuracy of typically 99% for the words in the training set, and 96% for new words. Also, HIERtalker used considerably less computer time than NETtalk did.

2. How HIERtalker works Data are obtained from a phonetic dictionary. Suppose we have the word "afternoon" followed by its phonetic transcription "@ftR_n_un". The letters of each word must correspond roughly one-to-one with the phonemes in its transcription to allow the system to pick up the regularities. Most phoneme representations have fewer characters than the word they represent, therefore a "-" character is inserted as a filler character wherever necessary. From this information we can infer a number of correlations between a string of letters and a phoneme. First we look at the word using a window of one letter ($L = 1$) to correspond to a phoneme. Those correspondences are as follows:

* Present Address: NYNEX Science and Technology, 500 Westchester Ave., White Plains, NY 10604.

$$a \longrightarrow @$$
$$f \longrightarrow f$$
$$t \longrightarrow t$$
$$e \longrightarrow R$$
$$r \longrightarrow \_$$
$$n \longrightarrow n$$
$$o \longrightarrow \_$$
$$o \longrightarrow u$$
$$n \longrightarrow n$$

Next we look at the word using a two letter window $(L = 2)$ and get the following correspondences.

$$\_a \longrightarrow \_$$
$$af \longrightarrow @$$
$$ft \longrightarrow f$$
$$te \longrightarrow t$$
$$er \longrightarrow R$$
$$rn \longrightarrow \_$$
$$no \longrightarrow n$$
$$oo \longrightarrow \_$$
$$on \longrightarrow u$$
$$n\_ \longrightarrow n$$

We then use a three letter window $(L = 3)$ to obtain the following correspondences .

$$\_a \longrightarrow \_$$
$$\_af \longrightarrow @$$
$$aft \longrightarrow f$$
$$fte \longrightarrow t$$
$$ter \longrightarrow R$$
$$ern \longrightarrow \_$$
$$rno \longrightarrow n$$
$$noo \longrightarrow \_$$
$$oon \longrightarrow u$$
$$on\_ \longrightarrow n$$
$$n\_ \longrightarrow \_$$

We continue in this fashion, looking at the word with a five, and a seven letter window $(L = 1, 2, 3, 5, 7)$. A window with smaller L is a sub-window of windows with larger L. If context is discarded or replaced by a wildcard from a window with $L > 1$ then the window will reduce to one of its sub-windows, which is one of the windows with smaller L. The idea is to capture the correlation between the phoneme and the shortest string of letters possible from the word. More general rules will be captured at smaller L, while exceptions are captured at larger L.

Following the work of Lee et al,[4] we use a "higher order" neural network approach. Each of these context window of L letters corresponds to a high order correlation neural network of order L for level L. The L context window items from a level L rule are correlated by multiplication or concatenation, depending on whether the input stream is numerical or symbolic. A symbolic rule cannot fire unless all context window items are present together. In particular, the patterns produced by the window of L letters are used as inputs to train a neural network of L-th order. We denote the input string of letters by $I_i = (I_{i1}, I_{i2}, ..., I_{in})$, and the output phoneme by O_j. The mean square error is

$$< E^2 >=< (O - O')^2 >$$

where $O = WI$, W is a matrix to be determined, O' is the desired answer, and $<>$ indicates the ensemble average. To minimize the error we must have

$$W =< O'I >< II >^{-1}$$

where

$$< I_i I_{i'} >=< I_{i1}I_{i2}....I_{in}I_{i1'}I_{i2'}.....I_{in'} >.$$

Taking approximately $< II >= 1$, we have $W =< O'I >$. This is the Hebbian learning rule. Obviously, in order to optimize, the Hebbian rule is more appropriate for high order networks than for first order ones, because for high order network $< II >= 1$ is a better approximation. In the HIERtalker we simply use the Hebbian rule without any error corrections.

A default hierarchy is schematically depicted in Fig. 1. When we train the hierarchy we

Fig. 1

start from the level 1. After we finished training the levels 1,2,....,L , we proceed to the next level. At the level $L + 1$, with every pattern encountered, a check is made to see whether this information has been included in former levels. If so, the level $L + 1$ does nothing, otherwise the level $L+1$ learns this pattern. Thus the lower L levels deal with the more general rules while the higher L levels absorb the exceptions. After the completion of training, the hierarchy is ready to be tested. With any given testing pattern, the system first searches the largest L level for a match . If there is a match, then an answer at this level of specification is obtained, otherwise the system goes to the next lower L level to seek a less specified answer. This "top-down training" and "bottom-up testing" is the logic of the default hierarchy. Because of the inclusive hierarchical window structure only exceptions to existing rules are registered. Every rule learned is indispensable, and thus redundancy is eliminated.

Apparently, the capability of generalization of a default hierarchy stems from its logical structure and is quite independent of details. Another outstanding feature of a default hierarchy is that it is not plagued with the difficulty of combinatoric explosion. This latter point is illustrated in Fig.2., where the number of indispensable rules $n(L)$ learned in each level L is plotted as a function of L. The reason that the number $n(L)$ decreases so rapidly with L is that only a few exceptions are learned in very high levels.

A simple mathematical argument will show why capturing the building blocks brings about tremendous saving. Suppose we are looking at strings of L letters as our context windows. The total number of combinations in such a window is m^L, where m is the number of different letters in the alphabet, 26 for English. When L is large m^L is too huge a number to handle, and this characterizes the so called combinatoric explosion. However, if we can

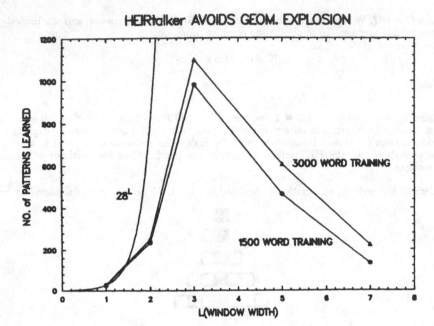

Fig. 2.

capture the building blocks, and assuming all the building blocks are of length less than l, ($l < L$), then the total number of patterns in a context window of length L is less than

$$L \sum_{i=1}^{l} n_i \le L \sum_{i=1}^{l} m^i \le Lm^{l+1}$$

where $n_i \le m^i$ is the number of possible building blocks of length i ($1 \le i \le l$). Therefore the factor of saving due to capturing the building blocks is at least (m^L/Lm^{l+1}), which is a very large number. This explains Fig.2.

The number of rules $n(L)$ has a maximum at $L = 3$, which is in accord with the fact that a significant amount of the information needed to correctly pronounce a letter is contained in the surrounding letters.[9]

3. The Training Sets

The training sets generated contained words with their associated phoneme representations. Articulatory features were not used. Later we may add this information for increased accuracy. A standard phoneme set was required as well as consistent allignment within word between letters and phonemes. The training data contained accurate and consistent data along with noise and inconsistencies.

The following is the phoneme set that was used for our training sets and test sets. It is based on the one letter phoneme translations used by Digital Equipment's DECtalk speech synthesizer. The symbols for the phonemes are shown in the first column and a word containing that sound is shown adjacent to it. The letter or letters in the word that translate to the phoneme are shown in italics.

PHONEME	SOUND
a	st*o*p
b	*b*et
c	t*ough*t
d	*d*one
e	la*k*e
f	*f*ill
g	*g*et
h	*h*int
i	s*ea*t
k	*K*en
l	*l*et
m	*m*en
n	*n*et
o	c*oa*t
p	*p*et
r	*r*un
s	*s*it
t	*t*est
u	l*u*te
v	*v*ase
w	*w*et
x	*a*bout
y	*y*ell
z	*z*oo
A	b*i*te
C	*ch*ill
D	*th*is
E	b*e*t
G	ri*ng*
I	s*i*t
J	*g*in
L	bott*le*
M	ranso*m*
N	butt*on*
O	b*oy*
Q	*o*ne
R	b*ir*d
S	*sh*ow
T	*th*in
U	b*oo*k
W	sh*ou*t
X	bo*x*
Y	*c*ute
Z	a*z*ure
@	*c*at
^	b*u*t

In assembling the training and test sets the question of allignment arises. Is a precise allignment between the letters in a word and the associated phonemes necessary? Consistency seems to be important for the allignment to allow general rules of text to phoneme translation to be captured. No allignment at all leads to the generation of a large number of inefficient rules. Regularity may be hard to see since there is not a one to one mapping between the letters and their phoneme representations.

Allignment is not an easy task. Most mappings are straight forward. For example, "top" $\longrightarrow$ "tap" is already one to one. Another example, "cake" $\longrightarrow$ "cek-" has a silent "e" which is shown by a "-". The word "there" $\longrightarrow$ "D_R-" becomes more complicated. We associate "th" $\longrightarrow$ "D-" and "er" $\longrightarrow$ "-R". The mapping "th" $\longrightarrow$ "D-" could as well have been "-D". The first mapping does appear to be more natural. It is necessary to decide on the type of letter associations as shown above and then to be consistent in applying them throughout the data set. When there are errors or noise in the data set due to inconsistent allignments or erroneous phoneme translations, extra exception rules will be generated to compensate.

We collected and assembled a number of different data sets for our work. Some contained different kind of allignment. Some contained errors in phoneme translation and allignment. One contained a very consistent allignment and translation. Most of these words were initially taken from a 250,000 word dictionary,[6] which had been translated using the phoneme set as described above. A rough allignment was done programmatically. This dictionary was used as a resource to selectively obtain the phoneme translations of sets of words. Clean up of the phoneme translation and allignment was done by hand. After the first training set was constructed and cleaned up it was also used as source data for a neural network to perform automatic allignment. It did a reasonably good job, but some allignment errors still remained.

Trainig set #1 consisted of 1017 words in alphabetical order from a book containing sets of words classified by phonetic sound.[7] The idea behind this training set was to collect a set of words which included many different letter representations of all the different phonemes. This set did not contain all possible combinations.

Training set #2 consisted of 1718 words, 1017 of which are from training set #1, with an additional 701 words randomly chosen from the dictionary. These 701 words were initially alligned using the allignment neural network and then corrected manually. This training set attempted to provide more examples to allow for more letter to phoneme combinations.

Training set #3 consisted of 1441 words from a book that contained a more systematic word set, from simple to difficult.[8] These were left in the order they were presented in the book. This training set attempted a systematic presentation of examples, starting with simple one syllable words, working up to larger words with suffixes such as -tion, -able, etc. A more efficient hierarchy with less rules was generated using this training set in its original order than when presented alphabetically.

Training set #4 consisted of 6219 words in alphabetical order. It contained training set #2 with an additional 4501 words randomly taken from the dictionary. The 4501 words were alligned using the allignment neural network and were not corrected manually. This training set basically contained a number of random words which contained a large number of letter to phoneme combinations than the above sets. It also contained errors or noise which bring about exception rules at the higher levels in the hierarchy.

Random words from the dictionary which are not in our training set were used for testing. We also generated a testing set which has been used by Stanfil and Waltz for MBRtalk.[3]

It contains 100 randomly chosen words. This set serves as a way to compare our results with MBRtalk.

4. Conclusion

We have developed a new, efficient learning algorithm, HIERtalker, based on the concepts of default hierarchy and high order neural network, which achieved a significantly higher accuracy in converting English words into phonemes and run considerably faster than NETtalk.

It is illustrated through this application that a default hierarchy is very efficient in learing the "building blocks". Building blocks in a stream of information are clusters of the basic symbols of the stream which appear repeatedly in the stream, and convey certain messages.

Default hierarchy captures and utilizes the building blocks, in a way similar to "cluster decomposition" in physics. English text happens to be one of those information streams in which there are definite building blocks. That explains why default hierarchy works so well for English text.

This algorithm generalizes readily to other important learning problems such as translating computer language compilers, and deciphering a genetic code[5]. We plan to report progess on the application of the default hierachy learning algorithm to these problems in a forthcoming paper.

References

1. T.J. Sejnowski, and C.R. Rosenberg, "Parallel Networks that Learn to Pronounce English Text", *Complex Systems 1*, (1987) 145-168.
2. D. E. Rumelhart, G. E. Hinton, and R. J. Williams, "Learning Internal Representations by Error Propagation", in *Parallel Distributed Processing: Explorations in the Microstructure of Cognition. Vol. 1: Foundations*, edited by D. E. Rumelhart and J. L. McClellend, (MIT Press, 1986) 318-362.
3. C. Stanfil and D. Waltz, "Toward Memory-based Reasoning", *Communications of the ACM*, (December, 1986) 1213-1228.
4. Y.C. Lee, G.D. Doolen, H.H. Chen, G.Z. Sun, T. Maxwell, H.T. Lee and C.L. Giles, "Machine Learning Using a Higher Order Correlation Network", *Physica 22D*, (1986) 276.
5. A. Lapedes, private communication.
6. J.E. Shoup, *American English Orthographic-Phonetic Dictionary*, Speech Communications Research Laboratory (May, 1973).
7. J. Griffith and L.E. Miner,*Phonetic Context Drillbook,* Prentice-Hall (1979).
8. A. Linksz, *On Writing, Reading, and Dyslexia*, Grune & Straton (1973).
9. J. M. Lucassen and R. L. Mercer, "An Information Theoretic Approach to the Automatic Determination of Phonemic Baseforms", *Proceedings of the IEEE International Conference on Acoustics, Speech, and Signal Processing*, (1984) 42.5.1-42.5.4.

Acknowledgments

The authors would like to acknowledge S. M. Mniszewski, and G. Papcun for many useful discussions, technical assistance, and for their comments on the manuscript.